*Searching
the Heart*

Searching the Heart

Women, Men, and Romantic Love in Nineteenth-Century America

Karen Lystra

New York   Oxford
OXFORD UNIVERSITY PRESS
1989

Oxford University Press

Oxford   New York   Toronto
Delhi   Bombay   Calcutta   Madras   Karachi
Petaling Jaya   Singapore   Hong Kong   Tokyo
Nairobi   Dar es Salaam   Cape Town
Melbourne   Auckland

and associated companies in
Berlin   Ibadan

Copyright © 1989 by Karen Lystra

Published by Oxford University Press, Inc.,
200 Madison Avenue, New York, New York 10016

Oxford is a registered trademark of Oxford University Press

All rights reserved. No part of this publication may be reproduced,
stored in a retrieval system, or transmitted, in any form or by any means,
electronic, mechanical, photocopying, recording, or otherwise,
without prior permission of Oxford University Press.

Library of Congress Cataloging-in-Publication Data
Lystra, Karen.
Searching the heart : women, men, and romantic love
in nineteenth-century America / Karen Lystra.
p. cm.   Bibliography: p.
Includes index.
ISBN 0-19-505817-8
1. Love—History—19th century.
2. Love-letters—United States—History—19th century.
3. Courtship—United States—History—19th century.
4. Sex customs—United States—History—19th century.
I. Title.
GT2703.L97   1990
392'.4—dc19   89-2890   CIP

2 4 6 8 9 7 5 3 1

Printed in the United States of America
on acid-free paper

For my parents
Lily Petersen Lystra and James Lystra
who know that to love another
is the beginning
of wisdom

ACKNOWLEDGMENTS

A CKNOWLEDGMENT is a minor recompense for all the debts one incurs in writing a book. I am grateful for the support of many colleagues and friends, as well as the financial and psychological boost of three summer research grants: a Summer Research Stipend from the National Endowment for the Humanities, a California State University Senior Faculty Research Grant, and a Henry E. Huntington Research Fellowship.

I owe a debt to the former director of the Huntington Library, Robert Middlekauff, who gave unstintingly of himself to the institution and its people. His kindness, gracious hospitality, and interest in my work was a source of considerable encouragement.

The Henry E. Huntington Library in San Marino, California, played a central role in the creation of this book. The Huntington has been my scholarly home for many years. Readers and staff alike have encouraged this project in countless ways. The people on the front line in Readers Services—Doris Smedes, Virginia Renner, Elsa Sink, Leona Schonfeld—have provided both personal and institutional support. Mary Wright, Fred Perez, and Jim Corwin made access to manuscript material efficient and pleasant. Brita Mack and Susan Naulty helped in the search for a cover design. Sue Hodson and Virginia Rust provided valuable archival information and suggested manuscript leads. I gratefully acknowledge permission from the Henry E. Huntington Library to quote from the manuscript letters in their collections.

I have made some very dear friends at the Library whose hearts and minds have enriched my own. At both lunch and coffee breaks, many readers, too numerous to mention by name, have entertained and enlightened. Because of their stimulating company, writing at the Huntington has never been lonely. In fact, the distractions of lively conversation sometimes proved an unconquerable temptation.

People in a number of fields and settings have nurtured this project: Doris Banks, Janet Brodie, Anne Feraru, Nancy Fitch, Pat Haley, David

Pivar, Judy Raftery, Carole Srole, Sarah Stage, Lucy Steiner, Catherine Turney, Ciji Ware, Paul Zall, and Betsy Truax, with whom I shared endless stories of Victorian romance. I thank them for their help and encouragement.

I am indebted to a group of gifted colleagues in the American Studies Department of California State University, Fullerton. Allan Axelrad, Jesse Battan, Wayne Hobson, John Ibson, Michael Steiner, Pamela Steinle, Earl James Weaver, and Leila Zenderland form a community of teachers and scholars who have supported and sustained me over the long years of researching and writing this book. Their intelligence, humor, integrity, humane values, and their skill and dedication to teaching and the life of the mind have nurtured every aspect of my life. The American Studies Department has given me both colleagues whom I respect and friends whom I love.

Also, special thanks to the American Studies students at California State University, Fullerton, whose faith in my work was so beguiling. I hope I have lived up to their trust.

I also acknowledge the support and good cheer of colleagues and friends in other departments at California State University, Fullerton. I want to thank a cooperative staff, including Norma Morris, Mary Castle, Joyce Griffith, Christine Quach, Helen Sims, and Jo Ann Robinson, who have cheerfully inquired about my progress and aided my effort in various material ways. Terry Spaise typed two early hard-to-read drafts of chapters before I discovered the joys of computers and mastered word processing. The services of Interlibrary Loan on the Fullerton campus have also been an invaluable research tool. Doug Temple never contributed directly to this book, but his sage aphorisms and good humored outlook on life were a source of personal satisfaction.

I am grateful to Heidi Sutton for her bibliographic assistance and her diligent labor in rechecking citations. Deborah Forczek generously helped with the time-consuming task of proofreading galleys. My manuscript also benefitted from the keen eyes of another generous friend, Leila Zenderland, whose passion for precision in language and thought has taught me much. Oxford University Press copyeditors Scott Lenz, Stephanie Sakson-Ford, and Leona Capeless each reviewed the manuscript at a different stage on its way to print. Special thanks to Rachel Toor of Oxford for her support and her confidence in the book.

Two individuals made a special contribution of time and intellectual energy to this work. Wayne Hobson gave the manuscript an exceptionally cogent reading. His intellectually rigorous critique challenged me to be bolder and resulted in a vital revision of the manuscript. He has been as fair, perceptive, and tough-minded a critic as one could ever hope to find.

Acknowledgments

Martin Ridge read the manuscript and made suggestions for improving my final draft. My writing benefitted from his sensitivity to the nuance of ideas and language. He has patiently offered counsel and reassurance and answered a stream of editorial questions with unfailing good humor. However hectic his schedule, he gave generously of assistance and support. His friendship has been a mainstay.

Though he did not live to see this book written, my teacher and friend, Gene Wise, left me a legacy of intellectual passion and creativity.

A note of appreciation to everyone who did *not* ask some variant of the question: "Haven't you finished your book *yet?*" I am especially indebted to my parents, Jim and Lily Lystra; my brother, Torrey Lystra; my sister, Gaylen Mollet; and their spouses, Carol Lystra and Ralph Mollet, who not only never asked that question but also inquired about my progress with tolerant affection.

Final thanks to my family and friends who kept me from faltering.

CONTENTS

*Searching
the Heart*

INTRODUCTION

I F EACH AGE nourishes its private rites and rituals, no era more carefully sequestered the rites of male-female intimacy than the Victorian. In a time before scientific polls and surveys, a formidable barrier of silence surrounded the intimate relationships between middle-class men and women.

In Victorian America, as in our own time, levels of intimacy were fixed by rules. Lovers shared certain experiences only with each other; witnesses would have contaminated their association, changing the confines of conversation and behavior. Total privacy was the foundation of romantic expression, and romantic relationships were guarded by a deliberate wall of secrecy.

Only an improbable set of circumstances would have permitted a nineteenth-century observer to penetrate this secret world. Consequently, many lovers believed that their private intimacies would not survive them. Nevertheless, some lovers left behind a means for twentieth-century readers to enter their hidden world: love letters.

This opportunity to understand lovers by reading their mail is neither as simple nor as straightforward as it sounds. Suspicious of strangers and jealous of their privacy, nineteenth-century letter writers or recipients censored, destroyed, and in general fiercely guarded their love letters.

The depth of the Victorian lovers' commitment to privacy and their stratagems of enforcement have made many scholars wary. Luckily, at the outset of this study, I was more optimistic and naively curious than informed. If love letters were treasured by nineteenth-century Americans, it seemed to me that they may have been specially preserved. Following this hunch, I guessed that surviving love letters, if they existed at all, were included in collections of nineteenth-century family papers—most of which were saved for other purposes. But this only pointed to sources; the best archival collections were not always accurately indexed—especially as styles of historical interest and relevance have changed over time. Thus I began what at times seemed almost a random search for love letters.

My quixotic search proved rewarding in several ways. First I discovered that caches of love letters are available to the patient researcher. In spite of their public reticence and active self-censorship, Victorian love letters survive in large numbers.

Second, I found that these letters, better than any other source, captured the experience of Victorian romantic love. Compared with love letters, traditional sources such as memoirs and other public documents were usually superficial reports on romance, ranging from the deliberately misleading to the self-consciously limited. Though private and solitary recollections, even diaries could be disappointing sources of self-revelation because they lacked the crucial dimension of social interaction.

Sources are, however, only as useful as the questions they provoke. One of my earliest questions was why middle-class love letters were so evocative and revelatory. What made them such an intimate experience even for an historian far removed in time? The answers to these questions required reconsideration of reading and writing letters as cultural acts whose rules and meanings change over time.

The mail of lovers was sacred. Family letters might be shared among a circle of relatives; business letters might be read by a network of associates; letters of inquiry, or recommendation, or introduction might be passed from hand to hand, but in Victorian America, letters exchanged in a serious romantic relationship were carefully guarded.

More than privacy was involved. As a romantic relationship deepened, middle-class correspondents experienced certain letters as actual visits of their lover. Insisting on seclusion, they often read and wrote love letters as if they were in a conversation that might be overheard. When alone, they kissed their love letters, carried them to bed, and even spoke to them. In both the act of reading and writing love letters, Victorians displayed an intimacy that affords the historian a remarkable opportunity to cross once forbidden boundaries without affecting behavior on the other side.

Although today's highly literate reader may not experience the easy flow of conversation, the spontaneity of dialogue, the spark of intimacy in their own correspondence, nineteenth-century lovers felt compelled to make love letters approximate their intimate face-to-face exchanges. By their own reports, their dialogue on paper felt akin to actual conversations. They told each other that love letters reflected the verbal intimacy of being alone together.

This evaluation is the most important measure of the quality of the evidence and the reliability of the conclusions in this book. Middle-class love letters, based upon the testimony of the only reliable witnesses, nineteenth-century men and women, were part of the fabric of their most intimate private interactions. Not intended for any other reader, these letters provide as genuine a record as possible of feelings, behaviors, and judgments as they occurred in romantic relations.

Scores of manuscript collections in the archives of the Henry E. Huntington Library in San Marino, California, provide the evidentiary base of this book. I read the letters of more than a hundred people, some of them coupled, others single; as well as parents, friends, and relatives. These letters, beginning in the 1830s, span a time period that roughly coincides with Queen Victoria's reign. The authors of the letters were a geographically dispersed lot, which included southerners, but perhaps best represented the West, Midwest, and Northeast.

The letter writers can be described loosely as middle- and upper-middle-class nineteenth-century native-born Americans. Most of the women became or were housewives and mothers; most of the men became or were doctors, lawyers, soldiers, teachers, ministers, engineers, businessmen, and farmers. As I use the term, "middle class" refers not only to economic and occupational level but also to cultural characteristics: concepts of privacy, the self, and standards of social behavior, such as the ability to demonstrate expressive skills and control in different social situations. These men and women used such concepts to separate themselves from those they considered inferior, to build social hierarchies, and to measure their own "class" performance.

This is an inductive study. I avoided, as much as possible, the imposition of potentially anachronistic psychological theories, striving instead to develop explanations directly from Victorian experience. Victorian lovers, in their correspondence, disclosed ideas and behaviors for which the historical literature had not prepared me; they regularly engrossed and challenged me. Therefore, their voices dominate this study because they played a large role in reshaping my thinking and forcing me to reassess conventional wisdom about the Victorian world.

Neither scholars nor readers come to any book without preconceptions. The reader may be caught up first in the stereotypes of Victorian culture. The most dominant of these—and in some ways the most beloved—is, of course, the sexually repressed Victorian. Thus the erotic intensity of the middle-class men and women in this study, as well as their passion for sexual expression, may take some readers by surprise. Clearly misunderstood in a contemporary context, the Victorian approbation of purity did not reflect a simplistic anti-sexual stance. Nineteenth-century middle-class Americans actually held an extremely high estimation of, indeed almost reverence for, sexual expression as the ultimate symbol of love and personal sharing.

The apparent contradiction between Victorian public repression and censorship and private sexual expression is perplexing. How can one integrate the public life of sexual condemnation and undeniable sexual repression with the private reality of sexual enthusiasm? How can one explain the relationship between public prudery and candid private correspondence? These questions encouraged me to read another kind of source—medical and moral advice books. With reputations as primers for prudery, these nineteenth-century public guides to sexual activity focused my efforts to relate public and private sexual values and behavior. Though this is a book about private experience and values, intimate reality is always a part of the world of public culture. The system of ideas and behavior commonly referred to as romantic love provided one significant means of integrating public and private worlds.

Romantic love was an intellectual and social force of premier significance in nineteenth-century America. Still it was not an experience readily understood by Victorian lovers. Not only did Victorians jealously secret their love from public scrutiny, but they also believed this love was essentially a mystery, at its essence beyond analysis. Such is the power of human belief that this ideology of love as an incomprehensible force meant that Victorians themselves often found the experience of love opaque. Historians have been equally bedeviled.

Romantic love is the largest area of discovery and significance in this book. It is also an ambiguous term. Many writers on the subject have thought of romantic love as a general term to encompass any historical reality that had some romance in it. This unwise practice has led to a lack of clarity and to ahistorical judgments. Historians must doubt a priori assumptions that romantic love was identical in fifth-century Athens, seventeenth-century England, Victorian America, or even late twentieth-century America. While late twentieth-century middle-class Americans are heirs of Victorian family culture and are in some respects still Victorian, the experience of romantic love must be understood in precise historical context. Unless it can be documented to be the same in different cultures and in different epochs, it is unwise to continue to assume that the phenomena referred to by the same word are necessarily identical.

While recognizing the importance of the companionate marriage ideal and the shift from property and social standing and other prudential factors to love, there have been few close historical examinations of the experience of love in the lives of ordinary Americans. Perhaps popular psychology, allied with the contemporary obsession with getting, keeping, recovering, or rediscovering love, has trivialized the topic. Curiously American historians have examined Victorian sexuality, family life, sex roles, domesticity, courtship, marriage, parenting, and birth control without really making an intensive analysis of romantic love.

Without denying the complexity of forces and the intricate web of experiences that are involved, love appears to be at the core of Victorian family culture. Basic themes of individualism, secularization, masculinity, femininity, and sexuality in the nineteenth-century American middle class constitute a constellation around romantic love. Understanding romantic love allows one better to integrate a surprisingly large range of Victorian experience.

Ideas of love formed the basis of middle-class standards of sexual purity, marital happiness, and emotional fulfillment in Victorian America. Constructs of love guided behavior in the public as well as private sphere, creating judgments of value in relationships and rules of openness or closure in expression. American Victorians cultivated romantic love during a special time of life—courtship—devoted to intensifying this emotional experience. Love, however, was important to many persons long after that period of life had passed. Cupid's arrows affected alignments of power within the family, and were especially influential in masculine and feminine responses to each other and the world. In short, ideas of romantic love suffused the world view of nineteenth-century middle-class Americans and guided their emotional experience and behavior.

In spite of the mystery with which Victorians shrouded romantic love, an old tradition that, to a certain extent, late twentieth-century Americans continue to foster, it is not some strange magnetic force or visitation upon the unsuspecting. The nineteenth-century Victorian experience of love was rooted in the concept of an ideal self. Not fully expressed in public roles, this ideal self was meant to be completely revealed to one person only. Individuals were taught to reserve their truest or best or most worthy expressions for a single beloved.

The feelings and behaviors of love were deeply shaped by the idea that completely unfettered self-revelation should be reserved for intimates and indeed defined intimacy. Thus the essential act of romantic love in nineteenth-century American culture, what gave it meaning and form, was free and open communication of the self to another. Middle-class Victorians encouraged marriages which originated in this view of love.

Romantic love in Victorian America was a complex alignment of idea, behavior, and emotion. Nonetheless the experience of love was anchored in this ideal of an essential self, what we today call a personality, and a closely allied cultural and social divide separating public and private expression. These concepts endured throughout the Victorian period (and persist to some extent today).

Nineteenth-century America underwent many diverse structural changes in economics, law, and social relations between 1830 and 1900, but one dimension crucial to romantic love was a fixture. Though structural conditions of industrialization were elaborated and spread throughout the period, the dependency of middle-class married women on their husbands' breadwinner role remained fairly steady. The continuity of the key ideas of romantic love and the constancy of a central structural factor in marriage resulted in at least a seven-decade period of relative stability in the experience of romantic love.

There were, however, changes that had some influence on romantic love. The first woman's movement introduced more doubt about women's role commitment in men's minds and some sense of potential role expansion to women; the Civil War probably affected southern courtship practices and may have created qualitative divides on other issues; the number of divorces began to increase; certainly other circles within circles may be found. But the basic framework of romantic love remained intact throughout this period.

Though the ideas of romantic love persisted, the experience of love had certain effects that altered, emphasized, and otherwise changed individual lives and through them, the culture itself. One such shift, at least in the inner world of some Americans, was particularly significant. Especially during courtship, it can be shown that romantic love contributed to the displacement of God by the lover as the central symbol of ultimate significance. Traditional religious influences were still strong, but nineteenth-century middle-class Americans were finding, sometimes against their own wishes, that their romantic relationships had become more powerful and meaningful than their institutional religious loyalties. They were making deities of each other in the new theology of romantic love.

Americans used religious language in talking about love. Infusing newer romantic meanings into older Christian symbols and structures, they saw love as rebirth—mysterious, uncontrollable. Older Judeo-Christian thought gave their romance form and substance. For example, Victorians believed that sexual expression should be a sacred act of worship, even more specifically a sacrament of love. Being in love was analogized repeatedly as being reborn. American romanticism and the ethos of romantic love gained power and popularity, ironically, through traditional religious thinking. Secular romantic love, strongly influenced by religious metaphors, contributed to an emotional and intellectual shift in world view toward the individual self.

One theme running through this study, as a result, is the contribution of romantic love to American individualism. Childrearing, the weakening bonds of community, and the works of American artists and intellectuals have all been assessed in terms of the themes of self-direction, self-expression, and autonomy. Though normally excluded from this litany of influence, the romances of ordinary middle-class Americans, and their romantic ideals, immeasurably strengthened American individualism. The goal of American middle-class courtship was to create some special experience within an individual before marriage that was not shared by others. Allowing considerable freedom in courtship, Victorians often succeeded in building unique, emotional bonds between lovers that emphasized their individuality, their distinctiveness, and their separateness.

Romantic love also contributed to male-female understanding and women's power in the Victorian household. Historians have paid serious attention to the power dimensions of nineteenth-century male-female relationships, but they have not analyzed the role of emotional power in private interaction. Having accepted Simone de Beauvoir's analysis of the dangers and costs of romantic love to European and American women, especially their loss of personal identity and social purpose through belief in a one-sided, self-sacrificing version of romantic love, I was surprised to find that women in Victorian America benefited in any fashion from romantic love.[1] The Victorian belief in romantic love demanded that men and women cross gender boundaries by disclosing and sharing what, from the romantic view, was their essence. Although this was a convention of relating that was neither perfectly attainable nor necessarily durable, middle-class American men and women in fact often developed intimate understandings of each other that allowed them to bridge some sex-role divisions while they were in love.

Furthermore, within the circle of romantic love, men and women shared their interior lives with such intensity that many felt as if they had merged part of their inner being. Contrary to the image of misunderstanding and distance between nineteenth-century men and women, both middle-class men and women shared romantic values that encouraged them to seek reciprocal understanding. This effort was not a fictional artifact but a behavioral reality that had important consequences. In an age when middle-class women had limited economic power, romantic love—not sex—gave women some emotional power over men. Men gained a similar advantage, but women's greater economic vulnerability and powerlessness in public realms meant that the emotional power of love was more important to them.

This is clearly illustrated in Victorian courtship. Courtship was a time of intense physical and verbal expression as men and women crossed sex-role boundaries by the bridge of romantic love and came to understand each other's perspectives on the world. Though love imparted vulnerability to each partner in the other, nineteenth-century women were considerably more vulnerable than men in the economic sphere. Both sexes recognized this. Men felt the burden of assuming economic responsibility for women, and women felt especially uneasy over their economic dependency. This resulted in the dominant motif of nineteenth-century American courtship: women setting and men passing tests of love. Not unexpectedly, women tested emotional ties more severely than men. In fact, middle-class Victorian courtship was a series of minor tests usually culminating in a major one, a crisis of doubt precipitated by the woman after the couple had made what often seemed to be a firm commitment, such as an engagement. Victorian men had to pass tests of love before marriage in order to prove their commitment to women whose limited opportunities in the economic realm contributed to their greater doubts and acute need for more intense emotional reassurance.

Love letters reveal dimensions of power, sex, and love in courtship and marriage. But spouses, after several years of cohabitation, generally expected less self-revelation from each other and paid more attention to the details of day-to-day living. They could assume a background of intimate knowledge that did not require verbalization. The most revealing letters between married partners were usually those in more troubled or dissatisfied relationships. These marital exchanges sometimes grieved over lost love. The most poignant of love letters, they disclosed the anger, frustration, and bitterness that accompanied the often futile struggle of one spouse to regain lost intimacy. The volatile emotional standard of romantic love, which offered no internal restorative, was joined to the long-standing ideas of continuous spousal duties and obligations. This union of necessary duty and spontaneous love created inherent conflict in the ideas and emotional standards of middle-class American marriage. Love letters permit insight into marital relationships which, whether they succeeded or failed, remained hidden from the public scrutiny of divorce.

Only by tracing particular couples through courtship and marriage— following their lives through separations, death, birth, war, quarrels, religious crises, everyday domestic details, pleasures of the body, and wounds of the heart—can an observer form an understanding of the patterns of behavior and meaning in intimate lives. While some individuals whose stories are recounted here make only brief cameo appearances, and others are featured occasionally, a number of other people's lives unfold throughout the book. The traditional biographical concern for the details of one life is redirected to a collective concern for the inner lives of many individuals—male and female.

The historical examination of gender and the family cannot reach its fullest potential by drawing only on the evidence of women. Family and gender are, in the deepest of ways, defined by relationships. It is essential to construct narratives that will chronicle both the fluidity and rigidity of emotional interaction between men and women. This will aid in recovering the most accurate picture of private reality in Victorian America.

Consequently, this study is concerned with the way ideas about men and women were translated into actual relationships. Sex roles are not flat one-dimensional entities of conformity or opposition. They are more fluid than any set of prescriptions about them. Nineteenth-century Victorians crossed gender boundaries, violated sex-role expectations, and yet still maintained ideologies of male and female differences. To look only at the end product of male-female interactions—summary statements, after-the-fact recollections and reports, and ideals and prescriptions—can lead one astray. Letters that record the way men and women acted on their values indicate that decision-making was itself a flexible process. Women and men were prone to reinterpret their interactions after the fact to fit traditional standards of male dominance and female submission that they did not always follow in the process itself.

A too-rigid view of separate spheres has led to a sense of male-female emotional segregation and distance in Victorian America that must be modified. No doubt nineteenth-century middle-class women had intense and emotionally fulfilling relationships with other women, but sisterhood did not preclude many of them from seeking and attaining deeply engaging and satisfying romantic relationships with men. The assumption that male-female intimacy was devalued or male-female mutuality debarred in middle-class America because of evidence of female-female intimacy, or because of the divergent conceptions and positions of Victorian men and women, cannot be sustained. Whatever the depth and intensity of Victorian women's relationships with their own sex, they had profoundly intimate emotional relationships with men.

Searching the hearts of women and men in Victorian America, the twentieth-century reader observes the pain and happiness, the self-deceit and truth, the loss and gain, of romantic love. Searching her own heart, one Victorian woman marveled: "Strange things hearts are."[2] Strange and wonderful indeed.

1

The "Pen Is the Tongue of the Absent"

Reading and Writing
Nineteenth-Century Love Letters

🐾 *Love Letters, billets-doux, are among the sweetest things which the whole career of love allows. By letters a lover can say a thousand extravagant things which he would blush to utter in the presence of his fair charmer. He heaps up mountains of epithets and hyperboles, expressing the inexpressible heights, and depths, and lengths, and breadths of his affection. Here he may at his pleasure revel and rave in eloquent nonsense about minutes lengthening into hours, hours into days, days into weeks, weeks into months, months into years, years into interminable ages. He inflates his heart into a balloon, which goes up and down through creation, with motions as light and easy as a thistle-blow drifts through the undulations of the atmosphere. He dissolves into sighs, and spreads himself out, on a sheet of gilt-edged note paper, into transparent thinness. He goes off into impalpable moonshine.* Dictionary of Love[1]

VICTORIAN LOVERS occasionally reveled in eloquent nonsense and felt as unsteady as a balloon in the wind, but the emotions of love were more substantial than moonshine and writing love letters was more serious introspection than light-hearted frivolity. With almost no other means of voicing themselves across even the smallest distance, nineteenth-century lovers bridged the silences with ink and pencil.

One courting couple described reading and writing love letters as "a consolation," "my only comfort," "delightful," and "my pleasure."[2] A young unmarried woman claimed that writing love letters was "a part of my *being.* Were it denied me I declare I dont know what I *should do.*"[3] She exulted in reading the letters of a long-absent lover: "they come winging their way many many miles, oftimes bringing a balm to my aching heart. Sweet are the hours spent in their perusal. How my heart and eyes read and reread the words of love and hope I find therein. Why they settle right down into my inmost soul and sweetly dwell and linger in my thoughts. I thank God for your letters. They are everything to me. . . ."[4] A married man contemplated his impulse to write love letters: "as crying relieves grief, so writing affection."[5]

Middle- to upper-middle-class American correspondents were very self-conscious about love-letter reading and writing. They distinguished love letters from other types of correspondence and commented upon the personal and social meaning of their intimate exchange of letters. One recently married man would carry his young wife's "last letter with me every where and look at it occasionally to see some particular sentence or expression."[6] This behavior, so typical of many correspondents, suggests that love letters were perused more intently and reread more frequently by ordinary Americans than any other form of written expression.

Since literacy was widespread by the early nineteenth century, and the postal service was reasonably efficient, both formal and intimate letter-writing flourished.[7] Love letters were an important part of the crescendo of verbal activity in nineteenth-century American life. "It may be doubted," one Victorian author observed, "whether as many reams of paper have ever been used in writing letters upon all other subjects, as have been consumed upon epistles of love . . . ."[8] Indisputably, love letters were highly prized by Americans who wrote to each other assiduously in matters of the heart. Whether or not there is any truth to the claim that "epistles of love" were numerically superior to other types of correspondence, nineteenth-century American love letters were a significant arena of cultural expression.[9]

Letter writing attracted many self-appointed guardians of cultural virtue. Both the general etiquette manual and the ladies magazine in Victorian America dispensed advice regarding personal correspondence, but a less familar source of counsel, called a letter-writer, was also widely available.[10] These specialized book-length guides provided their readers with sample letters to meet the various demands of business, friendship, and romance. Before the eighteenth century, the letter-writer was intended for the amusement of the sophisticated reader. But one scholar suggests that this genre of advice book evolved into a "utilitarian manual designed for the use of the partly literate in the eighteenth century and it has never been elevated again."[11]

The availability of model love letters in letter-writers and etiquette manuals poses an obvious question about the originality and reliability of love letters as scholarly sources. How much did native-born middle-class correspondents rely upon standardized book copy? The answer is: probably very little. The preface to *A New Letter-Writer, For the Use of Ladies* pointed to the obvious difficulty of attempting to use a letter-writer verbatim: "It is impossible that, even were ten thousand letters collected together, they would furnish epistles suited to *every* exigency, even during a single year of an ordinary person's life."[12] The number of model love letters was very small in general etiquette books, more copious in letter-writers. Nonetheless, sample letters in either type of book were often short—one page or less—and usually directed to a narrow and specific topic. The majority were not general expressions of love, but narrow rejoinders to particular problems of courtship and marriage, such as "A lady in answer to a letter in which her suitor intimates his wish to discontinue acquaintance"; "To a gentleman who had sent an absurdly romantic Letter"; "A negative, on the Grounds of a Pre-engagement"; "Another [negative], on the Grounds of Levity"; "From a Gentleman to a Lady whose friends are opposed to their union"; or "From a Lady to her Lover, whom she suspects of inconstancy."[13]

Interestingly, model letters on love in etiquette books were directed more to the difficulties and anxieties of persons in love than to celebrating the pleasures of the heart. Nineteenth-century advisers devoted more attention to losing than gaining love. Nonetheless, a plethora of positive love-letter models would neither have supplied enough material for a correspondence of any length and duration nor have readily translated to a wide range of individual circumstances.[14]

One reason may be that etiquette books placed the highest premium on sincerity in personal relationships. A study of nineteenth-century conduct manuals indicates that "mutual sincerity was regarded as the substance of the romantic contract."[15] The principle of sincerity was applied to all aspects of middle-class personal life, most especially to love and romance. In terms of explicit etiquette book instruction, self-expression was the highest act of epistolary etiquette in love letters.

Regarding letters of courtship and marriage, one editor of a letter-writer observed that if he had consulted his own inclinations, he would probably have excluded any advice on such matters "where befitting writing can only spring from the deepest recesses of the human heart . . . ."[16] Another advice book guiding ladies to perfect gentility apologized for making any ceremonial suggestions that could be construed as regulating the sentiments of the heart. This adviser urged readers not to use endearing epithets unless they were truthfully meant.[17] Another guidebook, which included model letters, gave some suggestions on how to introduce oneself to a lady, but after the initial meetings, the adviser noted: "There are, and ought to be, no set forms for inviting ladies to walk, to ride, or to go to balls and parties. The most simple, natural and unaffected way is the best way."[18] Letter-writers and etiquette books explicitly admonished familiar correspondents to write letters "not as a trial of skill, or a display of fine words and pompous expressions, but the communications of what we really think, and feel and wish."[19]

Victorian prescriptive literature in America, however, reflected an ambiguity in the treatment of self-expression and self-control. This ambiguity is illustrated by the inclusion of copy-book love letters alongside an emphatic insistence upon natural expression in intimate relationships. Dual and apparently incompatible messages were conveyed when guidebooks made love letters available for copying while they denied any formal or mechanical approach to matters of the heart.

Several letter-writers clearly offered their model love letters for the lower classes, whom they believed lacked the wherewithal to express themselves from the heart with a "natural" grace. One intoned: "But there is a class of people whose wants are likely to be greater and whose interest has been more immediately consulted in the following pages, viz: clerks, servants, sailors, and others, whose education may be very slight, but who may at the same time wish to be able to express themselves clearly and to prefer something like an English diction to the vulgarity which a little pain and care might avoid . . . ."[20] Model letters in this guide included "To a Young Lady from a Young Tradesman," and "A Man-Servant to the object of his affections."[21] Model letters in another included "Young Woman's reply to her Lover's (a Sailor) Letter," "A Servant to her Sweetheart," and "A Sailor to his Sweetheart."[22] A detailed letter-writer on matters of love was subtitled, "A Complete Guide for Ladies and Gentlemen, particularly those who have not enjoyed the advantages of fashionable life."[23] Thus it is possible that the overt affirmation of self-expression was a reflection of the "confidence" in middle-class upbringing in which gentility was supposed to be internalized through proper childrearing and education. The covert messages of self-control may have reflected the guidebook writers' estimate of the lack of breeding of the lower classes who sought admittance to a higher social rank. The implicit suggestion appears to be that only those with proper class background could "afford" to act naturally in the personal, private sphere. For example, Mrs. Eliza Farrar counsels young ladies to be "natural" with men, but only if the girls are "wellbred."[24]

By the early nineteenth century (if not before) copying from a letter book in matters pertaining to personal relationships was considered declassé by the educated classes in America.[25] Writing style and form, even in love letters, might be used to reflect social status. One nineteenth-century fiancée complained that her beau was not writing "proper" love letters. Her suitor's defense was to launch into a satirical attack upon various love-letter styles: the business man, the student, the farmer, and the piano tuner. He parodied the business man's approach as a stodgy, pragmatic calculus of benefits and costs. The student's style was glossed as overly literary and too self-consciously clever: "The student conjugates the verb 'Amo,' quotes poetry, and makes little sonnets of his own, and makes arithmetical calculations on the subject." By contrast, the farmer was portrayed as inarticulate, taciturn, and probably uneducated: "The Farmer takes his pen in hand to inform his Sally Ann of the perturbed state of his vest jacket, pleads his case in a very plaintive way and ends with the never failing 'Rose's red and violets blue Sugar's sweet and so are you.' " [26]

But this correspondent reserved his most dripping sarcasm and scorn for the Piano Tuner who "wears a monstrous moustache has a foreign air and accent" and sends his missives "to a young Boarding School Miss with her brain not being any stronger perhaps than the law requires . . . ." Both class and ethnic undertones reverberate in his parody of the Piano Tuner: "My most unbounded admiration of your . . . most educated and Lady like bearing induces me to embrace this auspicious moment to tell you of the feelings you have awakened in my inmost heart. It is love *love* true and enduring such love as never warmed this yearning heart before." He concluded this mock-serious illustration of the Piano Tuner's letter style by quoting various snippets of highly romantic poetry, such as " 'All thoughts all passions all delights/Whatever stirs this mortal flame/All are now ministers of love/And feed his sacred flame' . . ." or "Thy parting glance which fondly beams/All equal love may see/The tears that from these eyelids stream/Can weep no change in me/By day or night in weal or woe/This heart no longer free/Must bear the love it cannot show/And silent ache for thee." [27]

His specimens were not typical of nineteenth-century middle- to upper-middle-class courtship or marriage letters. More to the point, however, is his attitude of scorn and disrespect toward the Piano Tuner's copy-book rhetoric. Only persons of lesser social and cultural standing, he implied, would resort to such unnatural expression. In a middle-class context, copying a love letter was obviously beneath dignity, if not a little ridiculous, and was probably seen as the last resort of those whose disabilities of self-expression were themselves not only barriers to proper romance but also to joining the ranks of the genteel.

"We would impress upon you, in the beginning," one advice manual urged, "*that to be natural* is the great secret of success in love making. To disguise your character and study affection in courting, is the very error of the moon." [28] The advice to be natural and sincere, however, was not itself a universal standard of behavior, even when strictly applied to the middle classes. Nineteenth-century guides required an increasingly complex series of adjustments in gesture, tone, style and substance for different audiences and contexts. These adjustments of expression in nineteenth-century American dominant culture involved the extreme separation of public and private life.[29]

All expression in Victorian America must be located along a public-private continuum to understand its meaning and value. Sex, romantic love, and all the intimate communication of lovers (single or married) fall on the extreme end of privacy. Thus Victorian advisers of various stripes had no hesitation in asserting: "Make no public exhibition of your endearments."[30] Husbands and wives were warned "above everything, to avoid being personal" in society.[31] Private words were written and read in private places, reflecting the crucial relationship between words and space.[32]

The basic contours of the middle-class experience—work, residence, family, formal and informal groups—were shaped by the gulf separating public and private life worlds, and the expectations of expressive control or release within each sphere. More specifically, the expressive competence required to negotiate successfully the various demands of public versus private spheres was intended to differentiate social groups as well as individuals.[33]

This public-private division was a basic organizing principle of nineteenth-century middle-class culture. It may, however, be the point where twentieth-century culture is most at odds with the nineteenth century, and, consequently, a source of the deepest misunderstanding of Victorian culture. For the twentieth century, where the division has shrunk, too great a divergence between public and private persona and behavior equates with hypocrisy and deception. For the nineteenth century, by contrast, public reticence was a sign of good breeding and middle-class standing.

Since the public-private dichotomy was the very foundation of genteel Victorian culture in America, it was the compass that oriented individuals in courtship, marriage, gender role, and family as well as on the street and in the counting house. For the distinction between what could and could not be said and done in public defined the very essence of Victorian privacy. Privacy as a concept and a reality depended upon public prohibition and restricted expression. Without public prohibitions, there would have been no restrictive boundaries to give exclusive meaning to the private realm. Forbidden expression in Victorian public life, including feelings, vocalizations, and bodily gestures, attained a special and privileged meaning in private precisely because of the public prohibitions. This was also true of the nineteenth-century concept of romantic love.

The special aura of private communication, the heightened meaning of love itself resided in the contrast of public and private styles enforced by middle-class etiquette. Privacy was essential to nineteenth-century middle-class romantic love because the meaning of love was so deeply rooted in acts of protected and exclusive self-revelation.[34] But, for self-revelation to assume a special meaning, one must withhold responses from the world. And to hold back from the world in certain controlled ways was precisely how Victorians measured middle-class standing.

Middle-class Americans believed that intimate communication between individuals disclosed their "true" self and was an act of good breeding when conducted by the proper participants in private space. If romantic communication between men and women should be the ultimate in self-revelation, as Victorians agreed, then the act of writing a love letter either affirmed one's power with words or demonstrated one's ineptness with the language. This had class meaning for verbal skill indicated control over the self, and self-control was *the* basic indicator of middle-class standing.

Recognizing that the fullest and least inhibited expression of the self was dangerous and impractical in an increasingly anonymous society, Victorians designated the street and marketplace as regions of expressive control; the parlor as a middle landscape of self-restraint; and the inner sanctum of the home as the locus of freedom and the open heart.[35] Nineteenth-century letter types corresponded to this social geography: business letters were formal and unemotional, reflecting an expectation of the full exercise of expressive control in the marketplace. Letters to more distant relatives, acquaintances, colleagues, and social friends formed a middle landscape in letter writing; and letters to close relatives and friends shaded over into the ultimate unmasking of love letters. For example, one letter-writing manual for young men warned its readers to observe the rules of formal composition in correspondence with strangers; but in the intimacy of friendship it acknowledged that letters could be "unconstrained by forms."[36] Thus, advice books could urge expressive control and a tightly reined role performance in one social space, and the opposite in another. This differential tightening or relaxing of expressive control was one of the more crucial boundaries of Victorian culture and a guiding behavioral principle in all aspects of middle-class life.

🐦 Intimate correspondence may be studied from the perspective of social conventions, didactic forms, and rules of etiquette. But it is also important to understand how the act of reading and writing a love letter was actually practiced. Readers who have an image of nineteenth-century courtship and marriage as formalistic and distant may be surprised by the free-flowing associations and verbal intimacy of this type of correspondence. Nineteenth-century American love letters exchanged by middle-class women and men were often chatty and colloquial and ranged over a wide spectrum of topics from business to sex. Correspondents employed a conversational language, tone, and rhythm as well as more literary tropes, metaphors, and romantic figures of speech.

Though styles of love-letter writing varied, certain patterns of form and language usage were quite predictable. Most love letters began with a date. This practice of initial dating was an orienting device which allowed the receiver to locate a letter in terms of historical sequence and individual biography. The significance of time—even in an epistle of love— was underlined by the simple fact that most often the first act of writing a love letter was to record the month, day, and year. Following the date was the opening salutation. It added an important dimension of personalization and individuation to letters. For example, Lincoln Clark demonstrated the significance of the salutation when he inadvertently mailed a love letter without one: "I began my last letter with no title—I did not forget it at the beginning but I forgot it at the close—I wrote here in the office and did not like to expose too much what is delicate and sacred—I did not think of it again until after my letter went."[37]

In love letters the content of the opening form of address often indicated the level of intimacy between correspondents. Pet names were the most unambiguous emblem of a privileged relationship. After marriage, John Marquis addressed his wife as "Dear Pet Baby Wife," "My Darling Precious Wife," "My Darling Chickey," "My Darling Little Wifey," "Dear Dovey," and "My Dear Darling Chick."[38] Early in his courtship of Neeta Haile, John often used the all-purpose salutation: "Dear Neeta."[39] The evolution of emotion symbolized in these contrasting endearments is striking. In another example, Nathaniel Wheeler began his correspondence to Clara Bradley by addressing her as "Dear Friend." As their relationship progressed, his salutations changed to "My Dear Little 'Dot,'" "Dear Lonely Heart," "My Little Darling," " 'Hearts Ease.' "[40]

Closing forms in love letters also exhibited a similar evolutionary trend in that they metamorphized as the couple's relationship deepened and intimacy grew between them. For example, James Hague closed his love letters in an early stage of courting Mary Ward Foote with "Very truly yours," "I am reliably yours," "Yours truly and affectionately."[41] As the relationship developed more emotional intensity, his parting lines included: "A thousand kisses from Yours Forever," and "Darling, darling, *darling,* Oh! how I love you!! I want to *maul* you. Good bye, sweetie, Yours Forever."[42]

The sender and receiver were bound together by both the opening and closing conventions of love letters. The closing address, as well as the opening salutation, were often a highly ritualized display of the level of relational intimacy and commitment. Harriet Russell, obviously in love with Charles Strong, closed her letters to him, "Your 'No-No Hattie,'" "Your Pussy Hattie," "Devotedly Your Own Hattie. No bodys else."[43] Charles Strong concluded his letters to Harriet with such symbolically freighted good-byes as, "Your devoted," "From your own," and "Keep Your heart Warm for This Hombre."[44] Pet names and possessives—marking the beginnings and endings of a letter—symbolized an exclusive circle of sender and receiver, affirming a couple's emotional intimacy and mutual identification.

Though the language of emotions was sometimes conventional and the imagery hackneyed, within the body of the nineteenth-century love letter both women and men expressed elaborately detailed emotional states. Correspondents of both sexes energetically articulated their emotional ties with each other. They expressed their feelings through a rich variety of metaphors, employed a vivid array of adverbs and adjectives, and emphasized the subjective "I." Furthermore, they frequently established a context, explained their immediate surroundings—where they wrote and under what particular circumstances—and often elucidated events, settings, and non-romantic feelings in their everyday lives. Within the male-to-female and female-to-male exchanges of the love letter, few noticeable language differences divided the sexes.

This is particularly significant for the understanding of nineteenth-century middle-class gender roles. The range, depth, and intensity of masculine emotional expression in love letters challenges any unqualified generalization that nineteenth-century men were less emotional or less forthcoming about their feelings than nineteenth-century women. These men suggest that nineteenth-century masculine sex roles have been misunderstood.[45] In private love letters, middle- to upper-middle-class men fairly exploded with feeling, manifesting as much emotional intensity and range as nineteenth-century women. In the extreme privacy of romantic relationships men's emotional expression flourished. Contrary to the stereotype of the emotionally constricted Victorian male, the evidence indicates that middle- to upper-middle-class masculine role performance did not require men to be emotionally controlled and constricted at all times. In the protected romantic sphere, men led richly emotional lives.[46]

The cultural constraints that were yoked to the nineteenth-century masculine role have only dimly been appreciated. The male gender role reflected rules which controlled men's emotional expression in different social settings. The relative degree of privacy in an interaction gave a man greater or lesser freedom to be "himself." While everyone had constraints in public, women's role was essentially privatized. This reflected the fact that for middle-class women the home sphere was a place of work and a retreat from the world.

The nineteenth-century middle-class male role, however, was much more bifurcated between public and private spheres. The male role, in fact, demanded constant control of self in the public arena, which might include a strategy of emotional defense, and emotional expression and openness in the most private recesses of the private world: romantic relationships with women. This basic division between public and private life helped to mold a segmentation of self for both sexes, but most especially for men. Thus T.S. Arthur insisted that only withdrawal from the public eye allowed the revelation of a man's "true self."[47] In public forums, men's emotional expression was expected to be more constricted than women's. By contrast, in private romantic relationships, at least middle- to upper-middle-class men were allowed a range of expression that paralleled, if it did not precisely duplicate, women's. The evidence of love letters suggests that both sexes generally accepted the private range, depth, and intensity of men's feelings as normal.

A high value was placed on sincere, open, heart-felt expression by both etiquette books and actual nineteenth-century correspondents. In fact, letters were a closely watched gauge of the level of intimacy, trust, and love in a nineteenth-century middle-class courting relationship. Lovers constantly urged each other to "write freely" and love was often seen as both the cause and the effect of unfettered communication.[48]

These love letters were not formal, stilted exchanges full of self-conscious politesse. They were not even friendly parlor talk to those who wrote them. Correspondents reported to each other that reading or writing a love letter evoked powerful feelings of communion. The experience, they confessed, was very much like having an intimate face-to-face conversation. Eldred Simkins observed at the close of a long letter, "and it seems as if we were talking together and that you were indeed sitting by me . . . for as you will readily perceive I 'write without thinking.' "[49] After reading one of his girlfriend's letters, Nathaniel Wheeler confessed to her, "I had looked forward to Saturday with a half foreboding, for I thought even if I had a letter from you, it would only awaken associations and fill me with discontent and vain longing. But your letter did nothing of the kind. True, it was so natural that I felt as if you were not far away all the while . . . ."[50] "[I am] more *myself*," he reported, "as I sit here talking to you with the pen. . . ."[51] Emily Lovell referred to an interruption in writing to her husband as a distraction "during one of my greatest pleasures, that of talking to thee."[52] Eliza Trescot, Eldred Simkins's girl-friend, commented that reading his letters is "like conversation with my Eldred. Substitutes and I will not say poor ones either for nice little talks for which I trust they will be exchanged at some time . . . ."[53]

Correspondents repeatedly asserted that both reading and writing a love letter was an intimate experience. Nathaniel Wheeler underscored his feelings of intimacy while he was writing: "I'm just talking as I would if you and I were cuddled up on the dear old lounge in the corner . . . ."[54] Similar feelings were evoked by reading a love letter. As one man ex-claimed, "Oh! how I enjoyed it. It was a perfect luxury to read and reread it. I could almost fancy that I heard your sweet voice and could look up and see that 'quiet eye' gazing on me. Why I enjoyed it so very much I really don't know except that it all somehow seemed so like what you used to say when I was near you."[55] Augusta Hallock expressed practi-cally identical sentiments: "Reading your letters James seems like having a long conversation with you . . . ."[56] Referring to her own writing, she observed, "Sometimes I think—what *makes* me write such lots of silly things! and again, I think if you were *here* I would be very likely to *talk* them, and I have *presumed* that you would like to have me write as I would talk."[57] Reflecting the close identification of love letters with actual conversations, Eldred Simkins mused, "Somehow your letters seem to represent yourself . . . and when I look back upon the past your corre-spondence stands out in bold relief as my brightest happiness and seem to have been actual conversations with you."[58]

The representative power and symbolic transference between the love letter and the self of the beloved was even more graphically illustrated by Eldred nearly eight months later when he confessed to a behavior which was not unique: "Whenever I get one of your letters I always look to the end to see how long Lize has talked to me and then lock my door and have you all to myself. I cannot bear to read your letters near anybody but will keep it in my pocket until I am by myself. In fact I treat your letters as I would yourself. Have you all to myself."[59] The impulse to read a love letter behind closed doors underlined the tangible presence that the letter assumed for those who were bound by romantic love. Love letters for Eldred and others became so much like an intimate conversation that the receiver would isolate himself to read one. Clara Baker was "so eager . . . to know what they [his letters] bring me I can hardly wait to get to the quiet of my own room to read them, and I cannot bear to open them in the crowd."[60]

The symbolic interchangeability between the letter writer and the let-ter steadily increased, particularly as a couple grew more romantically involved. A mature woman, Clara Baker described her letters from Robert as actual visits, and acted accordingly. "No girl with her first love-letter was so eager as I am for my sweet hearts. The mail which brings them does not arrive until 8 p.m. so after dinner I wait down for the mail but when it has been distributed they see me no more—for as now, I spend the rest of the evening with my Lover—my Robert—my darling. When this little visit is over I shall ring for the bell-boy to post it that it [her letter to him] may go on its long trip at 6 a.m."[61] Samuel Francis Smith, carrying the transference even further, wrote before an imminent visit to his girlfriend Mary that he would probably *be* her next letter " 'written not with ink'—but coming happy and joyous, that we may rejoice together."[62]

As romantic love grew more intense, the nineteenth-century couple seemed more and more likely to anthropomorphize the letters of the loved one into the person of the absent lover. Reflecting the virtual merging of letter and the self of the beloved, correspondents reported a series of un-expected, emotionally charged responses. Robert Burdette confessed, "Dear Violet—your letters—they have so much of your own personality in them—they are so much *YOU*—that I sometimes find myself talking to them."[63] Albert Janin reported, when he received his girlfriend's letter, "I kissed your letter over and over again, regardless of the small-pox epidemic at New York, and gave myself up to a carnival of bliss before breaking the envelope."[64] On another occasion, Janin's response to an unopened letter was equally dramatic: "I entered my room and saw your letter on the table. I felt like falling upon my knees before it. The fact is, if I go on loving you as I do I shall run the risk of exploding."[65]

Nathaniel Hawthorne, receiving a letter from Sophia Peabody, whom he often called his "Dove," exclaimed, "So I pressed the Dove to my lips (turning my head away, so that nobody saw me) and then broke the seal."[66] Hawthorne candidly confessed: "(you will smile) I never read them [Sophia's letters] without first washing my hands!"[67] The symbolic equa-tion of the absent loved one with the letter was so physically intense that he must not only purify himself before touching her anthropomorphized letter but also seek protected social space. Retiring to the privacy of his room, he reported, "[I] locked my door, and threw myself on the bed, with your letter in my hand. I read it over slowly and peacefully, and then folding it up, I rested my heart upon it, and fell fast asleep."[68] On another occasion, he folded her letter next to his heart and felt "as if thy head were leaning against my breast."[69] Sophia's letters were "too sacred to be enjoyed, save in privacy."[70]

In the midst of preparations for a potential Union attack, the Confed-erate officer Eldred Simkins burned most of his personal correspondence, but he saved his lover's letters because they had become all too real a substitute for her self. As he explained, "I was afraid if they run us off here they might get my letters so I ran to the fire and threw them all in—not all either. I hadn't the heart to burn yours, but put them very poetically in my breast pocket till all the row was over. It looked like sending you away from me to part with them, so I kept them with me."[71]

Perhaps the most unusual and revealing anthropomorphic act was captured in Lincoln Clark's request to his wife Julia: "Pray write often as you can, and when you write put on your open black dress and beautiful white under dress."[72] Again, implicit in this request was the representational power of love letters to invoke the person of the beloved. James Bell observed quite simply and directly about the letters from Augusta Hallock: "They are mementos of 'your own dear selfe,' that I love, read and think of a great deal. They seem just like Gusta so I treasure them, way down in my heart, as a part of Gusta."[73]

James Hague also graphically personified his wife in letters: "I got your letter I just want to take you in my arms and put my face against yours and give you a sweet old fashioned hug. You see it was nice to get a *letter* after having so many postal cards. Postal cards are all open and before folks; and to get a letter, sealed and private, is something else. It was like going upstairs after being in company all day with other people, and finding ourselves alone and taking each other in arms and hugging and kissing to our hearts' content."[74]

Reading or writing a love letter was perceived as sequestered or bedroom communication by middle-class American correspondents. The emphasis on privacy was crucial in terms of the middle-class concern with loosening or tightening expressive control in different social space. The content of expression and behavior in Victorian culture was rigorously controlled according to its "place" or social location. Love letters were expected to be a "place" where communication was virtually unfettered by polite forms. In terms of social norms, love letters were the setting where one was permitted to be least constrained by and emotionally free of utilitarian goals and social strivings.

Though the enhanced technology of the late twentieth century requires the exercise of historical imagination to appreciate the power and impact of sending and receiving nineteenth-century love letters, it is nonetheless true that whatever their impact, letters did not compensate for the absence of the loved one. Correspondents also expressed frustrations with writing and reading love letters, particularly the impossibility of physical intimacy. While many seemed fairly content with the verbal intimacy of letters, many also wistfully yearned for the missing opportunity for nonverbal expression. For example, Eliza Trescot confessed: "I long to be with you that I may express my thoughts and feelings even without words;

that I may lay my head upon your ever willing shoulder and look into your eyes. . . . My heart craves the sound of your voice tonight and will not be satisfied."[75] After expressing intense satisfaction with the letter he had just read, one young man added, "Could I then have taken a delicious kiss from your lips I should have been almost too happy."[76]

Though never an adequate substitute, correspondents experienced letter writing as symbolically akin to personal presence. Paradoxically, as their romantic bonds tightened, their satisfactions with love letters increased. This is evident not only in letter content but also in how nineteenth-century couples behaved vis-à-vis the letter-writing process. Romantic love greatly accelerated the usual pace of their letter writing. The sheer number of letters and the rate of exchange gauged, in the minds of participants and onlookers alike, the romantic nature of a relationship and a couple's marital prospects. As one male correspondent observed: "They must think that 'your Johnny' is *dreadfully in love,* to write so often. But I suppose they consider you as badly off and that makes it even."[77]

Though the rate of exchange varied from couple to couple, as a general rule middle-class courting pairs as well as married couples who were separated wrote no less than once a week. However, Victorian lovers could be much more prolific. Robert Burdette, for example, in one passionate period of his courtship composed three letters a day to Clara Baker. The length of individual love letters also varied. Some were short, many averaged several pages, but as one record twenty-seven-page love letter attested, nineteenth-century correspondents might find considerably more to say.[78] Charles Strong expressed the longings of many lovers when he wrote: "While you are away from me I wish your every thought and *wish* could come to me in black and white every day."[79]

The practical reason for the length of letters and their accelerating rate of exchange in nineteenth-century romantic relationships is obvious: letters were the only vehicle for expression between people in love who were separated by any distance. The other reason is much more elusive but nonetheless equally important: the act of writing became progressively more satisfying to many correspondents as their romantic involvement deepened (sometimes much to their own surprise). Eldred Simkins, for example, observed, "I feel like doing nothing else—that is where you are concerned. . . ."[80] Furthermore, he contrasted his letter-writing experience before he fell in love—"such a bore"—with what it became as his love deepened: "I don't believe I ever took such delight in writing back before—."[81]

His fiancée, Eliza Trescot, expressed some initial discomfort and uneasiness in the letter-writing process: "I always feel as if my pen were a restraint upon me. I mean to say all that I want to but when I finish a letter I always feel that it is totally inadequate to express all that I wish to say."[82] Six months later she began a letter, "Nothing satisfies me so I have recourse to my cure-all, a letter to you. . . . My pride would not allow me to express the slightest shade of discontent to anyone else, but then I always feel like soothing my feelings by writing to the one who is dearest to me."[83] In another letter she confirmed that she had lost all trace of her former feelings of inadequacy and inhibition when taking up the pen to address him: "In writing I can express myself more easily than when I am with you, because I am not too happy to think."[84] And she added in a postscript: "I put up my letter very reluctantly last night. I felt as if I could have written all night. It is such a pleasure and compensation for your absence." Augusta Hallock also found that her writing had changed as her romantic relationship ripened: "But some how James when I write to you I feel just as tho' I was *talking* to you and were you here, we would not hesitate about speaking of this. I *never* felt such a freedom in writing to you as I do this winter."[85]

The progressive sense of ease and satisfaction in writing love letters can be accounted for in three ways. The first is simply that practice increased fluency in writing as in many other endeavors. These middle- to upper-middle-class men and women achieved an impressive level of expression in their private correspondence.

The second is the knowledge that reading one's letter gave pleasure to the beloved: the desire to please a lover was one important characteristic of romantic love. Thus, the thought that writing a letter gave pleasure to the person you wanted most to gratify was both a powerful emotional motivation and a reinforcement of romantic feelings. Samuel Francis Smith explained, "This letter, my love, is a pure gratuity, written just because I love you and love to make you happy."[86] John Marquis expressed similar motives: "You cannot think how much pleasure it affords me to get away by myself and write to you. It makes me feel like you were near me to sit down and call you pet names and think that you will soon read all the words I am writing here and that it will make your heart beat faster to read, that you will be glad. It is a proud feeling to think I can give pleasure and make anyone happy. Especially when that one is all the world to me."[87]

The third factor in the progressive facility and emotional fulfillment of love-letter writing is more complex. Writing love letters became easier as the self was brought into clearer focus through the communication process required by the conventions of romantic love. Within the nineteenth-century romantic relationship, both men and women were supposed to be able to abandon restraints and be their "natural" or "true" selves. Romantic love was the relational state in which the congruence of

outward expression and inward consciousness was expected and most encouraged. If to reveal one's "true self" was a central convention of
nineteenth-century romantic love, it was also a process of strengthening,
and sometimes even creating, a role best called the "romantic self." Though
a social and cultural construct, the self (now often referred to as personality) was and still is seen in the romantic tradition as beneath and behind
all social and public rituals of interaction. But what has come to be called
self-revelation or self-discovery can also be conceived (outside the romantic tradition) as a struggle to create a certain inner reality that is not innate
and therefore not inevitable.[88]

Personal expression in various forms was perhaps the most important
means by which this inner reality was developed and maintained. Given
the power of language to shape reality, little wonder that the act of in-
scribing one's life to another—both outward events and inner feelings—
should have cultural consequences in itself. The common acts of love-
letter writing and reading were processes of defining experience that fostered the development of a romantic consciousness of self in nineteenth-
century America. This included individual differentiation from the group,
empathy across gender divisions, and heightened emotional awareness.
Verbal communication aimed at self-expression and sharing "true" feelings was the central convention of Victorian romantic love. Thus, not only
the content of these love letters, but also the intimate acts of self-exami-
nation and self-disclosure they embody contributed to the individual
awareness that formed the foundation of American Victorian culture.

🎄 Love letters were fiercely guarded against the prying eyes of strangers,
but they were not discarded. Hidden in secret corners, tucked in trunks,
drawers, or even special boxes, perhaps eventually squirreled away in at
tics, these tender relics of the past survived housecleanings and moves to
distant cities. True, some people destroyed their love letters before they
died; probably others forgot them; but perhaps some left them untouched,
due not only to procrastination or senility, but because they could not
bring themselves to destroy what these letters represented about themselves.

Love letters "place us in the midst of past generations as if we lived
among them," a nineteenth-century magazine recognized in 1832. "They
lift the curtain which separates the illusive from the true . . . unlock the
most secret repositories, and give us a key to the most hidden thoughts."[89]

2

Falling in Love

Individualism and the Romantic Self

T HE RECREATION OF past emotion requires special sensitivity to the changing style and content of emotional expression.[1] Historical studies of the emotional expression known as romantic love, for example, have been plagued with difficulties of definition and context.[2] Romantic love has been identified in male-female relationships at almost every historical time and place. But as definitions of the concept expand or contract, so too does the measure of its existence in the past.[3] Perhaps what has been called by one name—romantic love—was actually a variety of emotional states within different historical situations. Recognizing that two people in any culture can fall in love under many circumstances, the widespread experience of romantic love cannot adequately be explained by luck, fate, or chemistry. The unique individual may have a unique experience, but collective patterns of behavior are constructed through cultural and social guidelines, limits, rules and the encouragement of the whole weight of the social fabric.

By 1830, romantic love was fast becoming the necessary condition for marriage in the American middle class. While the ratio of those who married for love versus those who did not can never be determined with certainty, evidence strongly suggests that American middle-class youth were selecting their own partners by at least 1800, with little interference from parents, and that "the heart" played an increasingly larger role in mating as the century progressed.[4] More than likely, therefore, a large number of middle-class native-born Americans experienced some degree of romantic love.

Falling in Love

✵ The question of what it meant to be in romantic love in nineteenth-century America is an intriguing one. Although couples saw love as an uncontrollable and baffling force, the evidence of their love experiences indicates that nineteenth-century lovers were involved in a *process*—initiated for a multitude of reasons—which led them to an identification of selves through an intensive sharing of their interior lives.[5] The beginnings of love are difficult to document for particular couples, but the pattern of the group suggests that only the decision to accept the possibility of romantic love was sudden. The actual identification with another person was usually a gradual process, influenced by an individual's social, economic, and psychological context. Moreover, this identification of selves was not an irreversible threshold. Some couples managed to maintain a high level of emotional involvement through years of marriage; in other cases romantic love eroded and sometimes disappeared.

Just how much opportunity for romantic love existed and how often it was expressed in America before the nineteenth century is a debatable question that depends as much upon definitions as evidence. Nevertheless, certain aspects of early American culture fostered a way of thinking that contributed to the widespread middle-class adoption of a particular construct of romantic love in the nineteenth century.

Nonconforming Protestants in colonial America were conditioned to emphasize self-disclosure and self-knowledge in the religious realm, especially to express their individual feelings in relationship to God. Personal religious expression was always seen as the sine qua non of Protestant religious observation. Introspection in order to discover and uncover one's true feelings was an integral part of one of the highest Protestant values: the direct confrontation of an individual soul with God.[6]

New England Protestants, especially Puritans, exhibited respect and enthusiasm for private feelings, and for the intimacies of family life.[7] Puritans, Quakers, and later evangelical Protestants encouraged private probing of the inner self and emphasized individual as much as communal responsibility. All this fostered self-consciousness in at least certain areas of everyday life and a tendency to be concerned with self-worth and self-justification.[8] This tendency confronted and eventually overcame countervailing forces, including communal structures of custom and social networks of ascribed status and position that held in check the private forms of experience in early American life.[9] These "checks" on personal expression and familial emotion, however, may have been stronger in the Anglican South than in the North.[10]

By the mid-eighteenth century the revivalism of the First Great Awakening catapulted emotional and personal feeling into the forefront of individual experience in both regions.[11] In general, the evangelical movement in eighteenth-century America, allied with ideas of equality, emphasized individual dignity and challenged in thought if not always in fact a hierarchical and role-bound sense of human order. In addition the ideas of the romantic movement began appearing in eighteenth-century periodicals and in popular novels.[12] These as well as other colonial social and cultural forces encouraged the development of romantic love.[13]

One idea of the romantic movement was especially crucial to the wide-scale enactment of romantic love in nineteenth-century middle-class relationships. Referred to by one scholar as the "anti-role," this was a concept of "a role that was different from all other roles in that it could *not* be integrated in the social structure of interlocking roles."[14] Though variously expressed, nineteenth-century lovers utilized an "anti-role" concept of the self to stress that the truest portion of their personal identity was concealed by social conventions. This sense of a hidden, but purer individual essence is the basis of the "romantic self."

The romantic self was constructed in nineteenth-century American culture upon the distinction between social roles and an inner identity that transcended all roles the individual filled, yet could be partially expressed in each. Though this romantic sense of personal identity was an historical creation, within its circle of belief, the self was regarded as something of an essence, antithetical to custom and society. Ironically, the romantic self, which was supposedly beyond social roles and above cultural conditioning, was an idea the romantic movement had intensely promoted.[15]

Romantic love flourished in nineteenth-century America in response to many factors, not least among them the growing popularity of this romantic concept of self. This concept of the self struck a resonant chord in the individualistic themes of early American Protestant teaching as romantic functions and forms sometimes replaced and sometimes were added to traditional Christian structures of meaning.[16]

The romantic concept of self was not just an idea but an experience as well. The distinction between the two should not be lost. One of the most important contributions of romantic love in nineteenth-century American life was to encourage the individual to experience and express a romantic self, to demonstrate significant subjective qualities not contained in the social conventions of his everyday roles. Thus the diffusion of romantic love must be linked to the wide-scale experience of a romantic self.

❧ Individualism is an important but often imprecise term used to describe and analyze American cultural development. For example, the breakdown of older, hierarchical, more communally oriented social orders is often taken as an adequate explanation of the creation of individualism. Yet, what is most often explained is the disintegration of the old order itself. Individualism is not only an absence but a presence. How was individualism created? What cultural templates helped to produce the experience it describes in individual lives? These questions have received less consideration.[17] Obviously, for "individualism" to be incorporated as part of a society, large numbers of historical actors must have some definable experience produced by specific cultural and social mechanisms. Romantic love is one of the specific mechanisms that gave form to the "individualism" of nineteenth-century men and women.

The experience of romantic love among the middle class in nineteenth-century American life was closely intertwined with a cultural commitment to individual differentiation from the group. Though romantic love was only one segment of a complex cultural and social matrix that nourished this individuality, evidence of nineteenth-century middle-class courtship clearly shows that romantic love was a powerful factor in the formation of identity distinct from social roles in young adulthood. While American middle-class childrearing and other cultural institutions such as the church were no doubt basic to the encouragement of individual self-expression, romantic love was a ready-made incubator of the romantic self in nineteenth-century American life. In America, romantic love was an active agent, not only a passive index, of the crucial social change that brought modern—meaning romantic—selfhood to a large group of men and women.

While in the throes of romantic love couples devoted obsessive attention to the identification of inner states. One of the dominant activities of nineteenth-century courtship was introspection. In courtship unchanges self-revelation and disclosure held the highest priority. Individuals in the formative stages of romantic love focused intensely on their own interior lives and those of their lovers.

Appropriately, disclosing and explaining the self formed the foundation of nineteenth-century American romantic love. Lovers, to strengthen their romantic feelings, repeatedly urged each other to reveal their "real" self and "true" feelings. "My darling," Albert Janin pleaded, "if you would not grieve and distress me, do not treat me with reserve and want of confidence—I mean, do not employ toward me the caution that you use in correspondence with others. . . . do write me your real sentiments, whether favorable to me or otherwise."[18] Vying for her affection, Janin appraised their level of intimacy through the extent of her uncensored personal revelations. He also emphasized the exclusivity of his own self-disclosure: "I write freely and frankly to you—you are the only person in the world with whom I am open and confidential—and I do it because I have always thought you desired it."[19]

In the nineteenth-century vocabulary of emotions, love was bestowed through access to the romantic self. "I am actually suffering for want of communion with you," Janin wrote. "One of the greatest pleasures of my life has been the free outpouring of your thought and feeling with which you have honored me . . . ."[20] During a crisis in their relationship, he implored: "Oh my beloved darling, do not refuse me your confidence."[21] To him, her love and expressive freedom were virtually identical.

In the vast majority of courtships, sharing the self was the distinctive feature of romantic love. As one young woman explained: "I want a *real live flesh* heart. One that is broad enough to take in all creation providing that creation is my *whole self*. All my wants, all my desires, satisfying all its yearnings . . . ."[22]

By the 1830s, the idea of the romantic self was already accepted by middle-class Americans. This was reflected in their insistence upon the central convention of romantic love: only when an individual un-masked—was somehow known beyond social conventions and roles—did a lover become a "true" self. "We don't fully understand each other as we should," James Bell explained to Gusta Hallock. "I've noticed one thing in people (perhaps its so with me). Its to often the case that they labor to keep a mask over their true carrachter. But I hold that if a man truly loves a woman, he will never try to deceive her in any way."[23]

To deceive, dissimulate, or hide behind words and thoughts that were insincere was taken as a sign of lovelessness and emotional distance. Charles Lummis, for example, accused his wife of wearing " 'intellectual masks!' "; and she responded to his accusation: "You insult me deeper than blood can wash out . . . ."[24] To most couples, revealing one's "true self" was both the ultimate ideal and the measure of romantic love.

Nathaniel Hawthorne claimed that Sophia's letters to him "introduce me deeper and deeper into your being, yet there is no sense of surprise at what I see, and feel, and know, therein. I am familiar with your inner heart, as with my home; but yet there is a sense of revelation—or perhaps a recovered intimacy with a dearest friend long hidden from me."[25] This urge to know the other person and be known beyond social conventions and polite sociability became itself a variety of social convention, but was usually considered "natural" by a middle-class culture already steeped in a romantic view of personal relations.

One of the most powerful and consequential assumptions of the ro-mantic idea of self was the belief that conventions, rules, and etiquette obstructed true communion in intimate relations. In the romantic framework of these Victorian men and women, free-form subjectivity was the only route to true understanding of others. Self-revelation became the primary symbol of intimacy, closeness, and sometimes even truth in nineteenth-century middle-class American culture. "Ceremony I think, should *never* exist among friends nor indeed can it where hearts are closely united," Mary Smith pressured Samuel in the early stages of their court-ship.[26] Ceremony and all the polite forms of society were seen as antithetical to intimate expression according to the conventions of the romantic self. Thus Samuel confessed to Mary that more than his mother or sister, "You know my weakness—my effeminancy—my folly."[27] Both Mary and Samuel interpreted self-disclosure as symbolic proof of the depth and sincerity of their love.

Another incident in their relationship demonstrates the significance of self-disclosure in nineteenth-century courtship. Mary reacted sharply when Samuel erected a barrier to their communication. Knowing that Mary's mother was perturbed by his interest in missionary work, Samuel wrote Mary of his decision to avoid the subject of missions in future correspondence.[28] Mary criticized what she considered his arbitrary and insensitive act of censorship. After venting her frustrations, she asked him to "promise me that you will write just what your feelings dictate—*Do* talk of missions, my love—tell me in your next all you was going to write in your last—Regard me once more as *one* with yourself—*Assure* me that you do—and we will *once more* be happy."[29] For Mary, being "one" with Samuel demanded complete self-disclosure; anything less was a harbinger of disaster for their relationship. Nineteenth-century middle-class couples, under the influence of romantic love, were obsessed with eliminating any barriers of communication between themselves.

Self-revelation in American middle-class culture had already become a far-reaching measure of intimacy and truth as well as the essence of the conventions that defined the romantic self. "This is but a scrawling letter," Nathaniel Wheeler wrote Clara Bradley, "but it is 'me' and what is the use of having friends if one cannot be careless with them, a doctrine you may think I've practiced before. So be good and write me a letter full of *yourself*."[30] Later, Nathaniel even quoted an unflattering journal entry in which he preserved reactions to his first courtship call on Clara: " 'I told her things of my inner experience that but one other ear ever heard. Strange that I should be honest, and to *her*. But really she showed more appreciation than I had thought possible, and more brains than her reputation gives her credit for . . . .' "[31] One might wonder why he chose to include such a candid remark, but he explained it when he admonished her: "But never say now that I have not trusted you."[32] Nathaniel offered self-revelation as proof of his love. Such self-disclosure obviously sealed the circle of intimacy for him because he closed this letter: "—at last, with no more doubts." For these correspondents personal revelation bore witness to romantic love and was central to the process that created it.

At the beginning of their extant courtship correspondence, Eldred Simkins asked Eliza Trescot to "write freely and when any thing bothers you let me know it."[33] Six months later Eldred was still closing his correspondence with "your affectionate cousin." The relationship was distant, reflected by such forms of address, but also by the fact that Eliza was more open in letters to her mother than to Eldred: "They told me in Florida that you were writing the most gloomy letters in the world to your Ma making her cry all the time . . . ," Eldred noted and then asked, "Why didn't you write to me and tell your troubles or are you afraid to confide in me? And I don't see for what reason. For your own Mother could not sympathise or advise you more warmly than I do. So for the future *please* let me be your adviser—and don't worry your Mother any more . . . ."[34] He harped on this theme, linking trust and confidence to unguarded communication: "Write to me plainly and I will endeavour to do the same. I have tried to do so today—I hope I have succeeded for really I have no thought or feeling which I care to hide from you ever. . . ."[35]

When Eliza finally unveiled her inner emotional life, Eldred was elated: "But Liza I am so selfish! While I *sympathize* with you, yet you, under the influence of the 'Blues' write me such long, nice letters without a thought of plaguing me, but lay bare your very soul that I am almost tempted to wish you a fit of blues every now and then."[36] At this moment of intimacy between them Eldred pressured Eliza for a more permanent commitment. Uncertain of her "true" feelings, Eliza had asked him not to press the marriage question until December. But in this letter, Eldred coaxed her for an answer: "All I want to know is will you write 'in full' which would be easier for you to do or wait for that distant and uncertain day 'to talk.' How easy twould be to say, 'Hope on, Eldred, the day *will* come when I shall be yours forever' tis all I ask; Liza, can you not say it? You said in your last speaking of *my* home, 'I have no right to it *yet*.' Did you really mean it: for the sentence implies a future certainty—or did you write it without thinking?"[37] Eldred seized the occasion of Eliza's self-revelation to press her for even more commitment. He recognized the connections between self-disclosure and romantic love: "the true reasons I merely wanted this from you . . . [was] that we should be perfectly open and candid, for there is nothing so adducive to perfect love as free and mutual confidence and not only to perfect love but also to true happiness."[38]

"Free and mutual confidence" grew during nineteenth-century middle-class courtship as lovers rewarded each other for their free-form subjectivity and thus were encouraged to reveal themselves ever more freely and fully. Sometime after they were engaged, Eliza wrote: "Are you not beginning to smile at my frankness tonight? Well smile on. I must write as I feel for as Florence expresses it I love you *so much* tonight."[39] When Eldred apologized for writing her a troubled letter, Eliza reassured him and encouraged even his mournful self-revelations: "let your sorrows be my sorrows; I love your confidence."[40]

Self-disclosure was heavily rewarded by nineteenth-century couples caught up in the process of romantic love. Working in concert with other social and cultural factors, romantic love was an incubator of self-consciousness in American life.[41] It helped to foster deeper and ever more intense levels of introspection and thus encouraged greater individual concentration upon inner states of consciousness and being.

The power of romantic love to cultivate a sense of individual identity is vividly illustrated in the relationship of James Bell and Augusta Hallock. Their expanding self-awareness was enhanced by the life and death pressure of the Civil War, and an unusually long separation which necessitated the discipline of writing out their thoughts and feelings. Receiving a photograph of James, Gusta asked him what he was thinking while his picture was being taken. "I thought," he revealed, "when I sat down that I must remember what I thought about for it would be just like *you* to ask me some time. I had just got things arranged for thinking when it was done. I thought of just 'nothing at all' . . . ."[42] This apparently inconsequential incident demonstrates that at this stage of their involvement thinking and feeling with enough self-consciousness to report to Gusta required almost herculean effort on James's part. This incident, the truncated correspondence that preceded it, as well as his reiteration in early letters of a dearth of things to say to Gusta suggest that James had not yet developed a romantic self; that he had not achieved a sense of self-definition that transcended the demands of his social roles. He hinted at this almost a decade later: "There has been a time when I could not write a letter to you without real trouble, or embarrisment . . . ."[43]

James made an even more telling observation on his lack of self-definition: "After waiting thus long for an oppertunity to answer your letter; I concluded to improve the present time. I hardly know what to write. I want to write a *good* letter to you. One that will express my great love for you; and will be an encouragement to you at all times. You used to say write just as you feal, well I do; but language is lame with me and I cant tell how I do feal."[44] Two factors were at work: the familiar one of deficiencies in language and expressive skills, but also the difficulty of explaining an amorphous inner life. His deficiencies as a writer were no doubt a central explanation of why he "cant tell how I do feal," but he also appears to be signaling some vagueness about his emotional experience.

Romantic love was a major causal factor in James's growing self-awareness. The dynamic through which nineteenth-century romantic love affected self-awareness was clearly delineated in the Bell-Hallock courtship: James was encouraged, cajoled, and summoned to understand what he thought and felt about himself by Gusta's desire to know his inner life. Her attention to his interior experience was urgent and expressive. "You felt bad in your letter . . . ," she exhorted, "James write it all out to me. Maybe twill be a relief to you. Let *me* share your troubles. Let me know them, I can *feel* for you."[45]

James reflected on how much Gusta's desire to know his inner life had affected changes in him: "But I am *so* glad you feal like writing freely to me. Your letters *do me lots* of good, and as you used to tell me, write just as you feel. I am shure there ought not to be to much reserve abou[t] us, and the better we understand each other the better for us both."[46] The complex process of identity-formation and self-creation were well under way in this romantic relationship. Their letter writing had been transformed since their courtship correspondence began. At the outset she could barely "fill the three sides of a tiny sheet of notepaper" and his first letter to her contained a mere three sentences.[47] Their letters, as Gusta pointed out, grew longer—more introspective, self-aware, and articulate—along with their maturing love. This was clear when James expressed a dramatic shift in attitude: "If I cant come to you, I will write often I *like* to write to *you*."[48]

One need only contrast his short infrequent letters, excusing the fact that he could think of nothing to say, with a later missive, *written without any intervening formal education:*

> I feal like writing to you this afternoon. My heart is so full of love for you. *All all for Gusta.* And I would write it to you in all its depth and fealing, but *words* fail to convey it in all its meaning. I find a love pure and beautiful, traced in *blue lines,* [she wrote in blue ink] and clothed in language expressive of a great and lasting love for *me.* This love finds a response in my own heart. It calls my own thoughts, fealings, and sentiments to you. . . . Love, how precious a gift it is, and how nessesary it is to me now, and I look back over the long years that I can claim this treasure, this priceless gift from you, and my heart reproaches me for the time I was careless of it. Now how wicked it seems to me, more so from the fact that there is so much *power, influence,* and *life* in this love of ours. I find my self really *astonished* that there is so much *real* happiness to be derived from the communings of a congenial spirit tho we are seperated,

tho we cannot see each others faces or see the expression of the counti-
nance or eyes, as our minds are expressed in letters to each other. There
has been a great change in me as regards this thing. There has been a
time when I could not write a letter to you without real trouble, or embarrisment: Now all that is gone. I feal *free* to write just as I feal and
think, and am *happy* in writing to you. tho I write a great deal that is not
sensible, I do not feel contented one bit if I am deprived of the time I
usuly . . . devote in writing to you. When I have nothing to do, my other
duties being all done, I feal like writing to you *every* day. Yes I do, and I
esteem it a precious privlege to write to you whome I love so much.
Perhaps you may think me extravigant in my expression of love to you.
Yet I find them existing in my heart, and you know you learned me to
write *just as I feal.*[49]

In this way he testified eloquently to the *"power,* and *influence,* and
life in this love of ours." Though he was not on a deliberate quest for
personality, the process of romantic love—along with other factors—propelled him into a self-development that expanded his conscious interior
life in conjunction with his powers of expression. Under the aegis of Gusta's questioning, probing demands to know all his experience, James was
"learned . . . to write *just as I feal."*[50] Through the dialectical pressure of
mutual self-disclosure and inner-directed attention, the existence of the
romantic self—a fairly amorphous inner reality after all—was confirmed,
defined, and given genuine content in an extremely close identification
with another human being.

Victorian lovers demanded self-expression of each other. Their obsession with self-expression directly contradicts what many believe to be the
prevailing concept of self in Victorian culture: the much vaunted ideal of
character. According to this theory Victorians enshrined the external demands of the social order in their concept of character, which promoted
the observance of social codes in every area of life regardless or in spite
of personal predilection, feeling, or individual taste. The assumption has
been that Victorians conceived of their code of character as absolutely
necessary social behaviors—thrift, hard work, honesty, and moral seriousness—which were the result of strenuous efforts of a self constantly straining against the force of its deepest inclination in both public and private
life. This culture of character was promoted in advice books of all kinds
and in the public images projected by nineteenth-century middle- to upper-
middle-class men and women.[51]

The culture of character, however, dominated only the public arena
of Victorian life. What has gone unnoticed is that in the protected realm

of private reality, a culture of personality was also taking form. While personality is sometimes seen as a twentieth-century concept of self, and character as the ruling concept of self in the nineteenth century, both ideals of individual identity coexisted in nineteenth-century culture.[52]

The Victorian concept of character based on analyses of public behavior fails to recognize that private life was *not* seen as an arena of self-constriction and restraint but of personal unmasking and freedom from etiquette itself. Even etiquette books prescribed a place—the inner sanctum of the home—and a relationship—intimate male-female—devoid of conventional etiquette.[53] Control was valued less in private relationships than expression of one's "true" self. As individuals moved closer to such a true self in communication with another, they became increasingly intimate. This romantic concept of a "true" self or personality—a set of qualities and characteristics that defined the essence of an individual—was an ideal that dominated Victorian private life.

Character was an ideal of self-control achieved by suppressing self-expression. But the nineteenth-century American middle-class concept of intimacy was an ideal of the fullest, most natural self-expression. There was a concentrated effort in middle-class courtship to comprehend the loved one. This most often involved an attempt by both parties to unmask, to abandon all outward forms of propriety, and to shed all normative social roles except the romantic self.[54] Courting couples pressured each other to greater self-awareness and self-definition in their efforts to know someone else as they sought to know themselves. Middle-class courtship by the 1830s or 1840s claimed this ideal of intimacy.

Nineteenth-century individuals in love discussed a wide variety of topics: the woman's movement, their division of labor after marriage, furniture, their moods and dispositions, character flaws, religion, economic matters, and anything of interest to them. The goal was learning as much as possible about the prospective mate.[55]

Self-disclosure, however, was not always positive in that faults, flaws, difficulties, and negative character traits were exposed. In fact, candid negative comments were common in courtship letters. A significant cultural pattern emerged. A prospective mate's emotional commitment was assessed repeatedly through negative self-images, presented to test the reaction of the potential partner and to establish a more realistic picture of their life together after marriage.[56] The implicit question usually posed by these self-criticisms was: "Will you love me even with all my faults and shortcomings?" If the courtship resulted in marriage, the response of the potential partner was almost always affirmative.

Criticism of character and behavior—negative estimation of personal qualities—was a significant part of nineteenth-century courtship. It was injected into the dialogue between lovers as self-disclosure. Self-criticism generally elicited positive response. The partners reassured each other that they were indeed capable, attractive, intelligent, and worthy. Although the self-criticism/affirmation response pattern was rarely violated, nineteenth-century couples did not appear consciously to offer self-criticism to elicit compliments. Positive assessments were tendered in response to what was mostly sincere self-criticism.

Self-denigration by one partner and affirmative response by the other could range from the sublime to the ridiculous. For example, Samuel Francis Smith insisted that his fiancée was heaven-bound, while he was on his way to Hell. She responded, au contraire: he was certainly saved, her destination in the afterlife was doubtful.[57] This seesaw was typical of nineteenth-century courtship letters. When Samuel bemoaned his lack of energy and piety, claiming that he was waiting for their marriage "to profit by your own example of devotedness and heavenly communion," Mary replied that the reverse was true and Samuel would be her spiritual guide because he was a better Christian.[58] The standard response pattern of courting couples was to hold up a roseate mirror to the loved one. By the implicit rules of courtship, criticism must be self-inflicted. For example, James Bell reassured Gusta and criticized himself: "Now you must not find any more fault; for *I know* that your letters *are good*. You think me to good, altogether, that will not do, for you will be *disappointed in me*. And as for my deserving a good wife. I *know* that I want one. And if I get you, I know I shall have one, That is the opinion I have of you."[59]

Negative assessments of self in romantic relationships were usually personal revelations. The motives behind such self-criticisms were instructive. "[W]hat I most want to be sure of is," James Hague explained, "that by and by when we get used to each other and the romance of courtship has disappeared, the charm of anticipation gone (which is now so delightful) and all the little illusion, that add so much to our prospects, have been dispelled—that then you will continue to *love* your old boy even when he is grouty."[60]

In a similar vein, John Marquis, astride the seesaw of positive affirmation of his lover and criticism of himself, insisted: "Do not call me *good* Neeta for *I am* not good. I am mean, sometimes wickedly mean and entirely unworthy of the pure love of yourself. . . . Do not deceive yourself. Do not place to high a value upon my worth for a future acquaintance may not fulfill your expectations."[61] Nineteenth-century courtship depicted a tension among lovers who wished to present the best possible image of self so as to nourish love but also needed to offer a realistic, even negative, self-image to test it.

Courting couples favored full and sometimes negative self-disclosure for several reasons, not least of which was a fear of future marital disappointment and discord without it. "[T]he simple, natural expression of our minds in all possible moods," James Hague suggested, "the less we shall have to learn or unlearn of each other by and by when all or at least some of our frauds will be discovered."[62]

Although love was not completely blind, lovers often praised, flattered, and generally reflected positively on the self-image of their beloved. Staring at a photograph on his desk, Nathaniel Wheeler described "the little blue eyed maiden that thus queens it over my study table, of her winning ways, her helpful hand and tender heart, her practical sense, her earnest womanliness, her happy, loving nature, her sweet little self. . . ."[63] Often induced by self-criticism, exchanges of reassurance helped contribute to the formation of personal identity. The flattery and praise of a romantic partner, however, had public as well as private implications. In a social order that increasingly emphasized individual motivation and achievement, the fostering of self-esteem and nurturing of individual ego strength served an important purpose.

This pattern of reassurance did not mean that nineteenth-century lovers lost their capability of estimating the physical, mental, or psychological weaknesses of their beloved. Gusta Hallock, for example, scoffed at a comment of Fanny Fern on romantic illusion: "but she sayes one thing that sounds queer to me. I dont believe it tho, Its that—'Whom one loves is *always* handsome.' I guess I should know it *quick* enough if I was loving a *homely* man . . . ."[64] John Marquis also expressed skepticism over the power of romantic illusion to hide individual faults: "The very fact of their being able to conceal their deformities so as to *appear* faultless is an evidence of deep deceit to me. I love to see a person knowing they have failings willing to acknowledge them. I am sure then that for every fault they have a virtue to counterbalance it. Such things may exist, but it is very seldom we find angels in human form."[65]

Albert Janin provides a striking example of someone who, in spite of a desperate love, retained the capacity to judge his lover's character. Janin, on the one hand, wondered without apparent irony whether other men "are so blinded by love as to entertain as high an opinion of their betrothed as I do of you."[66] When Violet warned him against her disposition, however, and insisted that there was danger in marrying her, he responded according to the rule of affirmative response by protesting that her disposition was glorious, but continued somewhat less normally, "Will you let me tell you, Vivy, what I consider to be the only defect in your charming and most remarkable character? They are a little want of charity for the infirmities and human errors of others, and a little too much love of flattery. But what a small offset these are to your many noble and striking qualities, which fill me with more sincere admiration than any mere words of flattery could express."[67] Though Janin was generally full of praise for Vivy, and was one of the most abjectly smitten of lovers, in this instance he violated one of the most sensitive rules of nineteenth-century courtship. Little wonder that Violet, possibly provoked by his uncharacteristic criticism of her, accused Albert of "putting on airs."[68]

While Janin recognized her faults, he repeatedly affirmed Violet's self-image at every turn, insisting she was "the most perfect and fascinating woman I have ever known," calling her beauty "superhuman," and praising her "wonderful example of goodness and self-denial."[69] He at times simply overflowed with enthusiastic paeans to her virtues, exclaiming "where on earth can your equal be ·found?"[70] Nathaniel Hawthorne also praised his fiancée: "Now good bye, dearest, sweetest, loveliest, holiest, truest, suitablest little wife. I worship thee. Thou art my type of womanly perfection. . . . Thou enablest me to interpret the riddle of life, and fillest me with faith in the unseen and better land, because thou leadest me thither continually."[71] Gusta Hallock, a more down-to-earth scribe, told James that she loved him because he was "good, noble, manly," because he had such a good heart, and "because—because I *cant help* it."[72] To the skeptic, Janin, Hawthorne, Hallock, and others in romantic love had simply taken leave of their rational faculties. True or not, the consequences of the assurances bestowed may not be too lightly dismissed.

The seesaw of self-criticism and reassurance that engaged nineteenth-century lovers was much more than idle flattery. This criticism and praise ritual involved attention to the details of personal behavior and inner states. It also involved owning up to certain qualities—a unifying set of characteristics—which cultural conventions taught the individual to identify as his essential nature or true self. It was a kind of personality testing that was an integral part of the larger cultural pattern of nineteenth-century courtship. Significantly, both sexes participated in and benefited from the self-enhancing, self-defining ritual of affirmation and praise so critical in the process of romantic love.

One of the more beneficial aspects of nineteenth-century romantic love was the strong dose of reassurance participants regularly administered to each other. Romantic love was an emotional experience in which self-doubts were openly expressed. Within this structure of feeling, romantic partners almost always responded by lavishly asserting the strengths of the beloved. Thus courtship exchanges of self-criticism and praise worked to enhance self-esteem, reinforcing the content of individual selves and simultaneously easing the burdens of individualism.

✄ Romantic love was a process of individual development that involved emotional expression, the reciprocal demands of personal disclosure, and the self-enhancing exchange of reassurance and praise. These introspective practices strengthened, shaped, and gave further definition to an individual's identity beyond or outside the social roles a person was expected to fill. Individuals in romantic love, by encouraging and nurturing each other's subjectively perceived qualities, contributed to the development of the romantic self.

In addition to self-definition and self-expression, nineteenth-century lovers described another common sensation: the subjective feeling of being immersed in and assimilating a portion of someone else's interior life. Romantically attached individuals repeatedly evoked the inner sense of sharing the identity of another. Nineteenth-century romantic partners believed that they had assimilated part of each other's subjectivity. There was mutual identification across gender boundaries in a nineteenth-century middle-class romance. This identification was a central experience of romantic love and required that lovers make their inner life comprehensible to a romantic partner of the opposite sex.

Nineteenth-century culture has been characterized as one of extreme separation between men's and women's life spheres. Historians have highlighted the divergent conceptions of the sexes as well as the gender-split ideas about the proper division of labor and social activity. Such portraits of gender differences, known as the doctrine of separate spheres, and the ideology of domesticity, emphasize the enormous gap between the sexes.[73] On this cultural foundation, historians have characterized same-sex friendships, particularly those between women, as closer and more intimate than women's relationships with men.

While same-sex friendships continued to flourish, as like responded to like, what has been less appreciated is that the gap between men and women closed dramatically under certain circumstances. Romantic love was the most significant circumstance in bridging the gender gap in Victorian America. For the experience of love created a mutual identification between women and men that was so intense that lovers repeatedly claimed to have incorporated a part of their partner's inner self into their own inner life. Men in the grip of romantic love swore that they shared a portion of their female partner's subjectivity and vice-versa. This extraordinary claim has profound implications for gender history because it suggests a popular means by which gender boundaries were crossed and mutual understanding and sympathy between the sexes was nurtured. After all, men experienced at least one woman, and women experienced at least one man, as part of themselves during love's tenure. This resulted in insight into differences as well as recognition of similarities between the two. Men and women in romantic love viewed each other not as useful performers in the drama of survival, not as antagonists in the battle of the sexes, but as two-in-one, a united being.

In struggling to analyze his relationship with Sophia, Nathaniel Hawthorne's observations are particularly valuable for his powers of interpretation and expression made him an acute interpreter of the heart.[74] Hawthorne was highly self-conscious about his emotional reactions to romantic love. In fact, he returned repeatedly to one theme of his love experience: the shared identity of two lovers. He was intellectually challenged and intrigued by his sense of unity with Sophia, and he worked to understand this facet of his romantic experience. He began an early courtship letter by addressing her as "Mine own Self," but later lost confidence in his feelings: "I feel as if my being were dissolved, and the idea of you were diffused throughout it. Am I writing nonsense?"[75] The phenomenon that Hawthorne could not deny was occasionally worrisome.

In Hawthorne's case, as in many others, the feeling of integration with another was deeply rooted and not simply a rhetorical flourish. Hawthorne, in particular, struggled to give this experience intellectual coherence. He was not always satisfied by his own efforts, but he continued to explore the consciousness of being "blended into one another. . . . Dearest, I do not express myself clearly on this matter; but what need?—wilt not thou know better what I mean than words could tell thee? Dost not thou too rejoice in everything that gives thee a more vivid consciousness that we are one?—even if it have somewhat like pain in it. The desire of my soul is to know thee continually, and to know that thou art mine; and absence, as well as presence, gives me this knowledge—and as long as I have it, I live. It is, indeed, impossible for us ever to be really absent from one another; the only absence, for those who love, is estrangement or forgetfulness—and we can never know what those words mean. Oh, dear me, my mind writes nonsense, because it is an insufficient interpreter for my heart."[76]

After learning that Sophia was ill, he elaborated on this theme of shared identity: "Partake of my health and strength, my beloved. Are they not your own, as well as mine? Yes—and your illness is mine as well as your's; and with all the pain it gives me, the whole world should not buy my right to share in it."[77] Hawthorne was speaking metaphorically, but not metaphorical nonsense, for he was describing the empathy that was commonly evoked by couples in this study who were bound by romantic love.[78]

Writing in the third person, Hawthorne, ever curious about his own psychological reactions to Sophia, observed that she was "transfused into his heart, and spread all round about it; and it is only once in a while that he himself is even imperfectly conscious of what a miracle has been wrought upon him." He was also fascinated by the fact that even in the midst of his custom house work, "I live only within myself; for thou art always there."[79] Hawthorne described the custom house as "that 'earthy cavern' . . . surrounded by all those brawling slang-whangers," and pictured himself at work amid "business and noise, and all sorts of wearisome babble."[80] Thus, when he claimed to live only in himself, he was referring to a romantic concept of self now partially defined by his lover, and clearly transcending his other social roles.

More than six months earlier, he remarked that if Sophia had been with him on his last trip to Boston, when he was alone in the train car, "What a blissful solitude would that have been, had my whole self been there! Then would we have flown through space like two disembodied spirits—two or one. Are we singular or plural, dearest? Has not each of us a right to use the first person singular, when speaking in behalf of our united being?"[81] Hawthorne's conclusion that he and Sophia were indeed singular—a united being—may appear hyperbolic, but his observations reflect the feelings of many nineteenth-century lovers. His insistence that "there is one good in absence; it makes me realize more adequately how much I love thee—and what an infinite portion of me thou art" was an evocative description of what it meant to experience the mutual identification of selves in romantic love.[82]

What was true for Hawthorne was also true for many others in romantic love. They became obsessive observers of their own inner states, and reflected upon the process of ever deepening identification between themselves and their lovers. Clara Baker felt such mental and physical union with Robert Burdette that she considered the possibility that he knew intuitively when she was menstruating—"And do you instinctively know now without referring to your calendar that I need loving, tenderly and restfully—but not playfully?"[83] Robert had been by her side for most of the preceding month, and she reflected: "It sometimes seems as if my 'other self' had gone—we are in such keen sympathy that when altitude was not good for one—no more was it for the other—so? wasn't it?"[84]

Many others expressed a similarly "keen sympathy" for their "other self." A young woman, who had orders from her mother not to write a letter on the Sabbath, protested to her lover, "but writing to *you* is just talking to myself . . . ."[85] Jane Burnett lamented, "I am so lost without

you my dearest Wellington—You are indeed a part of me . . . ."[86] Mary
Smith observed of herself, "I feel *intensely* feel, more than words can express every thrill of joy which bursts from your heart—and every sigh of
sadness which is breathed from your bosom."[87] Mary told her fiancé Samuel that he should write her whatever interested him, for "Are not our
interests one as well as our hopes and joys and fears?"[88] She also urged,
"Regard me once more as *one* with yourself—*Assure* me once more that
you do—and we will *once more* be happy."[89] Samuel responded that we
are a "wholly inseparable one, in feeling, interest, desire . . . ."[90] Gusta
Hallock observed, "James I write all this to you—my thoughts and feelings, because *our* hearts are knit together in bonds of sympathy, and somehow it is a comfort to me to, knowing that you feel for me."[91] James
confided: "I want to put my two arms around you, press my lips to yours,
and in a *low* whisper call you *Darling,* yes, call you *my own,* and in one
of those quick glances from your eyes, see your soul as it unites with mine
in one great love."[92]

Eldred Simkins mused on New Year's Day, 1864, about a love for his
cousin Eliza which began many years earlier: "that love has grown with
my life and now I feel as if every thought, wish and feeling of my soul
were centered in you."[93] Nathaniel Wheeler proclaimed, "My dear, I am
more than ever convinced that you were made for me—is it selfish to put
it that way?—and so, being the other half of me, it is no wonder I miss
you."[94] After spending time together at his family home in the summer
of 1877, Nathaniel told Clara, "You have become so mingled in my thoughts
and acts—my very life, that I am not myself now that I no longer have
you."[95]

Albert Janin expressed almost identical feelings: "My whole emotional
being seems merged in yours; robbed of you I should be poor indeed."[96]
Later he opined, "I have hitherto imagined that I could stand alone, that
I was sufficient unto myself for all the purposes of this life; but now I
seem to have yielded up a part of my strength to you and to need your
sympathy. I feel that you are absolutely indispensable to my comfort and
happiness, the necessary complement of my nature."[97] Janin had so identified with the woman he hoped to marry that "It seems to me almost at
times as if you were actually and bodily in my heart . . . ."[98] Janin was
in an advanced state of romantic love and, therefore, acutely conscious of
his interior life. In a moment of lover's anguish, he insisted: "I cannot
have a separate existence from you. I breathe by you; I live by you . . . ."[99]
Perhaps he summed up this aspect of romantic love when he said: "I have
no realization of self, of my personality, except in relation to you . . . ."[100]

This mutual identification within the nineteenth-century romantic bond
has significant implications for the historical study of gender and male-

female relationships. Men such as Albert Janin—whose "whole emotional being seems merged in you," that is in a woman's being—have indeed gained some measure of common ground with the opposite sex. On a subjective level, within the circumference of romantic love, the separate spheres of men and women were at least partially and intermittently bridged. Obviously nineteenth-century sex roles were barriers to such a bridge, but romantic love was deeply anti-role and predicated upon an ideological hostility to defining the person by his or her role alone. This is not to imply that bridging the gender gap was easy or always successful, even for those most desperately in love. But the ideology and, more important, the actual experience of romantic love required men and women to transcend their sex role differences at least to the extent of developing a reciprocal emotional relationship.[101] Under certain circumstances then, individuals in nineteenth-century America crossed interior gender boundaries and shared a portion of the other's perspective on the world. This was a critical factor in the history of male-female relationships. The evidence suggests that the experience of romantic love in nineteenth-century America produced a mutual sympathy between men and women that went beyond mere sentimentality.[102]

🕮 Nineteenth-century romantic love was not a single emotion, but a process of developing self-consciousness and mutual identification between two people. Once this identification process was under way, participants experienced an array of feelings. Yet the emotional fall-out of romantic love was not randomly scattered. Romantic love not only heightened and intensified an individual's general emotional life but it also elicited special emotions—jealousy, happiness, pain, and longing—in greater strength and frequency.

While nineteenth-century romantic partners did not always clearly understand their feelings, they usually experienced a predictable emotional response pattern. This response pattern included possessiveness, powerful feelings of dependence, obsessive thoughts of and extreme anxiety over the welfare of the loved one, and dramatic fluctuations between intense personal happiness and equally intense unhappiness and pain. In private, men and women ran the gamut of emotions from jealousy to intense sympathy, longing, and sharp pangs of joy and pain.[103] All of these feelings followed in the train of a deepening identification of selves between men and women in romantic love.

These emotions were a predictable consequence of romantic love. As such, they were also part of an emotional dialectic in which the effects of romantic love nurtured its growth and maintained its existence. The emotional highs and lows of romantic love contributed to an intensified concentration on the individual's interior life and added further impetus to the development of a personal identity separate from social obligations and public roles. The range and intensity of feelings that followed in the train of romantic love focused the individual's attention on himself/herself and one other who had become constituted as part of the self. Both sexes experienced a wide range of feelings and an intensity of emotion during romantic love which not only strengthened individual self-consciousness but also helped bridge gender divisions. After marriage, the emotional response patterns of romantic love were less intense or at least the range of emotional expression in letters was narrower. As long as romantic love survived, however, its characteristic response patterns might be activated within as well as outside a marital relationship.[104]

Possessiveness or jealousy was one of the least attractive emotional consequences of a shared identity in romantic love. Evidence of both masculine and feminine jealousy appeared in letters. Lincoln Clark responded to his wife's jealous suggestion "as to the influence of a 'certain lady': I do not blame you for making it for I would not blame you for expressing any thought that comes into your mind—but it surely was unnecessary as I have not spoke to the woman for years nor nodded my head at her except when I rubbed right by her."[105] Though "rubbed" was a less than felicitous choice of verbs, Clark apparently convinced Julia of his innocence for there was no further discussion of a "certain lady." In this exchange Lincoln expressed the romantic ideal of openness in all communication between him and his wife, including jealousy.

Albert Janin illustrates the lengths to which jealousy might drive an otherwise rational man. After his girlfriend Violet left for Europe, Albert responded to the possibility of cross-Atlantic rivals for her affections: "How often, I wonder, have I addressed you as my darling? Is it not a sweet word? You are my darling, sweet and dear and precious to me beyond everything; heaven's vengeance fall on any one who tries to take you from me!"[106] Albert's plans to carry out "heaven's vengeance" were tested almost immediately when Violet entered into a flirtation with a married man. She reported that Th. had asked her to go off with him.[107] "Half crazy with rage" and totally miserable, as he described himself, Albert desperately waited for letters from her, which did nothing to assuage his mounting anxiety.[108] Violet announced, with what can only be described as malice, that Th. told her he was primarily interested in her mind. Righteously indignant, Albert exclaimed: "In all your experience . . . have you ever, Vivy, heard or read of a man who was ready to sacrifice honor, duty, an innocent wife and a fellow-being who never harmed him for the sake of any woman's intellect?"[109]

Violet continued to fan the flames of Albert's jealousy and finally Albert, fearing the worst, went to New York, as he described it after the fact, to "seek out Th. [who had returned from Europe], read his character through and through, and if I found him to be worthy of your love, then release you and, if my fears were justified, say farewell to every hope of happiness in life and hasten away to some foreign land like Mexico where battles are constantly raging and I might find a speedy death."[110] Proving once again that life can imitate art, Albert found Th., wormed himself into his confidence, and plied him with questions about his private life over food and drink. Albert reported incredulously that Th. "acknowledges no restraints, and laughs at duty and virtue." But the *coup de grace* was Albert's discovery that Th. was strikingly unattractive. He described his competitor as sway-backed with small, close-set eyes.[111] Obviously concluding that his rival was not the threat he had envisioned, Albert becalmed himself.[112]

Violet's response to Albert's investigations was virtuous outrage. She indignantly accused him of thinking that she might actually have "given in" to a married man.[113] In response, he abjectly confessed his belief that under certain circumstances, neither men nor women could control their hearts. And he added, "I know that I cannot control my love for you."[114] But he insisted, without recognizing his inconsistency, that he never doubted her purity. Then, practically admitting his own guilt, he apologized profusely.[115] Somewhat more defiant the next day, he asked: "Oh why did you make me believe that you were so deeply interested in Th? . . . I could not suspect that you did it to make me jealous and tease me."[116] Falling back into a slough of despondency a forlorn Albert waited until forgiven.[117] Relieved and also reflective, Albert summed up his experience: "I have been a shuttle-cock of Cupid ever since you left—now writing in misery now exultant with happiness."[118] Indeed, nineteenth-century courting couples were often shuttle-cocks of Cupid, one day suffering intense emotional pain, the next enjoying overwhelming happiness. Both the pain and happiness of romantic love emanated from the same source: the feeling that lovers had merged some portion of their inner life with another.[119] Perhaps jealousy was also nurtured by greater attention to the nuances of the inner life in general.

Contrary to certain happily-ever-after images of bliss and unqualified happiness, nineteenth-century lovers displayed an awareness of both the unhappiness as well as happiness that love brought in its wake. Although courting couples took for granted that love's joys and pleasures outweighed its costs, they fully recognized the centrality of pain to their experience of love. From a more distant emotional calculus, the darker emotional hues of romantic love—anxiety, disappointment, suffering, depression—"educated" many people in their capacity for feeling and increased their self-awareness. For those in love, however, tolerance of this pain was rooted in the conviction expressed by John Marquis when he confessed to his lover, "I cannot *cannot* be happy without you."[120] He embraced a common romantic feeling: that the only possible condition of happiness in life resided within a relationship with the loved one. Participants often tolerated the emotional pain connected to a romantic relationship because they believed that there was "no hope nor calculation of happiness" without their lovers.[121]

The precipitates of genuine emotional pain in nineteenth-century romantic love were various, ranging from emotional coldness to temporary absence to death itself. Almost any "threat" of loss of the loved one created some discomfort. The duration and intensity of suffering varied, and depended somewhat on the source as well as the individual circumstances. Identifying and removing obstacles in the path of love—a standard feature of nineteenth-century courtship—always created some emotional suffering.[122] The risk of death in war, personality conflicts, financial impediments, medical problems, age-ethnic-religious-issues of compatibility, rivals for affection—the sources of pain were as varied as the life circumstances of nineteenth-century Americans.

Physical separation, whether for one day or many months, almost always produced some degree of unhappiness in romantic couples. Albert Janin, upon Violet's departure for France, described himself as sinking into "a condition of gloom and sadness. Everything seems to me so empty, flat and insipid without you. The only things valuable in life are the affections of the heart, and heaven knows mine are all centered upon you. Your absence is terrible to me . . . ."[123]

Feelings of gloom and sadness at the beginning of a long separation were common. Mary Granger worried: "if after you come home, you are obliged to leave me again, I should cry harder than before—for I know by experience how hard it is to be separated from you. I never was so utterly wretched in my life . . . fortunately I was able to overcome this dreadful feeling for I am sure I could not have lived if I had not . . . ."[124]

Certainly the pain of absence was much less dramatic for couples who lived very near. Nonetheless, almost all nineteenth-century courting couples endured some degree of separation, even if only in their housing arrangements.[125] Nathaniel Hawthorne, with some possibility of seeing Sophia the next day, bemoaned: "It is awful, almost (and yet I would not have it otherwise, for the world) to feel how necessary thou hast become to my well-being, and how my spirit is disturbed at a separation from thee, and stretches itself out through the dimness and distance to embrace its other self. Thou art my quiet and satisfaction—not only my chiefest joy, but the condition of all other enjoyments. When thou art away, vague fears and misgivings sometimes steal upon me; there are heart-quakes and spirit-sinkings for no real cause, and which never trouble me when thou art with me."[126]

The extent of the assimilation of two identities was felt and recognized when the couple was apart. Nathaniel Wheeler lamented, "[I am] not myself now that I no longer have you . . . I miss you every where and every moment. The house seems gloomy, the days are very long, the flowers are dull and faded. . . . I'm in danger of being very blue . . . ."[127] Eliza Trescot declared that her longing to see Eldred "was positive pain."[128]

Absence was painful, even for couples who saw each other regularly, such as Charles Strong and Harriet Russell. Harriet wrote to Charles, "My Charlie's 'Pussy' feels pretty well this morning but her heart was so full of sadness last night at parting with you. . . ." Later she explained why: "It did seem to me I could not let you go last night—I thought something might happen to you away from me . . . ."[129]

Fear for the well-being of the loved one was a constant theme in nineteenth-century romantic relationships and could become a source of extreme anxiety and pain. "I sit down to write you at the present time with a heavy, very heavy heart. . . ," Mary Smith wrote Samuel. "I conjured up a score of things which might have occurred—and thus have (foolishly you will say) made myself almost sick."[130] Nathaniel Wheeler confessed: "Darling, I never loved you, thought of you, planned for you—and *me,* longed for you, as I have the last few days. I have indeed been so full of tenderness, and anxiety for my absent dearest 'dear' that could I believe in presentments, I should fear something had happened to you."[131]

The psychological vulnerability of romantic love was summed up by Nathaniel Hawthorne: "where thou art not, there it is a sort of death."[132] When selves were merged in romantic love, even the potential loss of a loved one might feel like "a sort of death." This was the source of the feeling of special vulnerability to another that the experience of romantic love created in Victorian couples.

Pain and anxiety are perhaps the most misunderstood emotional consequences of romantic love. They were not abnormal feelings nor ones reserved for relationships in distress. Pain was an absolutely normal part of the nineteenth-century emotional response to sharing an identity with another human being. While the twentieth century seems to have progressively devalued emotional suffering, preferring to avert its gaze, the nineteenth century was more at home in the darker end of the emotional spectrum. This may indeed be a key difference in nineteenth- versus twentieth-century romantic styles and substance.

Contrast necessarily delineates and gives meaning, even reality, to its opposite: as cold defines heat so sorrow defines happiness. For nineteenth-century lovers, the pain and anxiety of either physical or emotional separation gave a sharper definition to the joys of love. Love brought anxiety in its wake, but as Eldred Simkins recognized, "anxiety tends to increase love."[133] To experience the negative potential in emotional life was more accurately to take the measure of the positive. Lovers testified that their pain heightened their experience of happiness with another. But pain did more than define love. It also helped to create love. This was not a masochistic pathology of a few romantically crazed individuals, but a normal component of Victorian romance. If love was, in some respects, a matter of paying attention, then pain often intensified and focused the concentration upon oneself and one other that was a necessary condition of romantic love.

Men and women in the nineteenth century often recognized that emotional pain was essential and even unavoidable in romantic love. They used their suffering to authenticate their love. Albert Janin claimed that his suffering proved his love: "This experience of suffering that I am now undergoing is something quite new to me, and that is the best evidence that you are my first real love."[134] Dorothea Lummis insightfully observed, "Our mutual pain has taught us that the love we each, perhaps foolishly, thought waning in the other's heart yet lives."[135] Pain taught nineteenth-century lovers as much about romantic love as pleasure and was just as fundamental to their experience of love.

Aside from marital prospects, the immediate results of romantic love were both intense feelings of joy and equally intense feelings of woe. Albert Janin testified eloquently to this paradox, "The bitterest moments of my life have grown out of my love for you—though through no fault of yours—but, also the most exquisite happiness, in fact the only real happiness, as I now understand that feeling, that I have experienced, I owe to you."[136]

Hawthorne, responding to Sophia's wounded feelings, wrote in some exasperation: "What misery (and what ridiculous misery too) would it be, if, because we love one another better than all the Universe besides, our only gain thereby were a more exquisite sensibility to pain from the beloved hand, and a more terrible power of inflicting it! Dearest, it never shall be so with us."[137] But Hawthorne quickly recognized that romantic love was a source not only of happiness, but also anxiety, depression, and disappointment.[138]

This is extremely significant in accurately assessing the power wielded in relationships between men and women. For romantic love—while it lasted—was a state of being that gave lovers emotional power over each other. One historian has suggested that "Love and power do not necessarily rise and fall together." [139] While true in purely economic and social structural terms, love and power do rise and fall together in certain dimensions of human interaction. This is why it is crucial to understand the emotional dynamics of nineteenth-century romantic love. To be able to inflict pain, to be vulnerable to intense misery, or to experience intense joy with another is to wield some interpersonal power. The dynamics of romantic love created interpersonal power through the pleasures and satisfactions as well as the vulnerability of a shared identity.

Perhaps the most persistent characteristic of nineteenth-century romantic love was obsessive thoughts of the loved one. Couples who were romantically involved sometimes used each other's thoughts to gauge the strength of their attachment. Gusta Hallock told James Bell that she thought of him "The *first* that morning brings and the last of nights are of *you*." [140] She thus effectively and economically communicated the depth of her emotional commitment to him. James corroborated the commonly available understanding of this symptom of romantic love when he remarked, "In all your hours, I see that I am a subject of your thoughts. I take this as an evidence of your deep, abiding love." [141]

One of the major pleasures and central activities of romantic love was simply *thinking* of the lover. "To night I feel very happy in thinking of you," James Bell mused, "in loving you and thinking how happy we may be in the future, if God wills it." [142] Albert Janin reflected: "I have been thinking constantly of you since I left Washington, and if you only knew what that implies—how much delight the mere thought of you affords me (when you are kind) you would assuredly envy me." [143] Albert Janin doubted "if it be possible for a man's mind to be more filled with one image than mine is with yours. It seems to me that I am always thinking of you, either waking or dreaming." [144]

Mary Smith told Samuel that she went on a fishing trip and "I was thinking so much of you and your letter that I was hardly talkative enough to be social. And thus it is my dearest one half of my time. I am so absorbed in *you* that I am at seasons almost unconscious of what is passing around me. A poor weak feeble child indeed!" [145] Samuel, in a similar predicament, admitted that he could hardly concentrate upon his ministerial duties "until absolute necessity drives me to it." [146]

Indeed, "the idea" of the loved one afforded the nineteenth-century lover great pleasure and satisfaction. Hawthorne sighed: "What should I do in this weary world, without the idea of you, dearest?" [147] Charles Strong said of Harriet Russell that "she is never from his thoughts a single conscious moment," and he confided that he "snuggle[s] her in my heart" when he thought of her.[148] "I did nothing yesterday—and the day before," Harriet wrote, "but think about you . . . look at you, read your letters etc."[149] Lyman Hodge informed Mary Granger in typical fashion, "I think of you very often, and hope and know that as often you think of me . . . ."[150] She confirmed that thinking of him was her "most agreeable occupation."[151]

Thinking of the loved one made those in romantic love happy, even in the midst of the most uncomfortable circumstances, such as those described by Eldred Simkins, a lieutenant in the Confederate army. "One great objection I have to this place is that we starve in the most respectable manner. We have very little for breakfast—pea soup for dinner and water for supper. . . . Another great attraction here are the mosquitos—who are beautifully drilled in 'Lancer' tactics . . . piercing through 3 blankets a coverlid and your uniform! Fact! Finally you eat, drink and sleep in sand—shut your eyes for a minute in your room, you have to dig the sand away before you can open them."[152] But amidst all these distractions, and sheer physical discomfort, Eldred exclaimed: "how can I employ this beautiful Sabbath evening better than in thinking and speaking to her whom I love more than all the world. Yes, my own! for when I think of you my thoughts rise finer and brighter and my heart feels happy and grateful to the Great Creator for his many, but most undeserved, blessings."[153]

Lovers testified repeatedly to such feelings of happiness, sometimes registering surprise at the excesses of joy they felt. "Sometimes I feel, deep, deep down in my heart," Hawthorne remarked, "how dearest above all things you are to me; and those are blissful moments."[154] After seeing Harriet Russell, Charles Strong recounted his homeward journey: "Just outside of Carson City [Nevada] it began to blow hail and snow. Could neither see the hills, lights, road or anything else from which to take 'bearings' as they say at sea, but my heart was so full of love for my Darling that neither heeded the dark, wind or weather and let the faithful ponies go whither they liked and I thanked the 'Bountiful Giver' for blessing me with such a love on earth and was warm and happy."[155] Harriet was surprised by her own feelings: "Such an arrangement as ours, I never thought could make one so perfectly happy."[156] James Bell was also surprised by what he experienced: "I find myself really *astonished* that there is so much *real* happiness to be derived from the communings of a congenial spirit."[157] Eldred Simkins summed up the feelings of almost all lovers when he said: "Liza! you cannot know how absolutely necessary you are to my happiness."[158]

The expectations of happiness, which courting couples fixed upon each other, had a basis in reality. While romantic love usually resulted in some level of emotional suffering, those who were romantically involved also experienced intense feelings of happiness. A most intriguing question then presents itself: why could even the idea of the loved one, much less their physical presence, command such passionate feelings of self-satisfaction and joyful exultation? Obviously, the culturally learned expectation that lovers will make each other happy played a part. When Albert Janin insisted: "I must and will be happy with you my darling, my treasure, my queen," his "must" and "will" indicate that he was straining to apply the standard romantic expectation.[159]

But more significantly, through the actual experience of romantic love, both men and women gained greater self-definition, validation, and mutual affirmation. Among native-born middle-class Victorians, who were beginning to expect less gratification from social roles and all the supporting structures of custom and institutions and more fulfillment from the individual psyche, romantic love provided enormously satisfying and fulfilling support for and sharing of the burdens of the self. Nathaniel Wheeler said as much to Clara Bradley: "Dear friend, you don't know how much I value those little evidences I saw and heard that you *think of me* daily. The world, even our part of it, is *so* selfish, it shuffles us out of the way so unceremoniously, that to know one dear heart is *ours*, as much absent as present, that in some ones thots our presence never dies and fades, is very pleasant—yes, it is the best thing this earth can give—."[160]

Such testimonies to a profound linkage of personalities and thus at least a temporary refuge from loneliness in romantic love can be instructively contrasted to the gloomy observation of one of the most respected students of nineteenth-century American life: "Thus not only does democracy make every man forget his ancestors, but it hides his descendants and separates his contemporaries from him; it throws him back forever upon himself alone and threatens in the end to confine him entirely within the solitude of his own heart."[161]

An acute and sensitive observer of Victorian public culture, Alexis de Tocqueville did not have access to the shielded realities of the heart. Exposure to the private realities of the period reveals that there was an antidote, albeit sometimes an ephemeral one, to the dangers of the solitary heart in nineteenth-century America. The antidote Tocqueville overlooked was romantic love. In many ways, of course, romantic love promoted the solitary self. It certainly helped to narrow social participation, inflate the importance of subjective feelings, and encourage individualism. But it also offered an escape from solitude through a paradoxical effect.

Two persons, by intensively sharing their interior lives, enriched and sharpened their separate subjectivities. Simultaneously such personal immersion in another often resulted in acute feelings of overlapping consciousness and subjective unity. Thus individualism—defined in terms of a romantic experience of the self—was strengthened in a process that made each participant feel less self-contained and alone in the world.

The implications of this complex emotional experience can hardly be overstated. Perhaps most significant for the history of gender, a gap was bridged between men and women by an involvement in romantic love that gave each sex greater insight into the nature and experience of the opposite sex.

Questions have been raised about exactly how much the ideals of love and companionship actually changed male-female relationships.[162] It has been difficult for some historians to believe that ideals of love and companionship had any real impact on middle-class Victorian marriages while men still maintained massive legal, economic, and physical bases of superiority. They are properly skeptical of the practical effects of a companionate ideal of marriage based only upon voluntary power-sharing, kindness, and a heavy reliance upon the whimsical favors of one's mate. Missing in that analysis—neither well understood nor appreciated—is that the glue of companionate marriage—romantic love—was an active force that identified men with women in an emotional process that was gripping and, while it lasted, had a compulsive drive. While romantic love was unstable in the sense of its duration or staying power, it had compelling effects on individual lives.

3

"Lie Still and Think of the Empire"

Sexuality in Victorian Courtship
and Marriage

Captain Frederick Marryat, an English novelist who toured this country in 1837, reportedly observed piano legs modestly covered by little trousers in a seminary for young ladies.[1] Although this story is often taken to epitomize Victorian prudery and excessive sexual repression, another incident in Marryat's life may be a more accurate symbol of Victorian sexuality.[2] When Marryat visited Louisville, he met a phrenologist, Dr. Collyer.[3] Suspicious of his wife's fidelity, Collyer decided to test her "bump of amativeness" against her "bump of discretion." Announcing that he would be gone all night, Collyer hid under his wife's bed "to await developments." According to Marryat's modern editor, "At one o'clock in the morning they materialized—the Captain [Marryat] in a shortish nightshirt, Mrs. Collyer in little more. The doctor [Collyer] after a diagnostic perusal, materialized too, from under the bed, shouting 'Fire,' 'Rape,' 'Treason' . . . . After Captain Marryat and Dr. Collier were separated, Marryat is supposed to have stammered something about having heard that Mrs. Collyer was proficient at easing *sprains,* and suffering such an ailment, he had come to her for treatment!"[4] This incident, unreported in Marryat's published diary, faded into oblivion, while the chastely covered piano legs live on in historical memory.

🌿 The "repressive hypothesis," symbolized so well by this popular image of prudishly sheathed piano legs, is now under increasing attack.[5] Long before this revisionist trend, however, a number of American historians were often skeptical about Victorian repression. Charles Rosenberg, Nancy Cott, Carroll Smith-Rosenberg, Daniel Scott Smith, and others characterized nineteenth-century America as a society of infinite sexual complexity, and carefully observed the distinctions between sexual ideology and behavior.[6]

Scholars working with diverse geographical samples of private expression have directly challenged the view of the passionless, prudish, and sexually inhibited American Victorian.[7] Elizabeth Hampsten concluded from North Dakota private writing that sexuality was discussed consistently and approvingly. Western historian Julie Roy Jeffrey, relying on diaries, letters, and memoirs, insisted that sexual passion had a place in women's behavior and expectations on the nineteenth-century American frontier. Ellen Rothman, utilizing an impressively large sample of New England letters and diaries—350 courting men and women—finds that they were very often comfortable with a wide range of sexually expressive behavior. Carl Degler, in several works, attacks the notion that anti-sexual ideology was ever put into practice by a large number of Victorians, men or women. And the Mosher survey—a questionnaire on family background, sexual behavior, and sexual attitudes, filled out by forty-five middle-to upper-class women who were born before 1870—indicates that sexual relations were often frank and sometimes enthusiastic. Peter Gay has also challenged the validity of the repressive hypothesis, both for bourgeois Americans and Europeans.

As many scholars have discovered, the discussion of sex in nineteenth-century letters and diaries is not exceptional. Diverse sampling in sources written by ordinary Americans of middle- to upper-middle-class social position confirms a pattern of sexual interest in both males and females, as well as testifies to sometimes passionate sexual behavior.[8]

Whatever the revisionist inroads, the historical examination of sexuality is still fraught with pitfalls.[9] The most serious problem of studying sex in the Victorian or any other historical period is the temptation to apply contemporary definitions and views to an erotic ethos that changes over time. For example, in contemporary American culture sex is often taken to mean intercourse; but this culturally constricted view of erotic activity must be expanded in order to examine the past sympathetically and critically. If erotic life is narrowly equated with coition, there may be little evidence of sexual activity before marriage in many Victorian courtships; but this would be a serious distortion of the historical record. Erotic activity has many forms, and the presumption that one dimension of sexual expression defines it or is superior to other forms is naively culture-bound.

It is misleading, but difficult to resist the assumption that contemporary standards of sexual gratification—which rely heavily upon orgasmic measures—should be applied to nineteenth-century erotic experiences.[10] Changes in human culture shape the biological sexual response into particular historical experiences. There are no sexual absolutes. Sexual experience is time-bound in that it both reflects and distorts culturally given models of reality. Questions about the personal experience, identity, and meanings both men and women attached to their erotic feelings can only be answered from direct evidence and not by analogy or public expression.

🎜 "Lie still and think of the Empire" is the wedding-night advice Queen Victoria was supposed to have given her daughter, thus representing the quintessential in Victorian repression and prudery.[11] Among American Victorians, passionlessness in female sexuality has been characterized as the dominant ideology.[12] It was an ideology propounded in ministerial and medical advice as well as in some feminist tracts, but it was not the ideal that guided men or women in the conduct of their private lives. Queen Victoria's daughters across the Atlantic did not take her advice to heart, if they heard it at all.

Middle-class American women gave no private indication that they believed in an ideal of female passionlessness. This does not mean that there was no sexual dysfunction among them, or that they had perfect sex lives by their standards or ours. But married women did not treat their sexual feelings as abnormalities. Many indicated that they accepted themselves as sexual beings. Individual personality factors influenced the way women expressed their physical desires, but it is clear that they did not consider themselves freaks, deviants, or even strange for having sexual needs or expressing sexual interest to men *in private.*

Married men showed no shock, horror, or even mild displeasure at their wives' physical interest. In fact, they seemed pleased by these private expressions of desire for them. There were surely passionless marriages, but passionlessness was not a dominant ideal within middle- to upper-middle-class American marriages. Moreover, romantically attached married men cared about their spouses' sexual responses. This did not mean that either husband or wife looked for identical sexual satisfaction. But under the spell of romantic love an ethic of mutuality seemed to operate in the Victorian bedroom. Nineteenth-century married men and women sometimes expressed a genuine interest in pleasing each other in physical as well as non-physical ways.

Although before marriage, Victorian sexuality was more ambiguous and potentially stressful, this should not be attributed to the belief in female passionlessness. Ironically, nineteenth-century couples might have found courtship less stressful if passionlessness had been a dominant Victorian courtship ideal. In fact, both men and women saw sexual desire as the natural physical accompaniment and distillation of romantic love. Some indeterminate level of sexual expression and satisfaction was acceptable in Victorian courtships when individuals were in love and the expectation of marriage was strong. Intercourse, however, was a physical boundary not to be crossed until after marriage. This was the crux of the courtship ambiguity. Sexual expression was approved as a symbol of love and typically accompanied nineteenth-century middle-class courtship. During a period in which love was intensively cultivated in a structure of privacy and minimal parental supervision, intercourse was supposedly postponed while some ambiguous level of erotic expression symbolizing love was condoned. For some couples, this created few noticeable problems, but for others the tensions were tangible.[13]

Nonetheless, married or unmarried, American Victorians recognized and expressed sexual desire, interest, and passion. Even humor surrounded sexual exchanges between nineteenth-century Victorian couples. There were earthy sexual comments by both sexes. Some women and men enjoyed the bawdiness of sexual experience unvarnished by sentimental rhetoric. Moreover, Victorians seemed to derive considerable pleasure from speaking of sex in private. Even the more reticent managed to convey that erotic activity was central to their view of male-female relationships. But if sex could be respectably "raunchy" in Victorian relationships, it could also be luxuriantly romantic. The mutual identification of two people "in love" was often symbolized by sexuality itself. Under the right circumstances, sex might be viewed as a romantically inspired religious experience, a sacrament of love.[14] The latter was perhaps the most culturally significant meaning attached to Victorian sexuality.

The fusion of love and sex, or rather the investment of sexuality with the meanings of romantic love, may seem unremarkable. Yet the nineteenth-century view of sex as the ultimate expression of love had remarkable cultural consequences. Imbued with romantic love, sex was seen as an act of self-disclosure, not so much in the sense of revealing one's body as one's essential identity. Sex was identified with the inner life and was perceived as part of the privileged revelation of an "authentic" self. Properly sanctioned by love, sexual expressions were read as symbolic communications of one's real and truest self, part of the hidden essence of the individual. This was the most salient meaning of sex to American Victorians. Thus any sexual behavior that was not the honest expression of an individual's "truest" self was deeply offensive to Victorian culture.[15]

Historians have concluded, on the basis of public exhortations, that nineteenth-century women downplayed their lower-status sexual/carnal side and emphasized their higher-status spiritual/moral capacities.[16] The implication is that the sexual was at odds with the moral or spiritual in nineteenth-century culture, especially with respect to women. This is a major stumbling block to understanding Victorian sexuality. Actually Victorians joined the sexual and the spiritual or moral in the concept of true love. Romantic ideology bridged the gap between purity and sex for Victorian women as well as men. For example, one nineteenth-century adviser defined *true love* as "Love in its truest, purest, highest form is that of strong, unselfish affection blended with desire—an honorable desire implanted by nature in the breast of men and women, and which is only to be condemned when it is perverted and seeks gratification in forbidden ways."[17] Sex could be sacred and sexuality might be spiritual, if affection were blended with desire.

In private discussions couples equated their love of one another with their sexual desire and pleasure. Desire was often cast as the sign and seal of romantic love. The son of a Presbyterian minister pined for his wife during the separation necessitated by his preparation for their move from Illinois to California: "When I lie down at night my mind is filled with thoughts of you, not *bad* thoughts but I do so long to have you beside me again and I find nothing to fill the vacant place in my arms and my heart. Nothing which I can closely fold up to my heart and feel blest in possessing. My dear wife you must come to me soon. I must have you with me."[18]

An unmarried woman, Betsey Meyers, also employed the common physical metaphors of love: "Oh if you knew how mutch I love you, you would have no occasion to fear that I would ever break my promises, for if you are as constant to me as I am to you, those vowes which we have made, will never be broken untill Death seperates us for ever, for my mind is constantly on you, you are my though[t]s by day and my dreams by night, some times I fancy in my dreams that you are by my side, and your arm around my waist and my hand in yours, and that you again lead me back to our old haunts of love and pleasures but when I awake I find it but a dream and the dear delusion flyes from· me, and I again sink back upon my bed and bedew the pillow with my tears. Oh that I had the wings of a dove that I might fly to your bosom and enjoy your embraces, but as I have not I must content myself as well as I can untill you come which I hope will be vary soon."[19] This unmarried woman's yearnings for love were inextricably bound in memories of physical expression and pleasure.

Celebrating eleven years of marriage, Lincoln Clark so intertwined sexuality with love, bodily pleasure with the metaphors of the heart, that love and sex were inseparable: "What would all this world be if we could not pour out the heart to those who care for us, and who have given us evidence that they care for us by smiles in health and angelic devotion in sickness. I want to put my arms around your waist—and kiss you and pat your black hair and say how sweet! It would do me just as much good as it did ten years and a half ago—and I have the vanity to believe that the pleasure would not all be on one side, and this idea enhances it to me: you say you can not 'say' you love: well I will say it, and you may express it as you please—the thing itself is the *sunshine* of earth." [20] Though the diction was uncommonly fine, the use of sex and physical desire as symbols of romantic love was not unusual. In private expression, romantic love and sex were intertwined in the everyday system of cultural meaning.

Although sexuality was romanticized, it could at the same time be funny and sensate. Alice Baldwin, a respectable army wife, was far from sentimental about her sexual feelings. Separated from her husband during his military campaigns against the Indians, she wrote after two years of marriage, "Oh how rejoiced I will be to see your dear face once more to feel myself clasped in your sheltering arms. Those dear strong arms have always loved and comforted me. . . ." [21] A traditional woman in Victorian terms, Alice Baldwin conceived of her own role as a strictly subordinate one in relationship to men. Alice was no feminist, which makes it all the more fascinating when in 1870 she reported to her husband on a flashing she received in a train station: "There was a man showed his 'conflumux' to me at one station where we stopped in Illinois while I was looking out of the window. I thought he might have saved himself the trouble because I had seen one before . . . ." [22] Her matter-of-fact reporting and earthy nonchalance were not what might be expected of a nineteenth-century lady if one read only certain medical and religious tracts on the female nature.

In later letters, she blatantly taunted her husband after learning that he had wavered between marrying her or another woman, Nellie Smith. "Aunt Mary," Alice wrote, "said you asked her advice about it you didnt know whether to marry her or me. Mr. Frank D. Baldwin, if I had known you was in such a quandry I would have settled the matter at once by giving you the mittens. I felt real queer and strange when I heard you had half a mind to marry another girl. I thought I held *undivided* you[r] love. Well its too late now. Nellie Smith dont know what she escaped. She would have been killed at one nab: of your old long Tom!!!" [23] Alice unabashedly referred to "old long Tom," a nineteenth-century slang term for a penis.

Alice Baldwin's feigned irritation at her husband was expressed in a humorous if vulgar attack on the recently revealed rival for her husband's love. The nature of the attack, however, turned the ideology of the Victorian lady around, for instead of deriding her rival as sexually loose and lacking in purity, she ridiculed her competitor's sexual inadequacy. In doing so, she conveyed a wholesome confidence in her husband's sexual appetite and her own physical response. Allie Baldwin was frustrated by many aspects of her sex role, but not her role in the marriage bed.[24] For example, she responded to a sexual tease by her husband with another of her own. "I got your letter last night . . . ," she wrote; "so you have been casting sly glances at Mrs. Sowters Bubbies [slang for breasts]. You ought to be ashamed. I intend to show mine to somebody before long."[25] Alice Baldwin's thinking about sex roles was fairly mundane. Her audacity with regard to sex was *not* tied to any overt inclination toward social change and cultural innovation.

Jane Cleveland Burnett, another nineteenth-century housewife, also thought about sex roles in traditional terms.[26] She yearned to be near her husband who was serving in the California legislature: "Could I but ley my head upon your bosom and whisper love to you I would be happy as I ask to be on earth. . . . My Dearest darling love how can I live away from you. This week has been so very long. I miss you in every thing— But the nights—to wake up and find you gone. I draw the baby close to my bosom but still—I grieve for you—our meeting will be sweet will it not darling. The baby wakes. A thousand kisses dear one."[27] The next day she wrote, "I am almost crazy to see you—This being away from you is torture. Do tell me everything darling. My own sweet Wellie. My dear precious love. How long are we to be seperated must it be much longer, I want to sleep in your arms again."[28] Eight days later she pined, "How I do long for your sweet kisses."[29]

Later in the year, during another separation, her passion fairly burst forth, "My brain is on fire what shall I do I can not live away from you three weeks, two almost killed me—Ikie yearns for you so, what will we both do—Wellington my own dearest best I love you with all my heart and soul—a thousand kisses from Ikie and Jennie . . . When will we rest in each others arms again. Darling I send you all the love that is so truely yours."[30] Two years later, she was still expressing the same passionate longing during his absence. "I could fly to you if it were possible, I wish to see you so very much tonight, when I do see you my darling husband what a meeting it will be . . . ."[31]

Jane Burnett, intensely aware of her physical passion, communicated her love in terms of physical embraces and inextricably fused her emotional and physical desires in anticipation of several conjugal reunions. Her letters again illustrate how sex and love intertwined in the everyday system of cultural meaning.

Emily Lovell was a housewife who observed many of the traditional nineteenth-century sex-role distinctions. She worried over her children, fussed over her husband's clothes, and conceived of her role as a behind-the-scenes one.[32] Yet as far as her sexuality was concerned, she expressed her passion openly and enthusiastically. She closed a Civil War letter by saying: "Now beloved—one of my heart—embrace me—and love me as I do desire—kiss me over and over again—while I say beloved one good bye—."[33]

"Dearest," she wrote two days later, "I have come to the end of my paper and have left no room for a nice long and loving embrace—would that it were such—so darling come and let me put my arms around thee and beg you to write so often as you can—one more look of affection—one more kiss—and a few tears—I'll go—your own poor loving wife."[34] She longed not just to be the object of his embraces and kisses but to be his passionate active lover: "I write you darling one a little messenger yesterday but on the arrival this morning of another song of Love, from thee, I could not help saying a few words . . . would that I could *kiss you all over*—and then *eat* you *up* . . . ."[35] Whether Emily meant her last phrase as a euphemism for oral sex or simply as general slang for physical enjoyment, she was not after all reporting on an actual event. What is significant is the intent she expressed—the open avowal of her sexual desires and the unselfconscious way she characterized herself as a sexual aggressor. Her husband Mansfield was equally enthusiastic: "Kiss me dear sweetheart, a thousand warm and loving kisses, take me to your beautiful arms and let me for a while enjoy a heaven upon earth."[36] He encouraged Emily to play the active sexual role and reveled in the prospect of *her* sexual aggression. This exchange took place after thirteen years of marriage.[37]

While women could be enthusiastic, and occasionally forceful, in expressing their physical desire, masculine sexual imagery tended to be more aggressive than female sexual imagery. One man, courting a young woman in 1838, teased her about their physical relationship by asking if she would give him "one kiss" for the gift he was bringing her. He joked that even her mother would approve, then continued: "I have written so much about kissing, don't be afraid I shall devour you for then I should have no one to kiss."[38] Albert Janin gave no such assurances to his lover: "May I confess that I am always wondering whether I shall find opportunities for folding you in my arms and covering you with kisses, as in the happy past."[39] Janin warned his fiancée that she should strengthen her ribs because he intended to "fold you in my arms à la boa constricter."[40]

John Marquis also expressed sexual longing and affection in a more aggressive masculine style. During his first year of marriage, while his wife was on a short visit away from home, John pined: "The nights are long and sad I do not sleep well without you. . . . expect to be smothered to death when I see you."[41] Anticipating a long-awaited reunion with his fiancée, James Hague also reflected an active masculine physical style: "Well, I hope I shall have my arms around the old girl herself before many weeks and if I don't behave with great impropriety then it will be for better reasons that I can now foresee. I'll just squeeze her and hug her, and kiss her forehead and eyes—yes I'll kiss them again and again, and when I have looked at them to my heart's content I'll kiss them again, and her cheeks and lips and throat, and I'll take liberties with her back hair and pull out her hair pins, and tousle and tumble her up generally until she boxes my old ears and goes up stairs to set herself straight. Won't that be nice, old Loveliness? glorified, exalted, ecstatic, radiant; and don't I wish I was there now . . . ."[42]

It would be a mistake, however, to draw the contrast between the expressive styles of Victorian men and women too starkly. Individual differences, combined with the difficulties of untangling the subtle signals of sexual desire, suggest caution. There was no subtlety, moreover, in the next invitation by Alice Baldwin, "How are you this hot day? I am most roasted and my chemise sticks to me and the sweat runs down my legs and I suppose I smell very sweet, dont you wish you could be around just now."[43] Alice was aware of her body and obviously enjoyed her erotic self. She savored her sexual appetite in this aggressively teasing invitation to her husband.

Sexual enthusiasm was expressed in the private correspondence of both married and unmarried nineteenth-century couples. With regard to sex, however, courtship exchanges differed from marital exchanges in at least one important respect. During courtship, correspondence offers glimpses of both physical and psychological tensions created by the traditional interdiction against premarital intercourse.[44] Nevertheless, there seemed to be little ambiguity in middle- to upper-middle class Victorian courtship over petting, caressing, kissing, and other non-genital forms of sexual interaction. In fact, the acceptable range of sexual conduct before marriage was wide.

Augusta Hallock wrote to James Bell that she went to a party and returned home "after 6 the next morning." This obviously reminded her of past history, because she exclaimed, "Oh! if I could only bite you over on the back of your neck. Does any one pinch or tickle your ears any more? I *did* cut up when you was here I'm sometimes ashamed when I think of it."[45] Gusta had an appealing vitality, but she was no radical nonconformist. Her sense of sex roles could be shrewd but also conventional for her time and place. She observed to James: "Most always when

I see men take care of horses and drive them, I form an opinion of their dispositions by it. If they are good to their horses, they are most always kind to their wives. I have heard that, and said it too a great many times."[46]

Bell, a farmhand and sometime sharecropper before he joined the Illinois Volunteer Cavalry to fight in the Civil War, wrote fondly of his own memories of their time together: "I am in just the right mood for teasing. O: if I was only where you are this afternoon wouldent you get fitts. I would pay you up for plagueing me so last winter. Biteing my neck and blowing in my ears for instance. Do you remember? I can say I do for the cold chills run over me whenever I think of it . . . . Do you remember going down to Hannahs one New Years eve? How I did tease you comeing home. I was thinking how many times I kissed you then. I would like to give you back some of them today if I could."[47] Gusta was a discriminating young lady and revealed a marked preference for quality over quantity in regard to kisses: "Now my dear Jimmie I will thank you for the kiss you sent me, and you can have one from me in return, Its a good *long* one, not one of those little sting ones we used to deal with."[48]

James and Gusta were separated for long periods as James struggled for an economic foothold on the agricultural frontier before the Civil War. Undimmed memories of their physical relationship gave them both pleasure and the reassurance of mutual affection: "I cant think of anything *very bad* you have done, unless its biting my neck, or plagueing me in the way you used to? Do you remember?"[49] Gusta remembered with unrepentant enthusiasm: "You speak as tho' you didnt have anything to forgive of me only biting you. Well I'm naughty enough to say that I *mean* to bite you the first chance I get, Now depend on it. I havent forgotten what a delight it was for me years ago and what a powerful effect this biting had on *you*. It makes me laugh even now to think of it."[50] James reacted with a mischievous gibe of his own: "Then you intend to have a good time do you, in biteing me, when I return. Let me give you a word of caution. You know that is a game two can play at. *My* teeth are a little poor but then I think they are a trifle *sharp*."[51] Continuing their good-natured banter, James shifted his attention from biting to blowing: "I wish I could manage it some how, to blow in *your ear* before I close this. . . ."[52] Gusta reacted with a playful challenge: she invited him to try his luck, but cautioned: "Dwight and Billy say they are afraid of me, and I guess you'd *better* be. I'm a powerful girl. Seems to me you would not make out much blowing—."[53]

Returning to the conceit of teeth, James chuckled to himself: "I had to laugh when you told me about loosing your teeth. Thinks I? You cant bite my neck now. But I have the start of you on that, for I lost part of mine ten years ago. So you see you have a sympathiser in me, on that point . . . ."[54] However, Gusta expressed an admirable determination to overcome all obstacles in her relationship to James: "—What! you losing your teeth too? Seems to me we will make a pretty looking couple—*gray* and *toothless* too. We both labor under the same difficulties you see, But I mean to bite your neck if I have to get false ones to do it."[55] Such playfulness illustrates the multiple dimensions of sexuality in nineteenth-century relationships.

Humorous sexual imagery might take a number of different forms. Ettie Elliott—who was a girlfriend of Gusta Hallock's—ventured boldly: "What a terrible warm time Dan and Em had of it for wedding, Probably Em has been obliged to squeal long ere this. (Please burn this) . . . ."[56] This undignified image of intercourse, perhaps drawn from the observation of farm animals, may have contained both a sly admiration of Dan and also a droll commentary on love-making. It too is a useful reminder that while intercourse was viewed reverentially, even religiously by devotees of romantic love, physical love could also be cast in a candid and earthy role by both men and women.

The humorous and serious sides of sex coexisted in the repertoire of Victorian courting couples. Sex was treated as *both* a serious spiritualized sacrament of romantic love, and the occasion for a good laugh. In fact, what may most surprise those who cherish the image of Victorian prudery and repression is that Victorians had a sense of humor about sex. Lyman Hodge, for example, recognized the incongruity between the reverential or sentimental and the sensate or pragmatic dimensions of sexuality. A Yale graduate with an exceptional literary flair, Lyman became a coffee and spice merchant in Minnesota. He and Mary Granger were unofficially engaged. She was teaching music in their hometown of Buffalo when he sent an unusual "mock" proposal of marriage: "Sunday night and five full months is it not—since I left you, —*five* months—each longer than the other—until I begin to think, that if one desires that very undesirable boon 'length of days,' here is about the place to get it. Days are long, nights are longer, for *mosquitoes*—those tormentors of every body who is in good eatable condition—" attack in squadrons at night. He suggested to Mary that they should write a "Mosquito Waltz" or "March" together and then proposed his own:

old Hodge is fat and his flesh is sweet,
and a Northerner is he, he, he,
I'll take his face, and you take his feet,
For he's just the game for me, me, me.

For we are Southern Skeeters,
Arksansians born and bred,
We'll eat the Northerners alive,
And pick his bones when dead.

He was then reminded of "an Old Arksansian's advice to his son on the boy's complaining of mosquitoes at night, 'Get married, my boy, get married, for if you have a woman to sleep with, the skeeters will eat her and let you be,—gal's flesh is tenderer than a man's'—flattering to the 'gal' isn't it?—Are you willing to try the experiment next summer and offer yourself sacrifice upon the altar of Mosquitoes as well as of Hymen?"[57]

That same year, after a recent reunion, Mary Granger waxed elegiac one Sunday evening: "This afternoon as I lay half asleep thinking of you, the meetings and walks that we took when you first began to love me came back oh so plainly, I thought of the words you said and the kisses you gave, until it seemed to me that I could feel your arms around me and your kisses on my lips, then it was all so new and strangly beautiful and I loved you so much from the first—that I did not for a long time think it would be a life blessing, but that I was to happy for this world, and you never would love me well enough to make me your wife, and when that thought was disipated, I felt that just to know that you loved was happiness enough, even if we were never married, but now I do not feel *exactly* so about our marriage, and know that choisest boon of love is yet to be ours."[58] In openly and self-consciously desiring "that choisest boon of love," Mary was not only referring to domesticity, but also to the physical consummation of her relationship with Lyman. The underlying tone of Mary's response was a kind of sensual nostalgia and joyful anticipation of their physical and emotional future together. Typically, Mary represented love through the physical dimension of their relationship: the emotional meaning of her interaction with Lyman was symbolized in sexual terms.

Lyman wrote her the following year from Minnesota, when it was cold, and he again used his discomfort to satirize the physical side of male-female relationships: "Married folks, whose better or worse halves are absent, shudder at the thoughts of cold sheets with no one to help warm them but their own shivering selves; or, if both are at home, decide by a game of euchre who shall go to bed first; while bachelors and maidens—half congealed already—wish and wonder—unutterable things,—and even some old maids declare for the first time for years that a husband would not be such an unmitigated evil after all. . . ."[59]

He also reflected the common nineteenth-century fusion of sex and love in a lyrical passage spanning the past, present, and future of their relationship: "and now, love, you with the warm heart and loving eyes, whose picture I kissed last night and whose lips I so often kiss in my dreams, whose love enriches me so bountifully with all pleasant memories and sweet anticipations, whose encircling arms shield me from so much evil and harm, whose caresses are so dear and so longed for awake and in slumber, making my heart beat faster, my flesh tremble and my brain giddy with delight,—whose feet I kiss and whose knees I embrace as a devotee kisses and embraces those of his idol,—my darling whose home is in my arms and whose resting place my bosom, who first came to them as a frightened bird but now loves to linger there till long after the midnight chimes have uttered their warning,—my life, with your generous womanly soul, my heart's keeper and my true lover,—Good night: a good night and a fair one to thy sleeping eyes and wearied limbs, the precurser of many bright, beautiful mornings when my kisses shall waken thee and my love shall greet thee."[60] Though Lyman and Mary had probably not yet consummated their physical relationship, they both eagerly anticipated the pleasures of a physical life together. Their courtship had an active sexual dimension which was sensate, humorous, and romantic.

During the Civil War, Eldred Simkins reminisced with his southern lady, Eliza Trescot. "Oh my precious darling! that bright moonlight night in Columbia when I held you to my heart and your beautiful eyes were looking up so sweetly and confidently into mine, I could say nothing! and yet we were talking in silent yet burning language of our sweet happiness, our enduring and engrossing love. . . . Daylight parted us and many days have come and gone since then but that night stands surrounded with a halo of enduring love and happiness beyond all the rest and I never think of it but as 'our night.' "[61] Though Victorian sexual expression was constrained by the taboo against intercourse before marriage, erotic pleasure nonetheless flourished in at least some segment of nineteenth-century courtship.

Eliza also enthusiastically recalled that night in Columbia: "it is getting late and I must . . . put my letter away until tomorrow morning. It is almost as hard to do that as it used to be to tear myself away from you those nights in Columbia. That last night, my own Eldred, was it not the happiest of our lives? An inexpressible sweetness lingers about the remembrance of it. I was so happy to have your head upon my shoulder and my arm around you and to feel as if you were entirely my own. Are you not beginning to smile at my frankness tonight?"[62] This courting couple spent long evenings alone. They were engaged, and it was wartime, both of which undoubtedly afforded them greater privileges than they might have enjoyed in the early stages of their courtship. But Eldred indicated that late nights together were a pattern in their relationship. "You know I always kept you up till 3 o'clock at night anyhow though you were so unwell and all the scolding I was in entreating to send you and while I would open my arms my bird would not leave— . . . ."[63]

The Civil War may have loosened the grip of parental supervision over southern courtship, though this couple indicated nothing remarkable about their freedom to be alone together. Nonetheless, they appear in the past to have operated under some umbrella of parental restriction, for in another letter Eldred looked forward to the time when he and Eliza could prolong their time alone together indefinitely: "I as usual stretched out on the sofa and you holding my head. Oh! how tired you will get sometimes for there will be no one else who can order us to bed and our talks can be indefinitely prolonged."[64] It is clear that they were doing more than talking when alone, however. He often wished she were physically present: "tis the wish of my heart at all times, but especially so tonight—for I feel very much like being patted and caressed."[65] Eliza expressed similar desires, imagining the time when Eldred would "put your arms round me and press me close to your heart and call me your own."[66] She projected this scene into the future of their relationship because, though she yearned to have him by her side, she was determined not to marry him until the war ended. Fearing that being married would render their separations even more painful, she rebuffed his entreaties.

The physical aspects of nineteenth-century courtship induced tensions as well as satisfactions. The romantic view of sexuality had the potential to create confusion over the meaning of purity in courtship. Though the taboo against intercourse before marriage remained powerful, the belief that love might sanctify sex could sometimes blur the legitimate boundaries of erotic behavior before marriage. But this potential tension might also remain dormant in Victorian courtship. Women such as Eliza Trescot had no doubts determining or trouble maintaining her sense of legitimate sexual expression during courtship. She clearly felt that some avenues of physical expression were closed to her before marriage. But she promised Eldred that, once married, she would demonstrate her love: "Never mind, one of these days when I have a right to be 'that sweet name' I will lose my reserve and let my words as well as actions, show how very dear you are to me."[67] She accepted the physical side of her love, but approved of its fullest expression only within the marriage vows. Nevertheless, outside matrimonial boundaries, she was willing to hold and be held by him, to touch and be touched. And she flirted suggestively: "How nice it would be if I could come in softly some evening when you were reading to kill time and take your book out of your hand. I would help you considerably to kill time then wouldn't I? Perhaps you wouldn't feel so fierce towards it in that case for twelve o'clock always comes too soon when we are together, you know. Vain dreams . . . but they are so pleasant, Eldred. I love to think of being with you."[68]

Mary Smith, more reserved about her sexual feelings than Eliza Trescot, had probably conducted a less physical courtship. She was less certain of the permissible range of sexual expression before marriage. Mary and her fiancé Samuel were also separated throughout most of their courtship. Apparently worried about her physical desires as the wedding day approached, Mary confessed: "though I would fain be all loveliness yet while I feel so much of evil rioting in my bosom how *can* I conceal it?"[69] Mary appeared anxious to reveal her physical feelings to Samuel, a Baptist minister, but she was worried about the reception she might receive from him. She remarked that she preferred physical converse to letters, adding suggestively, "I might and probably should appear *still more* naughty yet I would rather even prefer that—for then I could disguise nothing—now much is repressed."[70] Mary was a religious woman who lived in a world of high moral seriousness, yet she did not see herself as passionless. Her passion might be "naughty," and even "evil" before marriage, but she recognized it in herself and did not disown it. Respecting the "evil rioting" in her bosom, she asked Samuel: "how *can* I conceal it? I know not how, even if I could."[71] She clearly wanted him to recognize her physical desires; perhaps she was even seeking reassurance that her sexual feelings were acceptable to him. Her statements reflect, however euphemistically, the strength of her sexual feelings and her need to express them in spite of her sense of moral restraint.

Samuel responded to her revelations with a characteristic didactic piety, telling her that he believed their recognition of the warfare between spirit and flesh was encouraging because it proved they were trying to live more spiritual and godly lives. Yet he urged her to be careful "because I tremble lest the cup of pleasure should be dashed from our lips before it is tasted."[72] It seems unlikely that Samuel was referring simply to platonic love. Samuel hinted more than once at his physical interest in their relationship. He compared his relationship with God to his relationship with her: "The present has been truly an autumnal day. The fall of the leaf approaches. Well, we will fear neither change nor decay, if the encircling of the everlasting arms may only be our refuge. I long, my dear, for your presence with me. I need earthly soothers, as well as a friend in heaven."[73] Obviously, Samuel hoped to be encircled by less everlasting arms.

As their wedding day approached, Samuel revealed: "I am looking forward, my dear, with anxious longing to the day of our union. The thought that it is so near unsettles my thoughts and almost wholly unfits me for the performance of any duty whatever." His unsettled thoughts rested upon the imminent "consummation" of their marital vows: "I cherish the hope that all will be joyously and happily, yea, and to our eternal joy and happiness, consummated."[74]

Mary also anticipated the consummation of their wedding vows, using qualifiers that added parental overtones to her sexual meaning: "and might we not anticipate blessed results from the sweet union we are now so ardently longing to be consummated?"[75] Many couples gave evidence of courtships full of sexual exploration and playful physicality. Mary and Samuel appear to have been satisfied with, or forced by separation to settle for, less physical intimacy than the nineteenth-century courtship ethos allowed.

But indications from Mary, less than two years later, were that her relationship to Samuel had been successful in all respects, including the physical. She teased him confidently: "You seem always prospered truly—even when you made choice of a *wife*. Well you have one that adores you, if you have nothing more—and if you have a score of wives, who would all give you as many sons apiece, you would not after all I think find among them all any more affection or much more enjoyment. What think you my beloved?"[76] Mary had traveled to her family home to await the arrival of their second child, whom she described as a "love-token." Writing during this hiatus, she expressed a casual enthusiasm toward the sexual dimension of their life together: "I wish I could see you for one second—how delightful to have one sweet kiss—well we shall have a good number packed away and ready, when we meet."[77]

Marriage was a significant sexual threshold for courting couples and intercourse was supposed to wait until they crossed over to the "other side." Some couples found this sexual boundary definite and relatively untroubling, while others were deeply vexed. Though it is virtually impossible to quantify such a subjective division, evidence suggests that the symbolic equation of sex and romantic love left Victorian courting couples vulnerable to tensions on the physical side of courtship. In middle- to upper-middle-class courtship romantic love was cultivated during a commonly unchaperoned period of premarital relations. Yet until marriage, physical consummation, justified by "the heart," was forbidden. Ironically, those most deeply devoted to the feelings of romantic love were expected to forgo what their own culture defined as concomitant physical expression.

The difficulties of using the "rule" of romantic love to regulate premarital sexual conduct is explicit in the correspondence of Dorothea and Charles Lummis. During 1883–84, Dorothea and her husband were not living together. She was in Boston pursuing her medical degree; Charles was in Ohio editing the *Scioto Gazette*. Charles Lummis went on to become a prolific writer, journalist, and editor. He wrote voluminously on the Southwest, edited a leading western magazine, and did much to influence the cultural life of California as an archaeologist, librarian, preservationist, and literary figure.[78] Dorothea, his first wife, did not conform to the usual nineteenth-century female sex-role prescriptions. She graduated from the Boston University School of Medicine and practiced homeopathic medicine in Los Angeles.[79] Dorothea divorced Charles Lummis in 1891, also a less than typical pattern in late nineteenth-century America, but one that was becoming more frequent.[80]

Their marriage was already in deep trouble after only three years. While an exceptional woman to be sure, Dorothea was not separated from her husband purely for reasons of ambition or feminine autonomy. As she lamented in a letter to Charles begun on September 21, 1883: "If I thought we could drop right back into the old days, long ago, there isn't a thing in this world that would keep me from you, want of money, malaria, and loss of degree combined."[81] Dorothea was very much in love with her husband and desperately wanted their relationship to work. As if to underline this point, she wrote a few days later, "*I* am incomplete without you always and if its *sometimes* hell to be with you, its *mostly* heaven . . . ."[82]

An important part of Dorothea's heaven in her relationship with Charles was physical. She closed a letter written October 7, 1883: "I like you to want me, dear, and if I were only with you, I would embrace more than the back of your neck, be sure."[83] In a touching revelation of the vulnerability behind her physical yearning for Charles, she said, "I guess its the cherishing that I miss. You know I never had any real petting or tenderness until I had it from you, and I have missed it more than I can ever tell you."[84] She repeatedly urged him to visit her in Boston but was also willing to go west herself to see him.[85]

There was a hunger to her need to see him that was reined in by her pride and created deep emotional tension in her relationship to him. "I hope your heart will say 'come,' or 'go' at least, and that you won't think it best for us not to see one another until June. . . . But as I read over your letter again, where you say it is 'bad enough and extravagant enough' to spend $20 on a trip west . . . [I] feel sort 'o frozen up, and as if you had really come to think more of the money than of our seeing one another. If it [is] so, its another cause of our separation. As for me, I have saved the money from things I did without, yet I shall not spend it on the trip unless you wholly approve. I couldn't go to your heart or arms, if I thought you thought I had better have saved the money . . . ." Then she nervously interjected that she might be misinterpreting his remarks and was drawn away from her pride toward her need. "Love me sweet, and try to want me. . . . Now I won't think of anything but sweet things and will study a bit and then go to bed to dream that your arms is about me, and your nice sweet body sort 'o warm and close to me. . . ."[86]

In a letter she received the very next day, Charlie assured her that he wanted to see her. This seemed temporarily to release the tension in her, but Dorothea's objectivity was short-lived and very soon, she cried out in her painful craving for his affection and approval: "don't roar—I *am* quite bright once in a while. Really brighter than most ordinary women and I see them with the most devoted husbands, say and do far sillier things than I ever was capable of. So I think perhaps if I can be bright with you and sort 'o pleasant too, you will be an adoring husband."[87] There is genuine heartbreak in Dorothea's begging Charlie for some crumbs of affection.

However, this must be counterpointed to her pleasure in anticipating their planned reunion. Six weeks before she would see him again, she dwelt on her positive anticipations: "you said you wished I was there, so that when you got back from your work—and then there was an 'A—' and a dash, and yet it brought up a great wave of sweet memories and longings and the touch of clinging warm lips, and the still magnetic thrill of your warm body, and I was all lost to the present. Let us be very tender and careful of this physical bliss, which in its entirety means more—and not abuse it, but keep it sweet and clear through the years of our young strong life, at least if we can. Ah! love do you remember too its dearness, and with its sweetness to aid us and bind us together, we will try and not grow too 'morose and self-sufficing' and find still that our best of mind and body is still suited to, and completed in each other."[88]

Dorothea's anticipations for their meeting were concentrated as much if not more upon body than upon mind. She said in her next letter, "Pretty soon we can begin to think of nice naughty things cant we."[89] Dorothea's passion was unabashed and unashamed. "I want you tonight sort 'o awful badly! I want you to kiss me and say Thea! I love you. Would you if I was there eh? I shall like it a lot when the train get[s] into the depot a[t] C. and you come out of the dark to meet me, and then we go off up to your room a little minute and stop a bit for just one kiss, that makes us hungry for more, and then go up and see the maters' face brighten up to see her girl and even the pater beams, and yet under it all a sort o sweet impatience to be back with you alone, with just you and love."[90]

Dorothea was probably less inhibited than Charles. She hinted suggestively the week before: "if we only could be happy, not miserable once . . . *and not mind anything but love, and not think anything was naughty that made us happier.* Ah, I can't write of it. . . . How can I wait!"[91] As the time approached for Dorothea's reunion with her husband, she could hardly be contained, and asked him to write her a naughty letter: "for it makes me want you a lot to read one, and all the long ride toward you, the sweet words repeat themselves over and over to the noise of the wheels . . . ."[92] She closed the letter that contained this most unstereotypically Victorian request with: "I hope your heart and your lips and all of your sweet body will be warm with welcome and desire. Do you think so?"

After her trip to Ohio, she answered this question herself, when she broached the subject of their recent love-making. "Say, Carl dear, tell me something. Did you like me as well as you used to, and were you just as happy as you used to be—in bodily ways I mean—because I though[t] maybe you didnt so very much, and I dont want you to get too used to me, or tired of me, so soon. We were happy, but it didnt seem to me we were as nice as we used to be. Not that the happiest minute wasn't as nice, but we sort 'o hurried Love all up, didn't we. Don't let's us, will you. Because even in this, habit tells, and I want that to be the very last thing to become commonplace. I spect its partly because you are so busy, that you dont have time to be hungry for happiness, but its better to wait a long while, and then take time for love, isnt it, than to lose half the sweet from haste."[93] Such a frank and detailed critique of marital sex was noteworthy for its quality of sexual awareness as well as for the assertion of erotic taste by a married woman.

Dorothea was obviously confident of her sexual feelings, except in one significant respect. She had crossed a tabooed boundary in male-female relationships and her guilt and self-doubts about it were at times intense. She had consummated her physical relationship to Charles before they were legally married. Rumors had been circulating about her at school among the women of her class and she vacillated between proud defiance and tormented self-doubt. On February 6, 1884, she was defiant: "Nobody would or could believe, but our own hearts, how much of purity and passion was in our love then. . . . I shall never care or regret it, if someday we can hold our own before everybody . . . ."[94]

On February 22, however, her brave resolve wavered as she reported "What do you think they say of you and me. That the sheriff visited you and gave you the choice of three girls all needing marriage and you chose the one with the most money." She sought reassurance from him that he did not think less of her for having intercourse with him before marriage: "Ah tell me that you at least believe that it was only lack of wisdom not of purity that laid me open to all this insult." Then she confessed how deeply vulnerable she was to attacks on her character for this transgression of social norms. She observed, "A man can never know the deadening feeling of utter despair that a woman knows at the shadow on her fair name."[95] Dorothea was a strong confident woman, but the power of the cultural strictures against intercourse before marriage haunted even her sense of self-esteem and sexual identity. And she defensively demanded reassurance from Charlie: "I did not wisely, but too well, perhaps in not asking you to give your legal name before giving you myself, but *you* cannot hold me less pure, Carl, for that."[96]

It is practically a commonplace to note the emphasis on female purity in nineteenth-century public discourse.[97] However, historians have seriously misinterpreted its usage in Victorian America. Purity was a central theme in Victorian public life, often applied, it is true, to women who were supposed by nature to be purer than men in body as well as mind.[98] Purity was also regularly used in private correspondence to praise a woman's character, a relationship, or love itself. But in private correspondence purity was rarely made a referent to the asexual or passionless.

For example, Robert Burdette, popular lecturer and minister, believed in the purity of "True Womanhood," yet he gave no hint that the premarital consummation of his physical relationship with Clara (or her erotic interest) contradicted his sense that she was a "true" and "pure" woman. He waxed eloquent about her "true woman's heart, drawing me on to purer and nobler ideals, to higher and better work. . . . teaching me to love what is good; to hate what is evil; to do what is right, to forget what is wrong. I will have, from Violet, the best that true womanhood can give, which shall make me a better man in thought and in life . . . ."[99] He expected her to place his life on a higher moral plane, while at the same time drawing verbal images of her lying unclothed in his arms.[100] Neither Clara, a respectable Victorian clubwoman, nor Robert saw purity, moral uplift, or spiritual values as antithetical to their sexual activity. Robert characterized her sexual passion: "your love . . . so pure—womanly in its ardent passion . . . ."[101]

Clara, on her side, closed a letter written before marriage but after their second interval of physical intimacy with a guiltless sense of their previous love-making, which she also described as "pure." "Good night lover—mine—Good night Robin—a kiss for you in 'our own way'—and love, love—full and free and pure from your 'Sweet White Violet' . . . ."[102] Clara Burdette's physical exuberance is illustrated in a passage that survived her own censorship: "You haven't been kissed in a long, long time except 'on the side' as they say of some course dishes—and you just deserve to be now—kissed until you struggle for breath—kissed until you think you are Hobson—kissed every way, every where—every second until Violet yields to Robin and together they rest."[103] Neither Clara nor Robert discounted her moral virtue or purity on account of her frank sexual longing.

Another passage which survived in a heavily censored letter again illustrates that purity might have erotic content. "But it would be a greater luxury for me just now," John Wesley North wrote to his future wife, "to sit by the side of My Dear Ann, and enjoy the purity of her sweet kiss—."[104] In nineteenth-century American culture, purity did not mean the absence of all sexual activity; it meant legitimate sexual expression. Depending on their definition of "lawful" sex, nineteenth-century men and women could be sexually active and consider themselves pure.

What legitimated sexual expression from a Victorian American viewpoint? There were at least three answers to that question. The first was that nature, God, or both, had mandated sex. In each instance, the emphasis was on sex as a biological drive which must be satisfied or unnatural consequences might result. This was a traditional view when restricted to masculine sexuality but a radical view when applied to both sexes equally. From the traditional perspective, men, and from a more radical perspective, women, were fully sexed and the satisfaction of the sex drive was a healthy biological imperative, especially within but sometimes even outside the bounds of matrimony.[105]

The second answer also emphasized the natural side of sex but saw it as created (often in God's scheme) solely for the purpose of procreation. Nineteenth-century commentators on sexual matters often disagreed on what were natural sexual imperatives. But they all usually agreed on this: when nature or God-in-Nature was violated, disease, disability, or even death resulted. In the second view, only marital sex for the purpose of procreation was "pure" and thus escaped such punishments. A variety of evidence indicates that this view was not widely held, though its advocates have received considerable attention as spokesmen for Victorian prudery.[106]

The third and dominant view of sexual purity in Victorian America was romantic.[107] This means that for Victorians legitimate sexual activity always reflected the "fullest" expression of the individual personality. The Victorian mainstream continued to maintain that intercourse should be reserved for the marital relation, but sexual intimacies such as kissing, touching, and other varieties of petting before marriage might be pure if they were expressions of romantic love. The ambiguity surrounding coition and the marriage boundary, however, was always a potential source of tension in courtship.

In the romantic view, sex was as much an emotional as a physical activity, and the emotional actually determined the purity of the physical. Of course, a long tradition stood in the way of applying a pure romantic logic to physical affection. The taboo against intercourse before marriage was backed by traditional religious teaching against fornication and adultery, the double standard, as well as the fear of pregnancy. In fact, quantitative studies indicate the nineteenth century had a comparatively low premarital pregnancy rate. At least in a sample taken from New England towns, demographers have found the premarital pregnancy rate to be low in both the seventeenth and nineteenth centuries, and by contrast much higher in the last half of the eighteenth century.[108] Though the premarital pregnancy rate declined from an eighteenth-century high, premarital erotic expression was not banished from nineteenth-century courtship, and the problem of distinguishing pure from impure sex before marriage remained a vexing one for some Victorians.

Romantic love in nineteenth-century culture offered a spiritual logic for those who wished to regulate their sexuality according to more individual needs. Obviously disturbed by self-doubts, it is nonetheless revealing that Dorothea Lummis insisted that she and Charles might rightfully consider their premarital love-making "pure," because they had obeyed a higher law of love. There is no doubt that many nineteenth-century men and women, including authors of advice books, were afraid of this logic, but it also appears to be true that the investment of sexuality with the meanings of romantic love was widespread, at least among the middle- to upper-middle classes.

Romantic views of sexual expression predominated in a number of different areas of Victorian behavior and belief, including private attitudes toward children and birth control. One significant way couples saw children was as love tokens. Children were treated in private discussions between husband and wife as a symbol of their romantic love. Nathaniel Hawthorne wrote that his heart "yearns, and throbs, and burns with a hot fire, for thee, and for the children that have grown out of our loves."[109] Dorothea Lummis described a child she wished to conceive in the future as a "tie to both our hearts."[110] Mary Smith saw her daughter as a "little pledge" of her love for Samuel.[111] Jane Burnett wrote of her first baby, "We are happy, as long as we love each other—and surely that will be as long as we both shall live—and we have still another love, that will drew us closer to each other, if that can be . . . ."[112] Her husband, Wellington, claimed even more clearly that the source of the love for his first baby boy originated in his love for his wife. "I have never appreciated the blessing of having our baby boy so fully as since I have been separated from you this time. He is our common consolation and object of continued protection and care. His bright face comes to me in the night, protected by his Mother's—and my Wife's—care. I *see* you together; I *see* you separated but near each other. You are my life Darling, *and* I cannot look upon our baby, except through you. My love for him does not arise from consanguinity alone, roots deepest in the soil of the love I bear my wife."[113]

Love, however, could commingle with fear. Alice Baldwin's husband very badly wanted another child, and Alice was sympathetic with his desire, but she was also firm in her own resolve not to get pregnant: "Poor fellow you do want a boy so bad and so do I but oh I dread it so I dont know what to say hardly—I want to satisfy you and I want your name perpetuated as well as you do, and I feel as if there was something wanting yet to our happiness. I *do* want a son *just as bad as you do*—and I feel a sort of yearning come over me some times to feel a pair of baby lips close to me once more and a little pair of soft baby arms and hands groping in my bosom once more. I know I would be happier than I am now if I had already a little *baby boy.* But I feel so frightened when I think of it—I am afraid to risk my life. I am afraid I will die and I am afraid it would be a girl after all and then I would have nothing to pay me for my long suffering and suspense. But Oh dear I *do* wish I would have a boy. Would you love me any better Frank for being the mother of a son? than you do now? I have thought perhaps I wouldent be so sick again as I was the first time. I tell you what I will do Frank when Nita gets to be five years old I will consent to have another baby."[114] She wrote in the same vein a year earlier, commenting on her first pregnancy: "I used to feel then as if I never should be well again. Aint I glad I aint going to have another one? I think I will wait until Nita is about four years old and then I will have a little boy. Will that please you?"[115] Alice Baldwin expressed conflicting desires regarding having another child. Though she feared the sickness, pain, and the risk to her own life of another pregnancy, she also worried that her husband might love her less for refusing him the possibility of having a son. In spite of her intense fear of another pregnancy, Alice wavered, motivated by her love for her husband and her sense that having another child might demonstrate her love for him.

The unprecedented and unmatched decline in the native-born white birth rate in the nineteenth century combined with the evidence that passionlessness was not a dominant ideal much less a practice in middle-class romantic relationships suggests that sex was separated from procreation in the Victorian marriage.[116] The idea, based upon the repressive hypothesis, that abstinence was *the* Victorian choice of birth control is false. A number of birth control techniques were available to nineteenth-century couples who saw sex as something important in their lives outside of the desire to have children. While couples may have practiced intermittent abstinence in the form of the rhythm method or safe period in a woman's cycle, or abstinence may have resulted from physical disabilities or emotional estrangement, the predominant attitude of married couples in this study toward sex (within a continuum of course) was as a positive and highly valued expression of love.

After marriage the purity of loving physical relations was rarely questioned. Nathaniel Wheeler, Clara Burdette's first husband, sent out a veritable howl of unfulfilled physical desire while Clara was away on a visit to New York. Nathaniel, in some distress, asked her why she did not leave "me your lace 'robe de nuit' or a ruffled pair of 'draws' or something of that sort?" He reported being in a "dreadful state of mind (and body)" since he received her last letter, which he compared to a torpedo thrown into "such a powder-magazine as I'm discovering myself to be, especially when there's no waterworks or any patent extinguisher within reach."[117] Clara's primary role, in this metaphorical relationship, was to douse the flames. And indeed, Nathaniel appeared to take a rather one-sided albeit affectionate approach to their sexual relationship in this letter.

Openly yearning for her body he asked, "How much longer before I see you, feel you, look at you all over, Kiss you, hug you, put my arms around that soft little waist, draw you up close to me, closer, closer, forget everything but our two selves, then hold you off and look again, and so to sleep with my own wife in my arms, which ache with emptiness, these long, long nights?" At this juncture, with a mixture of self-doubt and smugness, he queried rhetorically, "Do you, *can* you miss me as I miss you?" His answer was no. "It would hardly be in nature for you to feel as I do. This long repression begets a wildness in me that I dare not even faintly hint on paper." Nathaniel was convinced that his sexual feelings were stronger than his wife's in this instance because of the "natural" differences between men and women. This was an attitude common in some medical advice literature of the day, though it was not completely dominant. It was an attitude of *degree* rather than absolutes. Women were seen to have more moderate sexual desire than men, but their sexuality itself was rarely in doubt.[118]

While in theory Nathaniel might question whether Clara experienced an equal measure of sexual longing during their separation, he revealed that it was Clara who encouraged the uninhibited expression of his sexual feelings: "This is not all—you know that this is only one phase of my heart, and you who first taught me not to be ashamed of my nature, will not blame me for talking in this way." Nathaniel ended his letter by describing the meaning of their physical relationship to him: "Oh the sacredness of wedded love. I begin to realize its depth and strength, and its purity, its frank surrender, its implicit confidence." Once again purity was used to describe a passionate sexuality. What made it pure? A loving marriage, not procreation. Victorians who believed that sex was legitimated as an act of love did not need to justify their conjugal passion by the possibility of conception. Therefore, they might practice birth control without a sense of impurity because sex was already purified as an expression of love.

Most couples appeared to separate sex and procreation. Imbuing sexual relationships with the symbolic value of romantic love, sex was seen, not primarily as the ultimate gift of heirs, but as the ultimate gift of themselves. Perhaps as a result of the dominance of romantic ideals of marital relations, and the decreasing practical value of children in an industrialized and bureaucratized economy, spousal relationships were sometimes given emotional priority over parental ones. Nathaniel Hawthorne told Sophia after seven years of marriage that he needed her "above all," even above the children he yearned to see.[119] Mary Smith, who had a young daughter, wrote to Samuel, "O that I had wings to fly to Waterville this very moment that I might enjoy one hours intercourse with him whom I love above everything. Yea *every thing* else on earth."[120]

Mary Walker told her husband, "Have wept many times to think how soon and easily my husband may be taken and I left a widow. It seems to me I could not be consoled. If son were taken away it would seem so light a stroke compared with this I should hardly feel it."[121] Mary Smith's position was even more forceful, because she insisted that her child's value stemmed from her deep love for her husband. "[S]o much for *our* child— how sweet that we can call her *ours*. What value would she be to me if you my Francis was not her father? The thoughts of her father makes her sacred—yes, that endears her to me a thousand fold."[122]

Mary Smith's feelings for her husband enhanced and intensified those for her child. This sense of children as emblems of the love between man and wife reflects the intense romanticization of sexuality in the nineteenth century. In children, Mary and others saw a confirmation of the romantic identity they shared with their spouses.

The romantic view of sexuality as self-disclosure and children as love-tokens profoundly affected women's power to control their own sexuality. Viewed as self-expression, women's sexuality was less easily seen as someone else's property or possession. The conception of female chastity as an economic asset of a woman's father and then her husband, and the view that a woman's sexuality was property to be transferred from one to the other was a long-standing basis of the double standard. Both English Puritans and the rising bourgeoisie challenged the "deeply entrenched" double standard in the seventeenth century. Certainly Puritans in America upheld the value of premarital purity for both sexes on religious grounds, though the double standard persisted nonetheless.[123]

In the nineteenth century the conception of female sexuality as vested in the woman herself gained credibility. Historians essentially agree that there was an increasing tilt (not a complete revolution) in the view of female sexuality away from something a woman's father owned, and then her husband, toward something that only she "possessed," but the trend has been to associate female control of sexuality with the Victorian woman's right to say "no" to her husband in the bedroom.[124] Historians have argued that the emphasis on female control of sexuality often depended upon a rationale of negative trust: woman's sexual desire was thought to be "naturally" weaker than a man's. Therefore, husbands should wait upon their wives' physical needs. Denominated as "voluntary motherhood" by nineteenth-century feminists, women's right to say no in the bedroom was championed by the woman's movement as well as a host of public-spirited moral guardians.[125] What has been less noticed, however, is that female sexuality was also conceived more positively within the meaning of romantic love. The recognition of women's right to sexual self-possession included the equally important right to say "yes." That is, Victorian women controlled their sexuality, not just on the basis of an image of themselves as sexless or passionless but on the grounds of their individuality. Women (as well as men) identified physical expression as part of the gift of self, and thus indisputably "owned" by the woman herself.

Though moral reformers of various stripes hoped that men would control their sexual impulses out of chivalrous compassion or less sexually driven and purer wives would dampen male lust, this was not the only basis of male-female sexual relations in private life. Rather a more straightforward rationale was frequently at work. Victorian women theoretically had the "right" to give their bodies as they had the right to share any part of their personality or experience with another: as "free" and "autonomous" romantic selves who chose with whom and how often to reveal themselves. In other words, when sex became romanticized, it became part of the act of self-expression over which each individual was sovereign. Romantic ideology gave women a rationale for treating sexual expression as part of themselves and as compatible with their spiritual, moral capacities. In fact, both men and women identified sex as a component of the romantic ideal of free and open communication and saw physical expression as part of the gift of self, indisputably owned by the woman, and not immoral (except for coition before marriage) if motivated by love. The romantic ideology of self-expression completely overshadowed the ideology of passionlessness in private life.

Sex was progressively separated from procreation through cultural values as well as birth control techniques. Historians have agreed upon the availability of douches, diaphragms, condoms, and the knowledge of the rhythm method and coitus interruptus in nineteenth-century America.[126] This meant that a couple could, if motivated, rely upon something other than female passionlessness to restrict the size of their families. This is not to say that evidence of women who conceived of themselves as passionless cannot be found, nor is it meant to imply that women did not fear pregnancy. Indeed Alice Baldwin and Violet Janin illustrate that women used this fear to justify family limitation. Nonetheless, Victorian sexual relationships could be intensely erotic with no necessary sense of female sexuality as unnatural or impure when separated from reproduction.

Furthermore, men could cooperate in the family limitation process. While urging his wife to come to Washington, where he was serving a term in Congress, Lincoln Clark exclaimed, "How much I want to see you. *Cautious!* How are your *lady conditions?*—safe or dangerous—tell me . . . ."[127] Lincoln was inquiring about her period and appears worried about the effectiveness of their birth control practices: "I feared you had some such trouble as you mention—I think it must be the result of exposure and cold at Springfield—that it can be the existence of a 'condition' I do not believe—there has been in my opinion no *adequate cause;* neither do I believe it to be the result of age—you had best lose no time in consulting Dr. Collins—you might pay severely for delay or false modesty."[128] Apparently Julia had not menstruated since they last slept together, and she was afraid of another pregnancy. Though Lincoln's use of *"adequate cause"* (which he underlined) remains ambiguous, other exchanges indicate he was probably defending some form of birth control he or they had used during their last sexual encounter.

For example, after a visit home, Lincoln returned to Washington and wrote: "How is your special lady health? No harm I hope."[129] Five months later he inquired: "Hope your health is good—How is that? Are you in any danger?—Will you not be a good, healthy, fat merry old Lady after a while? Come as soon as you can."[130] Though the Clarks were not always confident of the success of their birth control efforts, he was committed to family limitation. Perhaps both Julia and Lincoln were looking forward to her menopause, when no risk of pregnancy would strain their sexual relations. This, at least, is one interpretation of his query about her being a "merry old Lady after a while?"

Albert Janin was another married man active in family planning. Several months after her marriage, Violet reported in a mixture of German and English that Magda, one of her friends, had inquired about the sexual details of their marriage. Magda wanted "to know about our arrangements—she says if we zusammen schlaffen nichts anders thun [sleep together and do nothing else], it is because one of us kann es nicht thun [cannot do it]."[131] Violet responded that she must hint to Magda that her health was bad. It is unclear whether this was meant to confound Violet's catty friend or was some reflection of this couple's real sexual abstinence. Whatever the reality, Magda's sexual curiosity and rather frank assumption were striking.

At the time of the next exchange, Violet and Albert were clearly sexually active. "I was awfully scared," Violet confessed, "because something was a few days late—."[132] As Violet revealed three days later, Albert was tracking her menstrual cycle himself: "You miscalculated a certain matter—it was due on the 16th but was as usual several days late—I became very uneasy and uncomfortable . . . when my mind was set at rest by the arrival—much as I fear something that we have often discussed—yet if I were told that I would never be strong enough to have *that* supreme happiness, I think it would crush my very heart with sorrow . . . I suppose it is cowardly of me to shrink from any suffering or death, but I know that you would think the penalty too great—I do think that you value my life more than any other that is or that might be,—don't you, dear?"[133] Albert was not only an active and cooperative partner in the couple's family limitation practices but also probably his wife's chief source of birth control information. Magda, the snoopy girlfriend, inquired again about their sleeping arrangements. Violet wrote her husband, "If she asks me any questions I will refer her to you for information—As you have taught me almost everything I know on some subjects, I guess you could teach her ."[134]

Their next exchange indicates that marital sex could be not only publicly seen in terms of female power but privately transacted on those grounds. Violet's threat to withhold her sexual favors from Albert was made in partial jest, but she was serious about her future willingness to carry out the threat: "Even if you were disposed to be mean, I think I could manage you . . . , if you make me angry why I won't schlaffen mit Tiv capisci?! [sleep with you, do you understand?] I would not like to have to be so unkind but if I thought it my duty I would not hesitate you know—."[135]

Violet consulted a midwife, who advised her that she had a vaginal inflammation and must stop having intercourse with Albert for a while. In spite of her threats, Violet revealed a lover's confidence in her husband's willingness to cooperate in the bedroom: "However I know that you are ready to make any sacrifice for the sake of my health,—so it is not necessary to me to change our arrangements." [136] Violet was willing to sleep with Albert because she was sure he would be sexually cooperative. When Albert inquired about the length of this sexual interdict, she replied, "As for what you ask about in your last letter, I think that this month it would be dangerous to both of us, but after the lst of January . . . ." [137]

Not all nineteenth-century couples practiced family planning, however. Mary and Elkanah Walker seem to have left the size and spacing of their family to "God's will." Mary commented in her diary: "I find my children occupy so much of my time that if their maker should see fit to withhold from me any more till they require less of my time and attentions I think I shall at least be reconciled to such an allotment." [138] By contrast, a friend of Augusta Hallock wrote about her first baby, "Twas not a mistake Gusta, but because we wanted one, dont let anyone see this letter because as you say I am going to write just as I would talk, I have got a man that is pretty well posted about such affairs . . . ." [139] This young mother wanted her friend to know that her pregnancy was *not* an accident. This is a striking disclaimer because it is based upon a normative assumption, at least between these two friends, of family limitation in marriage.

🖉 Every aspect of middle-class sex life was touched by the ideas of romantic love. From courtship petting to the practice of birth control after marriage, from the cultural values attached to having children to the weakening of the double standard, sex was viewed as an emotional expression of the romantic self. Victorians reveled in the physicality of sex when they believed that the flesh was an expression of the spirit. They literally saw the ephemeral, hidden self made manifest in physical relations.

Sex was seen as the ultimate expression of the individual's inner self, and the romantic ethos joined the sensate to the most protected recesses of personal thought and feeling. Within this set of meanings, one gave one's self in giving one's body. Sex was wholeheartedly approved as an act of love and wholeheartedly condemned by the Victorian mainstream when bodily pleasures were not privileged acts of self-disclosure. The evil of sexual promiscuity, seduction, or betrayal was not found as much in bodily sensations as in the "spiritual" violation of the personality. Sexual expression needed to be protected and guarded, just as the inner recesses of the self should be shared only with the most carefully selected individuals.

When Victorians laughed about sex in private, it was usually the self-mocking juxtaposition of the ever serious business of love with the physical reality of cold, heat, sweat, squeals—the awkwardness and imminence of sheer bodily sensation contrasted with the reverential colors of romantic love. But the serious did not displace the humorous. Rather they worked together—love infused the physical and this juxtaposition itself tickled the Victorian funny bone.

Romantic love defined the boundaries of pure and impure sex in Victorian America. Already in the nineteenth century, observing the boundary between intercourse and marriage was sometimes a strain for courting couples. But the ambiguity of physical expression in courtship was a reflection not of repression and prudery so much as the Victorian approval of sex when associated with love. The highest values of individual expression and autonomous selfhood were heaped upon the erotic. Victorians did not denigrate sex; they *guarded* it. That sex had a place of honor and preeminence in Victorian culture is attested to by the fact that making love was the ultimate act of the individual personality.

One day in July of 1867 Lyman Hodge composed a letter in the form of a long rapturous poem to his fiancée Mary Granger. He began by telling Mary he missed her, especially on the cold nights, for "the days are warm,—no embraces needed then,—but when the sun goes down, the night rapidly grows cool, and I have often thought that if I had your arms around me or mine around you it would be better than whiskey to keep me warm."

> What will make man's heart beat quicker
> Than Ale or any other liquor?
>   Do you wish to know?
>   Two breasts as white as snow
> Warming cold hands with their amorous glow
>   Twin roses pressing
>   Out from their laces,—
>   Ripe for caressing—
>   Bed of the graces—
>   Love thoughts confessing,
>   Lover's hands blessing,
>   Warming embraces.—
> As delaying hand and finger
> O'er the swelling hillocks linger
>   Linger—all so loth to go.—

. . . .
How they redden and tremble and swell
 When a lover's warm lips
 Are pressing their tips,
 And the honey he sips
From each pearl tinted shell
But adds to the fire,
Till each rose fully blown
Stands erect and alone
. . . .
Mad with desire

. . . .
I could not leave them if I could
I could not leave them if I could,
—This dazzling "feast of Roses."
 List to their song,

The poem now changes perspectives as her breasts speak to him:

 To us you belong;
Yield us thy love fearless and strong,
 Kiss us, caress us,
 Handle us, press us,
Toy with us, play with us, do as you will,
 Bite us, unlace us
 Squeeze us, embrace us,
Drink from us, suck from us, drink up your fill.
Press us together till we touch each other,
Press us apart till the valley lies bare,
Bind us with ringlets from our common mother,
Part us with rivulets made from her hair
 How our mistress pants and sighs,
 How the love lights lights her eyes!
 We have betrayed her,
What a willing captive she,
And what willing captives we
 To the invader!

Lyman's poem is less arresting for its literary skill than for its luxuriant eroticism. Beyond mere attention to physical details, the poet empathized with his lover, imaginatively portraying her breasts speaking to him. His dramatization of the woman's arousal was probably based upon the sensual details of his physical relationship to Mary. While impossible to prove, it is unlikely that Lyman would have sent this poem to Mary if they had shared none of the sexual experience together as a couple that the poem so vividly portrayed. Lyman's passion was too bold and explicit to serve as the insinuating suggestion of a hopeful lover. Yet the very boldness of his passion again illustrates the unobjectionable sexual dimension of nineteenth-century romantic relationships. Lyman and Mary expected sexual passion in each other, welcomed it, appreciated its humorous, erotic, and spiritualized dimension in some combination, and enjoyed talking about it to each other.

Lyman closed his letter with a coda:

> Why need I further write,
> My head becomes unsteady,—
> *That* lover will be warm tonight;—
> He's *burning hot* already.[140]

4

Secrecy, Sin, and Sexual Enticement

The Integration of Public and Private Life Worlds

I F IT IS TRUE that erotic activity was thriving in Victorian American life, it is equally true that Victorians committed huge amounts of energy to hide that fact. One of the more interesting conundrums of nineteenth-century culture originates in the spectacle of secrecy and the veil of denial that surrounded sexuality among middle-class Americans. Does such reticence make sense, except as a reflection of hypocrisy and repression? Perhaps. But this does not fully explain the frank, teasing, sometimes celebratory descriptions of erotic experience in an array of private expressions. The well-documented Victorian penchant for secrecy, censorship, and restrictive public pronouncements must be explained in conjunction with the sexual appreciation found in private life.

Integration is at the heart of the problem of comprehending American Victorianism. One obvious integrative framework for understanding Victorian culture is the socially defined and enforced separation of public and private morality. What was acceptable, enjoyed, and commonly assumed in private life was shielded from the public realm. This is reflected in the evidence that much private correspondence was cut, clipped, and burned. Serious efforts were made to keep personal letters out of the hands of strangers. This behavior by ordinary Americans reflected a culturally prescribed separation of life worlds. The public-private codes of nineteenth-century expression were not expected to coincide or be harmonious.

Another and more complex dynamic was also evident. Secrecy and censorship may sometimes actually enhance rather than dampen sexual excitement. As Michel Foucault observes, "There is pleasure on the censor's side in exercising a power that questions, monitors, watches, spies, searches out, palpates, brings to light . . . ." On the side of the censored, there is a "pleasure that kindles at having to evade this power, flee from it, fool it, or travesty it." The relationship of the censor to what is being censored is a symbiotic one. Foucault insightfully describes that synergetic relationship: "The power that lets itself be invaded by the pleasure it is pursuing; and opposite it, power asserting itself in the pleasure of showing off, scandalizing, or resisting. Capture and seduction, confrontation and mutual reinforcement, parents and children, adults and adolescents, educators and students, doctors and patients, the psychiatrist with his hysteric and his perverts, all have played this game continually since the nineteenth century. These attractions, these evasions, these circular incitements have traced around bodies and sexes, not boundaries to be crossed, but *perpetual spirals of power and pleasure.*"[1]

On the private side of nineteenth-century culture, "spirals of power and pleasure" are found in both famous and obscure acts of censorship and secrecy. For example, before Sophia Hawthorne censored her illustrious husband's letters, she reaped the pleasure of receiving those "forbidden" words of sexual enticement and physical endearment; later, she exercised the power to make her sexuality an object of secrecy. Using her scissors, Sophia excised material from fifty-two of her husband's love letters, while she heavily inked out approximately 156 other passages. Obviously, the excised material was irrecoverable, but the inked-out passages have been read by scholars with a microscope, filtered light, and much patience.[2] Examples of material Sophia censored in Hawthorne's courtship letters include, as shown in the brackets: "Often, [while holding you in my arms,] I have silently given myself to you . . . ."; "[Oh, my heart is thirsty for your kisses; they are the dew which should restore its freshness every night, when the hot sunshiny day has parched it.]"; "[He is going to bed. Will not his wife come and rest in his bosom? Oh, blessed wife, how sweet would be my sleep, and how sweet my waking, when I should find your breathing self in my arms, as if my soul's most blissful dream had become a reality!]"[3]

After marriage other censored lines clearly reflected strong physical desire on Sophia's part. Hawthorne teased her: "Thou shouldst not ask me to come to Boston, [for the purpose of sleeping a night in thy dearest arms;] because I can hardly resist setting off this minute—and I have no right to spend money for such luxuries." Again after marriage, "[It is misery to go to bed without thee.]" Hawthorne lamented: "Oh, my love, it is a desperate thing that I cannot [embrace] thee this very instant."[4] In spite of Sophia's rather dogged efforts to keep the physical side of their relationship secret, she failed. For even without the recovery of the inked-over portions of material, Nathaniel and Sophia's sexuality was apparent. In fact, by censoring the letters so heavily, Sophia underlined her physical relationship to Nathaniel, accentuating its importance by the very act of trying to hide it.

What Sophia left untouched in these letters was fully as suggestive as what she altered. For example, these lines remain intact: "How happy were Adam and Eve! There was no third person to come between them, and all the infinity around them only served to press their hearts closer together. We love one another as well as they; but there is no silent and lovely garden of Eden for us. Mine own, wilt thou sail away with me to discover some summer island?—dost thou not think that God has reserved one for us, ever since the beginning of the world? Ah, foolish husband that I am, to raise a question of it, when we have found such an Eden, such an island sacred to us two, whenever, whether in Mrs. Quincy's boudoir, or anywhere else, we have been clasped in one another's arms!"[5] Sophia's eagerness to excise passages with reference to their holding, embracing, kissing, and general sexual desire actually emphasized the uncensored image of them clasped together like Adam and Eve, in Mrs. Quincy's boudoir.

Sophia's censorship was not effective in eliminating the impression of their physicality. When Hawthorne wrote that he intended to see her on Saturday and that he "yearn[ed] to . . . ," by censoring the rest, she in effect called attention to his yearning.[6] Sophia appears to be guarding the secret of her sexuality (or Nathaniel's) at least in her own eyes, by an act of censorship that only served to highlight what she was hiding.[7] The pleasure of speaking of sex appears, at least partially, to stem from the fact that it was considered "dangerous"; the fun appears in part to reside in the game of hide and seek.

Many less famous correspondents played this adult version of hide and seek. Eliza Trescot expressed disappointment that her secret engagement to Eldred Simkins "is no longer one." She enjoyed the secrecy: "Perhaps you will think me strange when I confess that it was so pleasant to know that our secret belonged only to you and me . . . ."[8] Alice Baldwin ordered her husband to destroy all her letters because: "I have written in most every one something I wouldnt want any one else to see . . . ."[9] Lincoln Clark gave instructions to his wife: "Now dont read this last to any body."[10] Gusta Hallock asked James Bell to send his letters to a neighboring post office in an attempt to escape the prying eyes of a small rural community. At one point, she not only ordered him to change post offices, but to address the outside of the envelope to her sister.[11]

In the private arena a correspondent could evade public authority simply by writing a "naughty" letter. Clara Baker reflected this sense of pride in private naughtiness when she remarked, after Robert Burdette lost a trunk on one of his lecture tours, "If those southerners get a chance at the love-letters it carries I'm afraid you'll never see it more. They like 'hot stuff.' "[12] Robert enjoyed the thought of Clara shocking the ladies of her women's clubs by reading his love letters in one of their public meetings. "You are tempted to 'bind my letters into a book?' " he chuckled; "Oh do, Santa Clara—do! Wouldn't they make what Horace Greeley used to call 'mighty interesting reading' for your monthly assemblies! Wouldn't I like to see you reading some of them?"[13] The irony may be that some of these same ladies took pleasure in similar thoughts. In a comparable vein, he told her that "you'll set the Book trade on fire, unless you edit—your—xxxxxxx's letters with an ice machine!"[14]

Certainly there was a pleasure in these letters—both explicit and implicit—at evading the public "no" to wayward sexualities, a pleasure not so much in actually scandalizing anyone, but in thinking of how scandalized people would be if they could read the letters. This, of course, made the speaking of sex in private "nicely naughty," in Dorothea Lummis's phrase, for breaking a secret every time one spoke of sex made the speaking of it that much more worthwhile. This may help to explain the extraordinary compulsion of some Victorians to verbalize their sexual feelings and experiences. Talk of sex was a hidden (and thus all the more exciting) secret in Victorian America that might forge special bonds of intimacy.

The older view of Victorian repression connected public and private Victorian culture either through the relationship of hypocrisy—when private sexuality was exposed—or the straightforward dominance of antisexual public values and behavior in the private sphere—when Victorian silence went unbroken. However, the integration of public-private life worlds involved a more subtle relationship: the public constraints on sexual expression actually encouraged the growth and intensified the experience of private eroticism. Fortunately for the historian, a central component of the Victorian public taboo surrounding sex is recoverable in the written word. Forbidding sexual expression in public helped to create and also to explain a veritable explosion of private sexual expression whose pleasure was partially enhanced by the thought that speaking of sex was forbidden. Victorians gleefully censored their own private correspondence and worried about being found out. But the idea of someone reading their letters hovered like a ghost of lasciviousness over correspondence, making their missives almost a secret erotic art.

The correspondence between Robert Burdette and Clara Baker illustrates the intense eroticism that might flourish not just in spite of but

because of the public-private dichotomy in sexual expression. Robert Burdette was fifty-three and Clara Baker forty-two when they commenced
their correspondence in earnest. Robert's wife had died many years earlier,
and Clara was a widow of two previous marriages.[15] Their courtship, in
spite of Clara's censorship, is both erotic and easily documented. It was
also an extraordinary seduction by a man who represented all the public
Victorian values of self-cultivation, didactic piety, and good-natured interest in moral order.

Burdette was one of the most popular lecturers in America during the
1890s. He traveled on the Redpath lecture circuit to many small towns,
delivering humorous talks on growing up in America, home life, and his
special forte according to his wife, moral advice to young men. After his
marriage to Clara in 1899, he became minister at the First Presbyterian
Church of Pasadena and then permanent pastor of the Temple Baptist
Church, which he and Clara built from the ground up.[16]

Clara, a prominent clubwoman at the time of her courtship, held key
leadership positions in Ebell, the California Federation, and General Federation of Women's Clubs. She actively participated in a wide range of
clubs, charities, and civic groups. A woman of spirit and vitality, Clara
led her public life within a framework of conventional Victorian values—
moral education, home, domesticity, time-thrift, honesty, service to God
and humanity.[17]

Both Robert and Clara espoused strong ethical and moral codes which
they themselves defined as lifelong tests of personal character. Neither was
a cultural renegade: both conformed to all the public expectations of their
cultural milieu. They lived intensely public lives, as they did private lives.
To suggest that somehow they were Victorian in public and not in private
may be convenient, but is not only ultimately inadequate, it also misrep-
resents the nature of their life together.

A casual acquaintance for many years, Burdette launched his romance
of Clara Baker innocently enough in April 1897 by wishing he could kiss
her fingertips.[18] They saw each other in May, before Clara's three-month
trip abroad, but their approach in the following months was tentative and
the courtship unfolded with a shy reticence until a letter reached Robert
on December 7, informing him that Clara was gravely ill. Clara recovered
fully, but her illness became an opportunity and an excuse for Robert's
increasing boldness. After December 10, 1897, he deluged Clara with letters, testing the extent of her romantic interest in him.[19] In early January,
encouraged by Clara's response, Robert made an unblushing pass: "Good
night, my darling. God bless you, and keep you, and rest you in His arms
of infinite love. So near to me your letter has brought you. I am to have
[quoting from her] 'What I will?' I will stoop at your side to take it then,

dear; I will hold the dear face close to my breast and close in my arms, and tenderly take from My Lady's lips what only her lover can take—."[20] A few days later, obviously anxious to act as well as dream, he set plans in motion to see her as soon as his lecturing schedule would permit.[21]

Robert boldly replied to her next letter with obvious physical innuendo: "Of course, while you are weak and helpless, I can take all manner of rough liberties with you. I can hold you in my arms, and press your tired head down to its throbbing pillow with my great square hand. . . . You are not at your best, yet, and my arms are iron, dear. Iron, to hold. But velvet to caress."[22] Clara's flirtations had opened the floodgates of Robert's physical desire. His response to another of her encouraging letters was an invitation to come to his arms for "always they are open, and waiting, and hungry for you. Climb into my lap when you are tired, and I will sing you to sleep with kisses. Lie on my breast when the sweetness of love shall draw you; you are not safer on your mother's."[23]

Robert's romantic seduction technique was to take great pains to reassure Clara that she was safe in his arms—by dramatizing scenes in his letters of her resting innocently in his embrace: "Tired little Girl, dear tired Little Girl, tired of too much work, tired of the big, dry books, tired of the hard slate and the gritty pencil and the long, tangled sums; tired of too much play in the hot sun, tired of the close rooms and the chattering crowd, go take off the beautiful bonnet and the regal gown you've been playing president in, put on some-[thing] light and easy and dainty, that clings to your little figure and falls away from it and coquets with it, and makes you look tenfold more bewitching every time one looks at you, and come here, and climb into the lap of a Boy who loves you so dearly he doesn't know how to tell you so. And he will rock you, and tell you stories about yourself, and kiss the tears away from the tired eyes, and push the hair back from the throbbing temples, and tell you how dearly he loves you. . . . And you shall fall asleep there, rocking in his arms, And more than likely then, the boy will fall asleep too, nodding over the dainty head cuddled down in his arms . . . ."[24]

Robert created elaborate word pictures of physical intimacy. Through these tableaux, Clara was meant to identify with the action as she read the part he created for her. Robert's role was both leading actor and the benevolent, wise, and omniscient stage director. That he was self-conscious about staging this unfolding physical and psychological drama for her, there can be no doubt: "What fair, sweet pictures do I draw of you in my visions, Dear heart—my own dear Heart—loving and tender, pure and sweet, and true; I kiss your hair, dear—that's for I don't know what, save that your hair has always held a strange, sweet fascination for me. Do you remember in my earlier letters that was the only think I dared to kiss? I kiss your cheeks, dear Little Girl, and that's for friendship. I kiss your brow, sweet Lady Violet, and that's for reverence. . . . And I stoop to cover the soft warm lips with a lover's kisses—many and long—and lingering—a Lover's kisses, dear. And you can only get them of one man. Only one. No one in all the world—no one else, can kiss you as I do. My darling; My darling." [25]

Robert, conscious of his rhetorical success and sensing he had won a measure of her trust, pressed for greater intimacy. In a March letter, for example, he undressed her, but carefully portrayed himself as the fond parent preparing the tired child for bed. [26]

Robert created several characters to personify different aspects of Clara's personality—Lady Violet, Mrs. Baker, Sweet White Violet, Sweetheart, and his favorite, "Little Girl." [27] In loving "Little Girl" Robert celebrated the childlike part of Clara. He asked: "What do I see in my darling's eyes? I see in their tender depths, a child's innocence, sweet and touching; I see a woman's sincerity, direct and true as a light-ray; I see the loving nature of a girl, to whom this new love is as new, and strange and sweet, as her first kiss; I see affection, stronger than passion, deep as the sea; I see an intellect, strong, well-poised, just and generous—a wisdom that ennobles love; I see this intellectual strength bending, with lovely grace, its beautiful strength to the gentle power of love, so that the brain and the heart are yoked together for the best that life can do; I see the tender rapture of a loving woman, whose love is a joy to herself, a blessing and a happiness to her beloved; I see a woman's truth, clear, shining, transparent in its crystal depths as one of her own mountain lakes; I see in her dear blue eyes all womanly grace and perfection . . . . " [28] Although Robert delighted in her childlike innocence, sweetness, and girlish charms, he also appreciated her strength, the power of her intellect, her passion, and her womanly poise. The "Little Girl" conceit allowed him the symbolic control of a father and a confident manner in which to approach her, but it did not deny the powers of her maturity as a woman, which he also recognized.

Clara, however, worried whether he loved the real woman or the idealized version of his own creation. But she was more than responsive to his verbal seductions, apparently even encouraging him with tableaux of her own. [29] Almost as if to fan the flames of his enthusiasm, Clara posed the obviously tantalizing question for them both: what were his intentions regarding their real meeting in June? Never at a loss for words, Robert responded that he only knew for certain what he would like to do. [30] His letters during this period, sometimes twenty pages in length, combined doubtful hesitation with brash anticipation of their June encounter. [31]

Both feared they would disappoint the other, but their anxiety was often swept away in highly charged anticipation of their long-awaited rendezvous.[32] Robert probably revealed what they had both been thinking: "And then as our love grows, as our hands and hearts become acquainted, there will come the sweet freedom of 'Robert and Clara Land' in our real hand-and-heart lives, as now we find in our letters. And these sweet pen pictures dear—oh, they are but 'dress rehearsals,' after all! What will the play itself be like?"[33]

Calling his letters "dress rehearsals," Robert disclosed that he had wooed Clara ardently and carefully as well. Taking her through several stages of physical intimacy in his letters, intimacy became familiar and non-threatening. The response pattern he so carefully cultivated in his "dress rehearsals" was a type of verbal conditioning that prepared them both for genuine intimacy when they met.

When he learned a few days later that his visit would be delayed, he thought Clara would be troubled and pained. In response, he staged a dress rehearsal whose intent was comfort and sympathy, but whose erotic intensity was barely concealed. "Good night, my darling," he whispered, and added seductively, "You shall have your pillow to night, dear love. And I? I whose heart always beats madly at the thought of this hour, I will not even kiss the snowy breasts that my lips so often seek in dreams. For I can only see your dear, dear grieving face to night, my darling. See, I cover the white breasts tenderly, while I kiss the dear dear lips, and fondle you with ever increasing tenderness. The soft outlines of your dear body melt against my own, but there is no ardor of passion to night, My own Sweet Violet. Only tenderness. I feel the carressing touch of your white limbs, dear, but my eyes, my thought, my heart are on your dear, sweet face. Rest then, my darling. All your matchless charms to night, the dimpled knees, the curving sweep of the ivory thighs, the snowy breasts, and the waist that my arm is clasping, only make you more tenderly, sweetly, sacredly dearer to me, sweet love, and all my love to night is turned to comfort. . . ."[34]

The next day Robert created another scene in which "Little Girl" rested in his arms: "And this broad collar with its frills of lace may be open a little bit? For I have one free hand, you see, and it wants to play hide-and-seek with two sweet, soft, snowy play fellows now and again. 'And you have a hand?' Well, my Violet, it has its own hiding places . . . ."[35]

Sensing even before this letter arrived that her control was faltering, Clara tried to slake the fire of Robert's intensifying passion, asking him to write "only" three times a week.[36] By contrast, he had been sending her two, sometimes three letters a day, seven days a week. From May 2 to June 1, 1898, he obeyed her edict, but not its spirit. Composing "on" day letters that often reached fifteen pages, he wrote "off" day letters that he eventually mailed to her. In one "legal" letter, he asked her to be the sexual aggressor in one of his "dress rehersals." He aimed to be the passive partner and just take love: "Just now, my darling I want to *be* kissed. Am I heavy, dear? . . .Can you hold me—so—just for one little minute, while you rain your sweet honey-kisses on my face, and cover me with dearest, tenderest caresses? . . . And I, the happiest, happiest, happiest man am lying in the shower of your caresses, giving you not one in return. Just taking all the love and petting you have to give. Do you mind, Violet dear? Does it bring a flush of protest to your woman's cheek, when I 'make' you do this for me?"[37]

Offering an insight into masculine sexual preferences, Robert explained why he wanted to receive caresses as well as give them: "Don't you feel that way some time, My Sweet President of Ebell—when you are tired, and just want to lie in my arms in languorous ease, and not move, and just let me pet you, and comfort you, soothe and caress you—JUST DO AS I PLEASE WITH YOU? ANYTHING I WANT TO? Just as long as I let you lie still and lose yourself in a waking dream of love and a lover!"

Robert asserted that his sexuality was much like hers—sometimes passive, just as sometimes she wanted to lie in "languorous ease." Contrary to stereotype, Victorian men might be more than receptive to a "respectable" woman's sexual passions and enjoyed a passive sexual role, at least occasionally. Burdette urged Clara to become the sexual aggressor because he enjoyed "just taking all the love and petting you have to give."[38]

Robert's expectations of a woman's sexual passion was not unusual in American middle-class males. What set him apart was the power and intensity he attributed to Clara's erotic self. In deepest sincerity, he gallantly assured her that he would not take advantage of "the roused and throbbing nature of your woman's heart, which in its sweetly awakened passion might feel tempted at some wild, exultant loving climax, to throw all its 'reasons' to the wind and give life, love, thought, being, destiny— everything—to the lover who holds you in his arms . . . ."[39] Such a romantic notion might in part be wishful thinking, but it did not mask the attribution of a driving sexual passion to her or to women generally.

Clara, withdrawing her edict, freed him from all restrictions on his writing. Confident, he staged a "dress rehearsal" that was more daring than any so far: "Come here then, sweet. No, your head isn't in the way— you know I love the scent of your hair—I love to touch it with my lips, and feel it upon my face. See, I kiss it here on the moonbeam that marks its parting; and I lay my face into its coiled masses as one might smell a mass of clustered violets. And my lips follow it down to the dear white neck, which again and again I kiss—can you feel my breath playing upon it, dear heart? Is it that your hand on my cheek draws my face from your hair to your lips, or does it just meet my face as I seek for your own? . . . How warm and soft your lips are, dear. There is no laughter on them now—not with that love-light softly glowing in the dear blue eyes. A soft love-born dew is on the tender lips, like the honey-moisture on a dawn-kissed rose. . . . Holding you closely and more closely still, yet with all gentleness, I smooth the silken hair back from the temples, with caressing hand and many kisses. Softly I press the dear, white, beautiful breasts that rise like snowy mounts above the heart I love—the heart that throbs for me. Sweet and white, balmy and fragrant—my lips in loving homage lay their loyal service upon them so tenderly, softly pressing their swelling grace, lingering in the sweet warm valley that divides these twins of snow and warmth, and clinging with loving compression on the dainty tips of soft October tint. And all the time I am whispering your own dear name . . . ."[40]

Robert finally fulfilled his promise to visit her in Pasadena and became her living epistle.[41] After the visit Clara wrote in her most coquettish fashion: "Want to see Violet? Want to kiss your sweetheart? Want to play with your Little Girl—she's spoiling for a little romp—and much kissing." She closed on a note of wistful yearning, "Good night, sweetheart—oh! for a good night such as has been—but the memory of it is very sweet."[42] Though Clara acknowledged that Robert thrilled her to her very being, she remained undecided about marriage.[43] Perhaps because she was older, had already been married, and was financially secure, she did not immediately accept his proposal, although they had already consummated their physical relationship.[44]

Since she later censored her letters more heavily than his, only his version of their real sexual intimacy survives. In his view, her caresses were so "child-like" at first that he said to himself, "I have been dreaming. . . . There can be no woman's passion slumbering in such a tenderly affectionate nature as this—this is pure child-love. But by and by, when my own caresses and kisses—given at first, under this child-influence of your own . . . as they quickened into the fire of passionate desire by their repetition, as they became amorous and eager, as I forced my way into your arms, upon your breast, stifling your protests with my kisses, struggling into the snowy smother of your skirts,—I cannot go on dear—I saw the new light dawn in your blue eyes and transfigure your beautiful face, and lo, My Little Girl was gone, and My Lady—My Lady Violet . . . lay panting in my enraptured arms."[45] In spite of her two previous marriages, and in spite of his own oft-repeated rehearsals of their mutual passion, he confessed, "[I] half-expected, when my wanton hand followed the guiding of my hot desire, that it would rest upon a dimple-grotto, sealed with the tender veil of virginity, and smooth as the child's own cheek. And lo, the silken vines that meshed and fringed the throbbing fountain of a Woman's Love! My Little Girl—My 'Sweet White Violet.' "[46]

Their correspondence emphasized the mutual expression and pleasure of physical love: " 'Violet wants to be loved,' " Clara wrote enticingly, " 'and kissed, and petted—' " and he responded, "Violet must let me take off some of these things then. And this that is left I will gently raise out of my way, till its fleecy folds lie above the milk-white breasts. How beautiful you are, my Violet! How delicately graceful are all the curves in your dear white body! How smooth and satin-like. How ravishing the perfume of your breasts. And this soft brown silken thicket that discloses and yet conceals your—how strong are your clasping limbs, my Violet— so beautiful and so strong. Is that right, my darling? There, dear! Do I hurt you, Sweet? Hold me close, close, close then my darling. And now, Sweet White Violet—my love phrases, my pettings and my kisses are raining on your silent face, and you are my sweet heart and I am your lover."[47]

Robert enjoyed a luxuriant eroticism in another passage that, had it caught her censorious eye, Clara would doubtlessly have suppressed: "Which wants the most? We'll make a little 'match' some day. I will draw your dear white warm throbbing body into my arms and nestle you up to me until swell and dimple, muscle and hollow fit into each other, and every breath and heart throb feels the other—and I will crush you to myself and fasten my lips to yours, and see who can lie there the longest and do the most petting. Your petting Sweet Violet, will be prettiest, with the cooing of my Dove, but mine will be the stronger. From your sweet brown hair to your dainty instep I will kiss you, love, and I'll murmur a thousand impassioned words with every kiss. I will cling to your perfumed breasts with my passionate lips till mother and sweet-heart love will throb together in your heart as it upholds the weight of your lover-husband. I will bury my face in your fragrant hair and praise its silken beauty till your heart is aflame with love. I will clasp my arms around the dear rounded waist and belt it with a girdle of kisses. The swelling thighs shall taste my lips until you can stand no more of it, and then I will kiss your dimpled knees to rest you, while I laugh with an overflow of joy and love. My Little One—ten kisses to your one I will give you—long—sweet— lingering—passionate kisses. We will love each other My Beautiful Betrothed, until love itself cries out in an agony of bliss, in the very pain of ecstasy—'My Violet!' 'Oh, Robert!' . . . ."[48]

These passages illuminate a most intimate area of Victorian life; but beyond their physical details, they reveal an unabashed sexuality and an enjoyment of bodily pleasure that had no sense of sly amusement, no hint of unnatural desire, but rather celebrated a joyful sensuality. Robert expressed a genuine appreciation of Clara's erotic feelings. While it might be tempting to laugh at lines such as "the throbbing fountain of a woman's love!," to do so would be to risk missing the unselfconscious acceptance on Robert's part of the pleasure of their physical coupling. Though many correspondents expressed erotic interest and enjoyment, the Burdettes' highly respectable social standing and generally strong involvement in the didactic and moral code of Victorian America call attention to their attitudes and behavior in the area of sexual expression.

Unfortunately, Clara's diligent censorship limited the extent that her perspective can be read directly, but her sexual enthusiasm was unmistakable: "Such a lover! And such a fancy! I wonder sometimes how he dares to pen it all: what if other eyes should read? If your fancy must run riot—and I'm not saying it isn't delicious to have it—pen the fancies by themselves . . . ."[49] Clara urged Robert to compose his sexual fancies on separate sheets of his letters so that she could destroy only those pages and spare the rest. Quite obviously Clara missed a few pages, revealing more than she might have wished by what she called "delicious."

As she prepared to leave for the East to see Robert again, another passage survived which unambiguously and directly conveyed Clara's attitude toward their physical relationship. "Tho we shall begin where we left off at parting—tho it will seem the most natural as well as sweetest thing in the world for you to take me in your arms and love me and kiss me and talk to me as only my Robin can. . . ," she said, "my heart beat[s] a little faster and I wish it was to be tomorrow—oh to night and *now*."[50] Clara and Robert were reunited on the East Coast the first of October. After two weeks, they separated, planning to meet again in Rochester, New York, at which point she finally agreed to marry him.[51]

Robert Burdette was a man who opposed card-playing, spoke out publicly against bicycle-riding on Sunday, and received newspaper notices such as: "but here is a man who, if he could make a million men laugh at the cost of a single hidden tear from some poor stung one's eye, that tear would never fall. And his language is as clean as his thoughts are pure!!"[52] This man, however, could joke slyly with Clara: "But my arm is all right, dear love. I'll go to the dentists and have it pulled, and send it to you. Then you can tie it around your waist—'cinch it'. . .Tie it on tight dear, or sure as love it will slip down. I know that arm."[53]

If Robert and Clara had actually built a bonfire for all their letters, as Robert suggested, historians would have little more than pious newspaper reviews or reverential comments on their relationship in Clara's autobiography, such as "with frank acknowledgment of loyalty and devotion to the memory of those we had 'loved and lost,' we pledged our future together to building a companionship which should glorify a home for our boys and render a service to God and humanity."[54] To call such public piety "hypocrisy," a common practice in the past, is a superficial way to characterize the relationship of public and private life worlds in Victorian culture. In keeping their sexuality out of the public domain, the Burdettes were observing the rules of expression that defined sincerity and full disclosure of emotion as a privilege of the private sphere.

But Robert and Clara's correspondence is an illustration of more than the separation of public and private life worlds. It also reflects the extraordinary compulsion of some Victorians to express their sexual dreams and acts in words.[55] Robert and Clara obviously relished the telling almost as much as the doing. No doubt this stemmed partially from an intimacy that their shared secrecy reinforced. The common stratagems of letter censorship suggest that the "hidden" nature of sexuality was itself a pleasurable enticement to written erotic exchanges: more spirals of "power and pleasure."

Social constraints that impose secrecy may not always function to extinguish or even "depress" the excitements they seek to control. Powers that constrain may also be powers that enhance or excite. A more obvious example of this paradox is the attention the censor lavishes upon suppressed material, bestowing glamour as well as publicity on experiences that would otherwise have remained humdrum and obscure.[56] This enhancement process was at work in Victorian self-censorship. Thus secrecy, including anticipatory as well as actual acts of self-censorship, doubtless heightened the erotic experience of some people in Victorian America.

Secrecy was a special force in middle-class experience as public restriction shaped the contours of private intimacy. For example, the Rev. John Bayley's *Marriage As It Is and As It Should Be* urged the "true-hearted wife" to be "a faithful keeper of her husband's secrets." He insisted that "The sacred seal of secrecy is upon her lips" by the very fact that she was in a social relation of marriage.[57] For many Victorian advisers, only withdrawal from the public eye sanctioned the revelations of the true self.[58] Victorians cultivated an aura of secrecy in their male-female relationships, believing they must shield their "true" feelings from the world. Their concept of privacy, a symbol of negation, received its special aura from public prohibition. Thus the middle classes could take pleasure in verbalizing their erotic experience precisely because such talk was forbidden in public life.

Composing and later censoring nineteenth-century love letters afforded Victorians the pleasure of talking about sex: capturing, dissecting, and describing it; whispering fantasies, confiding personal secrets, and hearing confessions about it; but still maintaining an unblemished public image. To do both was to integrate the twin demands of Victorian America.

🐝 The evidence that private sexual attitudes and behavior did not fit the stereotype of Victorian sexual repression raises significant questions. Especially pressing is the disjunction between expressions of sexuality in private correspondence such as the Burdettes' and the large body of anti-erotic public pronouncements. Though public sexual discourse assumed many forms, medical and moral advice books have often captured both scholarly and popular attention. Given their reputations as primers for prudery and repression, Victorian prescriptive advice on sex sharply focuses the problem of reconciling the seemingly contradictory evidence of public versus private experience. Thus, advice books serve as a convenient focus in the examination of the relationships of public and private life worlds.

The current historical view of Victorian sexual advice literature is that it was varied rather than a monolithic repository of repressed, luridly anti-sexual ideals and values.[59] Emphasizing variety, this interpretation often stresses industrialization, social disorder, and the need for social control.[60] It offers insightful explanations for the extremely negative stances toward sexual practices taken by some Victorian sexual guides. Establishing the fact of variety within the advice literature and analyzing the connections of anti-sexual attitudes to general social developments, however, does not fully explain the relationship between the negative response of some advisers to birth control, masturbation, and intercourse, and the evidence that Victorians in private enjoyed sexuality, rejected the ideals of passionlessness, and sometimes even relished the erotic. Did the private world of middle-class Victorians exist at such variance from the public world of sexual advice literature written by their own kind? No; but the relationship of public advice to private values and behavior requires even more refinement. For example, which sexual advice was most and least representative of private values and behavior? Did the different authorial roles that Victorian sexual advisers assumed in addressing their audiences affect their ability or even interest in reflecting the cultural beliefs and behavior of their time?

Victorian sexual advisers can be divided into three camps: restrictionists, moderates, and enthusiasts.[61] The enthusiasts, viewing sex as the key to health and happiness, encouraged sexual expression. Though some enthusiasts placed limits on sexual practices, they saw sex as a healthy biological drive that in an ideal world should be satisfied untrammeled by the artificial restraints of civilization. This was a radical stance in nineteenth-century culture.

Generally representing the mainstream of Victorian thinking, moderate views were indeed fairly integrated with private reality. Though differing among themselves on many issues, moderates separated sex from reproduction and generally approved of the fullest sexual expression only as an act of love.[62] They tended to associate sexuality with romantic love. Moderates always disagreed with their restrictionist colleagues about the purpose of sex and were less exercised overall about what were deemed sexual transgressions. They advocated moderate sexual indulgence and gave more weight to pleasure than punishment. The moderate viewpoint was consonant with the private attitudes toward sexuality expressed by Victorian couples.

Finally, restrictionists published ringing scientific jeremiads, calling for sexual limitations and restraint. Presenting lurid visions of sexual excess and indulgence, restrictionists urged that sexual activity be limited to one purpose: procreation.[63] Restrictionist advisers offered extreme examples of stereotyped anti-sexual Victorian public pronouncements. If the relation of restrictionist values and beliefs to the private correspondence can be understood, then one of the most difficult roadblocks to integrating Victorian public and private life will be removed.

꙰ Contrary to stereotype, American advisers from enthusiasts to restrictionists recognized the sexual appetites of women as well as men. Though post-Civil War sexual and moral advisers were much more explicit than their antebellum colleagues, there was considerable unanimity in guidebooks throughout the Victorian period that both men and women had a natural sexual urge.[64] While some restrictionists and moderates argued that men had a stronger sex drive than women, no American adviser saw women as completely asexual.

Several examples illustrate these attitudes. Edward Foote, from the perspective of a sexual enthusiast, claimed that women were "the greater sufferers from sexual starvation" than males in this society. He believed that passion was an "integral part of the individual" of either sex.[65] The moderate H. R. Storer portrayed men's sexual drive as overwhelming and characterized chaste unmarried men as martyrs. He believed that women's sexual instincts were subordinate to their maternal instincts, but he noted that "both sexes are impelled, the male by far the more strongly, towards bodily union."[66] He granted that women also experienced physical desire. It was a matter of degree, not kind. George Napheys, also a moderate, described the wedding night as "attended with more or less suffering" for women. But with the loss of virginity, he believed, a woman's distress should disappear. He defended the dignity and propriety of the sexual instinct in women, arguing that sexual passion for both sexes was a "natural and healthful impulse. Its influence is salutary."[67] Mrs. E. B. Duffey, a moderate, characterized men's passion as "much stronger and more easily inflamed" than women's. She nonetheless acknowledged that women could be carried away by sexual desire.[68]

An anonymous restrictionist physician addressed men: "You may talk of the instincts of nature, but in you these instincts are brutalized; in her they are artificially suppressed. You have the double task of curbing the former and of developing the latter." And this adviser continued, "You should never forget that this passion is ordinarily slower of growth and more tardy of excitation in women than in men, but when fairly aroused in them it is incomparably stronger and more lasting. This, of course, with due allowance for differences of individual temperaments." Moreover, this physician worried that "a strongly passionate woman may well-nigh ruin a man of feebler sexual organization than her own . . . ."[69]

Perhaps the most oft-quoted observation on women by a nineteenth-century medical adviser is found in an English medical advice manual that was popular in America. William Acton's statement that "the majority of women (happily for them) are not very much troubled with sexual feelings of any kind" has been widely cited; the fact that it occurred in a long section on male impotence has received no notice.[70] Yet, its context is crucial to a more subtle interpretation of its import. Acton revealed a nervousness and anxiety about female sexuality that was at least as dominant as the substance of the remark. He perceived a serious problem in contemporary masculine fears about the overpowering nature of women's sexual urges, noting that some men believed "that the marital duties they will have to undertake are beyond their exhausted strength, and from this reason dread and avoid marriage."[71] His widely cited view of female sexuality was offered in a section on "false" male impotence, by which he meant impotence without a physical origin, and his purpose was to reassure men who had no physiological disorders that "No nervous or feeble young man need, therefore, be deterred from marriage by an exaggerated notion of the duties required of him. The married woman has no wish to be treated on the footing of a mistress." His claim that "a modest woman seldom desires any sexual gratification for herself" has to be juxtaposed with the image of imperious female sexual desire that it was intended to mitigate. The irony is that throughout his famous passage proclaiming the absence of female sexual feelings, the ghost of feminine lasciviousness hovered. What has been missed is Acton's attempt to allay men's anxieties through a strategy of reassurance that paradoxically betrayed fears of a strong feminine sex drive.

Though American sexual advisers did not single out women's sexuality as a special threat, the fear of sexual anarchy loomed large for the restrictionists who most often insisted that intercourse was intended, by God or nature, solely for reproduction. For example, both John Cowan and John Kellogg argued that the natural end of sexual intercourse was reproduction. Following this logic, both recommended continence unless reproduction was desired.[72] Kellogg, however, was less rigid and more pragmatic. Recognizing the need for some compromise, he accepted intercourse during a woman's "safe" period—a time when pregnancy might be avoided "naturally." Restrictionists feared the dissipation of external controls over sex that the ethic of privacy embodied in middle-class lives. Worried about surrendering behavior of such importance to the individual conscience, they insisted on tying sex to reproduction because it established an absolute rule of practice which admitted no ambiguity. When reproduction was the only legitimate grounds for intercourse, sexual purity could be clearly defined and judged.

Restrictionists spurned a relativistic sexual ethic. They feared that the boundaries of sexual activity would be blurred if the morality of sexual behavior was at the bidding of unstable privatized feelings. They attacked this spectre of chaos with their own logic of sex for reproduction. Flatly declaring that the one object of marriage was the preservation of the species, John Kellogg insisted: "a genuine woman looks forward to the possibilities of motherhood with glad anticipations . . . ." But he also betrayed a disquieting anxiety: "it is not to be supposed that motives of so high and chaste a character are always the actuating ones. The passion denominated love might often be more properly termed lust."[73] So long as intercourse was separated from procreation, restrictionists feared that love would be the gateway to lust.[74]

The intensity of the restrictionist condemnations of non-reproductive sex was revealing. The author of *Satan in Society* denounced a diaphragm as "the invention of hell."[75] The controlling assumptions behind such condemnations by late nineteenth-century restrictionists were as important as their rhetoric. Their advice reflected a powerful conviction that the vast majority of Americans, women as well as men, had little sexual restraint and vast sexual appetites, only held in check by the desire to avoid continual pregnancy. Only fear of pregnancy, they thought, imposed a limit on the "violence of sexual inclinations" abroad in the land. Artificial birth control, they contended, removed the only brake upon the lust of an unregenerate people. Ironically, the restrictionists actually held a highly sexed view of the world; and they exhibited a keen, perhaps even hypersensitive, appreciation for the "rising volume of sexual sins."

Restrictionists rivaled or exceeded enthusiasts in their view of the average person's sex drive. In fact, they certainly held the highest and strongest estimation of the Victorian libido, seeing it in the aggregate as barely containable and ready to escape into unbridled licentiousness unless restrained by a fortified system of social control. Their powerlessness left restrictionists frustrated because society afforded them few resources other than moral suasion to bridle this force. Their escalating rhetoric was aimed at the individual conscience of a reader who must be driven or frightened into practicing sexual restraint.

Even restrictionists, however, did not present a monolithic body of advice. They expressed complex attitudes which often mixed positive and negative values toward sex. The anonymous physician who called the diaphragm an "invention of hell" also insisted that women should share with men in sexual pleasure during intercourse. The husband, he claimed, often killed a woman's passion on their honeymoon. Openly blaming the husband, he warned that another man might easily awaken in a "cold" wife "the slumbering spark, and the flame be all the wilder for the rights it has been denied. If this does happen, alas for poor mortal frailty, if only natural virtue sustain her!" This restrictionist believed women were capable of great passion, and, openly sympathetic to female sexual needs, he urged men to cultivate womanly passion as one of their primary conjugal duties.[76]

The restrictionist wing of the Victorian medical and moral establishment, which achieved high visibility, has mistakenly been treated as representative of the nineteenth-century mainstream's sexual attitudes and values. Restrictionists had a complex relationship to their own society. They offered their advice in a jeremiad tradition: they bewailed and dissected the sins of the American community. Urging reform upon the people, they warned that if their listeners did not mend their ways, retribution would follow. Restrictionists were especially perplexed by two developments whose conjunction seemed ominous to them and terribly threatening to American morals: the growing ethic of privacy with the physical segmentation of space that accompanied it; and the struggle over how to define sexual virtue or how to distinguish pure from impure sexual behavior.

Interestingly, restrictionists focused upon sexuality as the consummate expression of the new individual freedom. They feared the unbounded opportunity for sexual expression in the home, where the middle-class ethic of privacy prevented any outside control. James Foster Scott inveighed: "It is hardly necessary to say that improper sexual conduct is rife among us, and that it is polluting the sanctity of our homes to a degree only superficially appreciated." Moreover, "Unchastity, being a secret sin," he insisted, "is therefore all the more dangerous."[77] John Todd, noting the potential freedom of private space, resorted to a standard evangelical threat of divine judgment: "The eye of God reads all . . . . No darkness can conceal your deeds. . . . He will bring every secret thing into judgment."[78] The "secret sins," which so troubled Todd, were more than masturbation: "The repining thoughts, by which you secretly rebel against the providences of heaven, are secret sins,—but they cut you off from religion. The envy of the heart by which we covet what others have, but we have not—and the thought which is unholy and impure, is a secret sin;—but, indulged in they will destroy your hopes of eternal life."[79] Todd was upset by the fact that people might escape the prying eyes of neighbors and friends to an inner sanctum where there was no one to enforce social or religious codes.

The authors of moral and medical advice books were sometimes frightened by the total absence of supervision in the middle-class home. Enlarged and segmented room structure along with smaller families allowed for more potential freedom and thus more abuse. The notorious Victorian anti-masturbation mania itself may be interpreted from many angles, among them that it objectified and symbolized a deep fear of the implications of privacy.[80] Samuel Woodward's *Hints for the Young* indicts masturbation as the choice to be alone.[81] It was the asocial disease of privacy par excellence in Victorian America, and its vigorous condemnation must be read in light of the fact that it was the ultimate symbol of private freedom and atomistic individualism.

Early nineteenth-century restrictionists expressed horror at the opportunity for riotous living and unchecked indulgence that the growing ethic of privacy created. Their dilemma was in how to control a space where, following accepted social practice, outsiders were now excluded.[82] William Alcott's *The Young Wife* discussed the impurity hidden after marriage in the private sanctums of family life. Recognizing the high value already placed upon freedom of expression in private life, he worried about private indiscretion, which he defined as "freedom of manners, and a levity of conversation . . . ." He continued defensively: "And some are even so blinded as to make their married state an excuse for laying aside that delicacy which they regard as an unnecessary formality. No doubt, the artificial reserve of former times ought to be discarded. At all events, modesty is not stiffness. There is, however, no little danger of going to the contrary extreme. Odious as formality is, it is better—far better—to be deemed somewhat formal, than to be actually indiscreet."[83]

The early Victorian restrictionists (1830–60) shared with almost two generations of later restrictionists (1860–1900) a fear of secret space and the potential abuses of privacy. Both were chagrined by the actual loss of social control in the middle-class home and the absence of reserve it implied. The pre-Civil War generation relied upon what were time-worn threats of divine punishment. After 1860 restrictionist advisers more often substituted the judgment of Nature for divine retribution. Using threats of physical pain and deterioration rather than supernatural intervention, they thundered against masturbation and sexual excess. They continued to express moral judgments but, struggling for scientific legitimacy, warned the people of the punishments for violating inexorable physical laws.

Nature, in restrictionist advice books after 1860, became a surrogate for God, punishing and redeeming according to a dynamic of natural law which the authors steadfastly turned into moral judgment. Thus, they portrayed the punishment for sexual sin as a "natural" judgment: disease. Augustus K. Gardner explained, "Excess in lawful desire is subject to the same corporeal laws as in unlawful, and its penalty is disease and debility."[84] A restrictionist physician insisted that he was simply revealing the laws of science to his readers and if he was considered a severe "Interpreter of the decrees of Nature, we could not be otherwise."[85] James Foster Scott stated succinctly: "The laws of Nature and the Laws of morality which we have accepted for our standards will always be found to coincide . . . ."[86] Scott assured his readers that, though "Nature is leisurely," she grants no pardon for sins committed against the body, and that sooner or later, horrible diseases—sterility, venereal disease, invalidism in women, birth defects, blindness, insanity, and all manner of suffering—always result from "unlawful sexual indulgence."[87]

Though prone to less teleological explanations of organic processes, even moderates might characterize natural law as a physical punishment. R. T. Troll's *Sexual Physiology* reaffirmed the basic natural mechanism of judgment found in most of the later medical advice books: "A more pernicious doctrine was never taught then that of absolution from the penalties of our misdeeds. Causes and consequences are as unalterably related in the organic as in the inorganic world. Nature punishes always, and pardons never, when her laws are violated, or rather disregarded. In the vital domain, as in the moral, 'no good deed is ever lost' nor can any wrong act be performed without an evil result. When this great primary truth is recognized practically; when it is taught in our schools and exemplified in our lives, we shall have the true basis on which to prosecute our physiological redemption."[88]

Along with a more scientific rationale for the wages of sin, post-Civil War restrictionists as well as moderates became increasingly explicit in their discussion of sexual matters. While many earlier moral advisers resorted to extreme circumlocution if they did not avoid sexual references outright, their late nineteenth-century colleagues struck forth boldly in the public discussion of sexual concerns. Public discussion of sexuality in nineteenth-century America, however, was not a novel historical development. In the seventeenth century, before American public life and private life were so dichotomized, Puritan ministers, for example, used homely sexual metaphors in their sermons apparently without hesitation.[89] Historical memory is notoriously short. After the establishment (probably between 1730 and 1830) of the public-private dichotomy around family life in general and sexuality in particular, those boundaries appeared almost timeless.[90] Thus late Victorian medical advisers characterized themselves as brave pioneers of public sexual discussion.

The public-private sexual boundary did not fall at the first incursion of these writers into forbidden territory. As the entire spectrum of advisers realized, they were still on dangerous ground. H. R. Storer in *Is It I? A Book for Every Man* wrote defensively: "It may, perhaps, be alleged that the topics of which this book must treat are such as cannot possibly be discussed without offending good taste or transcending propriety." He disagreed: "This opinion, like many that are merely preconceived, may be found an erroneous one."[91] The author of *Satan in Society* worried that there was still some question as to whether sexual matters should be discussed in public forums. But, he reasoned, since this subject already was being "forced upon the public attention" by misguided authors, he intended to utilize "the ruthless invasion of the sanctity of private life" to educate "the masses in things which, if they know at all, they should know rightly."[92] He reflected an uneasy recognition that public discussion of sexuality was a violation of certain public-private boundaries of expression and that the walls were coming down, "the ruthless invasion of the sanctity of private life now become the fashion . . . ."[93] George Napheys, in a "Preface to the First Edition" of *The Physical Life of Woman,* noted: "The difficulty is to express one's self clearly and popularly on topics never referred to in ordinary social intercourse."[94] Even the bold Edward Foote, whose original edition of *Plain Home Talk* was published in 1870, observed nervously, "knowing the prejudices which frequently arise against those who dare to meddle with the delicate subject, I have myself felt many misgivings in giving publicity to my views . . . ."[95] Henry Guernsey, M.D., described himself as "emboldened" to speak what his title announced, *Plain Talks on Avoided Subjects.*[96] Late Victorian sexual advisers of all stripes contributed to the rising volume of public sexual discourse.

The public discussion of sexuality promoted interest in sex and accented sexual values, even in the case of restrictionists who overtly aimed to restrain sexual conduct. A more extreme restrictionist, John Cowan, illustrates how even public condemnations of Victorian sexual transgressions contributed to erotic stimulation. Cowan suggested that three years between successful acts of intercourse was ideal.[97] (For Cowan success meant conception.) Appropriately perhaps, given the rarity of coition in Cowan's scheme, he devoted thirty pages of detailed instructions to preparing a couple for the moment of conception. He spent two pages alone on instructions concerning the setting, time of day (11 a.m. to noon), clothing, when to breakfast, what to do between breakfast at 8 a.m. and 11 a.m., when that rare but precisely scheduled time to make love arrived.[98] Such lavish, albeit controlled, attention to one act of intercourse indicates that even this most restrictive medical writer was, however negatively, obsessed with sex.

Typically, restrictionists saw sex everywhere. John Cowan lamented, "That the prevalence of sensuality is widespread in this our day and generation is a fact sadly evident. . . . In the matrimonial bonds and out of them, rich and poor, high and low, learned and unlearned, sexually thwart the chief end and aim of their existence. The abuse of amativeness is the great crying wrong of the age."[99] What was most striking about those Victorian writers dedicated to restricting sexual freedom was how all-encompassing and overwhelming sexuality was to them. Even more significant was the fact that they despaired of effecting any change in American sexual behavior. Restrictionists railed against sexual excess, but in their own minds it was so linked to the conditions of modern life that only turning back the clock would have been an adequate antidote. Cowan revealed, "I here lay it down as an undeniable law, that a man or woman, living as men and women usually live—eating what they eat, drinking what they drink, cannot live a pure life, cannot *possibly* live other than a life of debauchery and licentiousness."[100] If the majority of his fellow Americans, living normal lives, could not possibly escape debauchery and licentiousness, then there was little hope, as Cowan himself admitted, that they would follow his advice. Realizing the extremity of his position, Cowan observed dryly that his three-year rule of continence would have many opponents.[101] He also acknowledged that his position was not a popular one, even among medical and reform writers.

Another restrictionist claimed to believe in moral progress, but he too found "such widely spread looseness of morals on the subject in particular [referring to birth control], that every day confirms us in the belief that it is verily and indeed a national curse." He predicted that "maidenly freshness and innocence . . . at the pace we are going, will soon cease to have real examples . . . ."[102]

The authors of restrictionist medical advice books believed that sex was out of control in nineteenth-century America. To them, passion was always on the verge of being ignited—like a spark to dry grass—in both men and women. Pure-minded girls, for example, were in constant danger because if the "spark" of passion is "once developed, it is uncontrollable in direct proportion to the strength of love and confidence."[103] Even ministers were suspect. John Kellogg in *Plain Facts for Old and Young* lamented, "to the majority of mankind, apparently, amativeness, or sexual love, means lust. The faculty has been lowered and debased until it might almost be considered practically synonymous with sensuality . . . . If one may judge from the facts which now and then come to the surface in society, it would appear that the opportunity for sensual gratification had come to be, in the world at large, the chief attraction between the sexes. . . . Even ministers, who ought to be 'ensamples to the flock,' are rather 'blind leaders of the blind,' and fall into the same ditch with the rest." Kellogg found that clerical lapses abounded, due in large part to the fact that the "indulgent hostess" served the unsuspecting pastor "the richest cake, the choicest jellies, the most pungent sauces, and the finest of fine-flour bread-stuffs. Little does the indulgent hostess dream that she is ministering to the inflammation of passions which may imperil the virtue of her daughter, or even her own. Salacity once aroused, even in a minister, allows no room for reason or for conscience."[104]

Obsessed is an extreme word, but restrictionist sexual advisers do not make its use an exaggeration. For Kellogg, the causes of unchastity were legion, including tobacco, bad books, idleness, fashionable dress, the Waltz, constipation, late suppers, and the most revealing of all, "modern modes of life." Kellogg elaborated: "It is not an exaggeration to say that for one conforming to modern modes of living, eating, sleeping, and drinking, absolute chastity [by which he meant lawful indulgence] is next to an absolute impossibility."[105] If modern modes of living made unlawful sexual indulgence almost a certainty, Kellogg and Cowan could hope to accomplish little, short of abolishing modern life. The extreme "primers for prudery" were actually one long wail against modern life, in the tradition of the American jeremiad.[106] In fact, though certainly transformed from their seventeenth-century relatives, they were nonetheless secular sermons, bewailing the sins of the people, denouncing American sexual declension, and calling for a general reformation.

Seventeenth-century Puritan jeremiads also bewailed the sin of sexual promiscuity, but for the seventeenth-century Synod, which attempted a definitive catalogue of American transgressions, sex was number eight among a typology of twelve major sins.[107] By contrast, for these latter-day Victorian jeremiahs, sex was almost the only sin, or perhaps more accurately, sexual transgression had become *the* sin of modern life. John Kellogg ex-

pressed the centrality of sex to these cultural critics: "the organs of reproduction may in a certain sense be said to rank higher than any other
organs of the human frame, since to them is intrusted the important duty
of performing that most marvelous of all vital processes, the production
of human beings. That this high rank in the vital economy is recognized
by nature, is shown by the fact that she has attached to the abuse of the
generative function the most terrible penalties which can be inflicted upon
a living being."[108] Late nineteenth-century moral advisers had dramatically narrowed their concentration upon sin to sex.

Victorian jeremiahs of sex conceived of the mechanisms for punishing
sin in more scientific terms than their predecessors. The doctrinal anchor
of this latter-day jeremiad form was that physical sin—defined as acting
against the laws of life and health—resulted in the natural punishments
of disease and death. The physical sins most often condemned by these
guardians of virtue were masturbation, "artificial" birth control, abortion,
and sexual excess. The premise was that to "thwart nature" was to court
disaster. Augustus Gardner threatened: "Death is not uncommonly an
immediate result. Decay and debility, followed by weary days and nights
of languishing on beds of sickness, are the penalties which many pay for
these transgressions against God's holy laws."[109] William Alcott noted,
"Some may object to this use of the word *sin,* but I know not how to
express the idea intended in a better way. Sin is the transgression of a
law, and there are laws of the human frame as fixed as any other and as
necessary to be obeyed."[110] Alcott summarized the moral shift from supernatural to organic judgment when he conflated the two: "In a word, to
be healthy, we need to be holy," which for him was to be "fully re-
deemed."[111]

Calls for reformation are an old formula in American life, and in
several important ways, restrictionist medical advice conformed to the tra
ditional conventions of the form. One of the most important of these
conventions involved the author's role in relation to his audience.[112] Restrictionists assumed the mantle of prophets scorned in their own time.
They saw themselves as lonely figures, crying out in a howling wilderness
of sexual excess. To interpret their stance as representative of Victorian
ideology is to misunderstand and misinterpret their role vis-à-vis their
audience and their social milieu. The severe restrictionist wanted to intim-
idate his auditors. As one admitted openly, "Those who shall seek in our
pages the gratification of a libidinous curiosity, will be disappointed, but,
better still, they will be scared!"[113] Thus, they often dwelt quite lovingly
on the "retribution" in store for readers who did not reform their ways,
just as Puritan ministers lingered over the possible manifestations of the
wrath of Jehovah visited upon an unrepentant people. For example, in the

time-honored tradition of American jeremiahs, Augustus Gardner closed his nineteenth-century medical advice book by threatening the destruction of America: "Let us hope that our present high state of civilization will not repeat the iniquities perpetrated in the corresponding Golden Age of barbarism, of which we read in our school books, *Roma fuit,* lest we may have inscribed, upon our natural ruin, for succeeding races to read in the obsolete tongue of a forgotten people, AMERICA WAS!"[114]

Defenders of sexual restraint sometimes reveled in their own persecution. John Kellogg was certain that any author who spoke against marital excess was "vilified, and his work is denounced and relegated to the ragman. Extremist, fanatic, ascetic, are the mildest terms employed concerning him, and he escapes with rare good fortune if his chastity or virility is not assailed."[115] J. R. Black did not allow himself to believe for a moment that his view was popular. Denouncing marital debauchery, for example, he acknowledged, "Perhaps the number is not small of those who think there is nothing wrong in an unlimited indulgence of the sexual propensity during married life. The marriage vow seems to be taken as equivalent to the freest license, about which there need be no restraint."[116]

Jeremiahs prospered in adversity. True reformation left them nothing to do, and no one to reform. Most restrictionists were, therefore, at least as pessimistic about their prospects of success as J. R. Black: "Experience, it is true, does not warrant the hope that as soon as mankind are duly impressed with what is their duty in regard to the laws regulating reproduction, they will immediately give a faithful and cheerful obedience to its requirements. No conquest in the restraint of an animal pleasure was ever made in this summary manner."[117]

John Kellogg scolded, "If young Ladies were brought up to work as their grandmothers were, there would be far less need for books of this character . . ."[118] Yet, he knew that increasing numbers of late nineteenth-century women and men were forced to cope with an expanding commercial, industrialized social order—a rural-to-urban transformation that was unlikely to reverse itself. Elizabeth Blackwell in a book aimed at the reformation of American life played on a similar theme: "There is a fearful amount of vice, misery, and disease existing under the fair outside of society, from which we should shrink with horror if we saw it in its naked truth. And this vice grows with our growth. The closer we come together in a dense population, the larger we build our fair cities, the more fearful becomes this pestilence of evil which spreads through every rank of society . . . ."[119] For Blackwell, neglect of the laws of the body resulted in sin which she told her readers was punished "naturally" by insanity, disorders of the stomach, nervous illnesses, and diseases of the "generative system."[120] Gardner, in his ringing jeremiad, claimed that modern living—dances, dress, parties, late suppers, the company of the other sex, music, fashionable society, "and various etceteras," most especially birth control—all cause disease which is a natural punishment for such sin.[121] In case his reader missed the point, Gardner titled his last chapter, "What may be done with Health in View, and the Fear of God before Us." In fact, though there were differences between the religious and secular strains of jeremiads, the ecstasy of Puritan denunciation of ordinary behavior applied to their nineteenth-century cousins.[122]

Nineteenth-century medical jeremiahs found sexual transgression everywhere: in modern courtship customs, in fashionable life, in modern diet, and even in staying in bed too long. John Cowan preached, "The morning is an important period in the life of the incontinent individual, and the plan all such should adopt is to leap out of bed as soon as they wake in the morning." Undaunted, Cowan admitted that the early morning leap to safety was much neglected in favor of the widespread "habit of early morning licentiousness."[123] Nineteenth-century jeremiads condemned many sexual sins so vigorously because they recognized them as endemic to "modern life" and thus incurable. The sexual sins they so vehemently denounced were a more accurate mirror of their times than the denunciations themselves, which they predicted would be greeted by their auditors with appropriate scorn and indifference.

Though restrictionists might praise and encourage the spiritual elevation of love, "that union of mind and communion of souls that lifts one above sensualism," they were unwilling to trust the nineteenth-century conscience to the moral law of love.[124] By contrast, moderates and enthusiasts usually had fewer fears of sexual excess than their restrictionist colleagues and were thus more relaxed about the subject of coital frequency and the practice of birth control. Moderate medical advisers, such as R. T. Trall, advocated "*temperate* sexual indulgence, even when off-spring are not desirable nor proper." Suggesting that once a week was a safe frequency, he also left much to individual discretion. "Surely, if sexual intercourse is worth doing at all," he admonished, "it is worth doing well. And it would not exalt its importance one iota above its real merits, if certain days were set apart, consecrated to the conjugal embrace. It might be one day in seven, or one day in twenty, or more or less . . . ." This moderate advised people to set aside whole days for making love. He was unambiguously positive: "whatever may be the object of sexual intercourse—whether intended as a love embrace merely, or as a generative act—it is very clear that it should be as pleasurable as possible to both parties."[125] George Napheys defended the physical passion of love as "an honorable and proper pleasure which none but the hypocrite or the ascetic will effect to contemn." He argued that sexual passion "is a natural and healthful impulse" when exercised in moderation.[126]

The association of intercourse and romantic love affected the meaning of both in profound ways. The "rule" of mutual love limited sexuality as an emotional expression, but theoretically freed it from legal and other non-emotional constraints. Restrictionist fears were not completely far-fetched. Accepting the premise that sexuality should be governed by the heart laid the groundwork for the antinomian acceptance of personal moral judgment—the conclusion, in other words, that only the individual might rightfully judge, on the basis of his emotional relationship, when and if an act of intercourse was pure.[127]

Mrs. E. B. Duffey, a moderate, aptly illustrated the antinomian propensities of sexual conduct when viewed as a "sacrament" of love. She boldly argued that love made a couple's sexual activity lawful, not marriage. "If a man has been guilty of persuading his affianced into improprieties, it ought to be the strongest reason for their marriage; whether she has been mutually carried away with him under the influence of passion or whether she has yielded from loving him too much, he is equally bound to her. . . . It is not likely she will make the less true and faithful wife for this indiscretion." Duffey's reasons for accepting premarital sex are crucial: "In the eyes of God she is his wife already, for God's laws were made before man's laws, and marriage consists in the exchange of vows of love and fidelity between a man and a woman, and a physical consummation of these vows. The legal form does not make the marriage, it only recognizes it."[128] Mrs. Duffey was not a wild-eyed radical. She was restive with the privilege of privacy routinely accorded to courting couples, and she urged caution concerning the "undue familiarity between the sexes."[129] She believed in holding passion in control outside as well as inside marriage.[130] Thus her concession to the purity of premarital intercourse when it was hallowed by love (and followed by marriage) is all the more significant.

From the moderate point of view, the essential element of sexual purity—in marriage or out—was emotional. While some Victorians were willing to act upon this romantic logic, many no doubt resisted the very thought. Nonetheless, distinguishing the boundaries of sexual morality within the emotional logic of romantic love was a peremptory agenda for many cultural authorities in the last half of the nineteenth century. Whatever their persuasion, Victorian medical, sex, and health manuals were often obsessed with the problem of dividing the pure from the impure physical act. In fact, whatever side they took, and their arguments were quite varied, they all labored strenuously to distinguish lawful and pure from unlawful and impure sexuality. But including erotic activity in definitions of purity and chastity created a deeply vexing question within American nineteenth-century dominant culture: by what logic could pure and impure sexual indulgences be distinguished? And even more perplexing, how could these distinctions be enforced?

For the restrictionists, what ultimately made sexual pleasure lawful was the desire for children. As one restrictionist adviser declared unequivocally: "Sexual intercourse is not intended for so trifling a purpose as that of giving a pleasurable sensation. . . . Sexual pleasure is merely an incident in the union of the sexes, which draws them together in order to ensure a result—birth . . . ."[131] But for moderates, sexual intercourse was intended not only as an act of procreation but for the expression of love.

Restrictionists hoped to lower the frequency of sexual contact and limit or reduce marital sex through fear of pregnancy. Moderates did not usually share the restrictionists' negative intentions. Their goal was a more egalitarian enjoyment, sometimes embodied in the right of refusal but sometimes in the right to say yes, even outside the legal boundaries of marriage. Believing that sexual intercourse was an emblem of love, Alice Stockham insisted that "it should never occur except when there is mutual participation on the part of both male and female" and procreation is "governed and guarded."[132]

For moderates physical desire was purified and elevated by love. They believed that every sexual act should be governed by the rule of love. Compared with restrictionists, moderates appear less troubled by the potential abuses of personal judgment and individual freedom that the ethic of love implied. Some moderates, however, were probably avoiding the dilemma of individual freedom which their rule of love espoused. Marion Harland in her book *Eve's Daughters; or, Common Sense for Maid, Wife and Mother* insisted: "A loveless marriage is an unchaste union." She continued, "The connection is unnatural, impure, and unsafe."[133] But she said nothing of the connection between sexual purity and love outside the marriage bond. The Reverend John Bayley, who wrote very little about sex, repeated almost identical sentiments about love. "Love is the soul of wedlock, without which the performance of the marriage ritual is an unmeaning ceremony."[134] By staying within the boundaries of marriage, Harland and others sidestepped the connection between sexual purity and love before marriage.

A moderate occasionally confronted the issue directly. The Reverend George Hudson defended young women who granted men "improper liberties" before marriage, arguing that if the woman loved the man "devotedly," "undue liberties" were no evidence of impurity for "She only yields because she loves you."[135] Hudson clearly distinguished between "liberties" and intercourse, urging men to avoid both, but betraying a certain tolerance of non-coital physical expression.[136]

Braving the question of purity and love before marriage, Hudson boldly exposed the moderate dilemma: "Some reason thus: 'We are engaged, we intend to marry, it is only a question of time, whose business is it, if we do choose to enjoy the privileges of married life beforehand: The marriage ceremony will not unite us any more sacredly; beside, what harm can it do?' "[137] Having admitted that "the marriage ceremony only sanctions the union consummated by the parties themselves," Hudson did not argue against the morality of premarital sex with love but rather paraded the attendant embarrassment, shame, and disgrace which might result from intercourse before marriage.[138] Basically his reasoning was pragmatic as he reminded his male readers of the potential for disaster if one's fiancée became pregnant or one of the parties wished to break the engagement. Hinting that the pleasures of married sex would be increased by delaying the "sweets of connubial life" and noting that it was against nature, but offering no rationale, he fell back upon bald admonition: "You can wait, you are a man; be master of yourself."[139]

Many moderate advisers were more cautious than Hudson, avoiding, if possible, a concurrent discussion of the ideals of love, morality, and premarital sex. Nonetheless the conclusion that the presence of love made a courting couple "morally" married according to a "higher" law and thus morally entitled to the full range of sexual expression, even without the legal entitlement, was difficult to avoid.[140]

The approval of sexual intimacy as a love act, with intercourse ambiguously contained within marriage, was probably widespread in Victorian culture. Moderate views of sex in advice books seem strikingly similar to the private views of nineteenth-century correspondents who regularly approved of physical desire as an expression of love. In the Mosher study, the one known sex survey of nineteenth-century women, many respondents separated sex from reproduction and saw its purposes in terms of furthering the companionate relationship between husband and wife. Several women in the Mosher study explicitly stated the belief that love was the sole legitimate purpose of intercourse. As one said, "I think it is only warranted as an expression of true and passionate love."[141]

The logic of this moderate position led in two cultural directions. The first is a common enough view of Victorian life, but with a new refinement: condemnation of sexual promiscuity and a profoundly censorious attitude toward anyone who was seen using love as a stalking horse for mere erotic seduction. Viewed from the moderate perspective, men like

Aaron Burr should be vilified for their sexual exploits, not because they took pleasure in the flesh per se but because they divorced their sensual pleasures from true love.[142] No Victorian adviser of any persuasion condemned what they defined as lawful pleasures of the flesh; they were disturbed about behavior that violated the proper boundaries of sexual morality; sexual behavior that stayed within those boundaries—however they were drawn—was affirmed.

The erotic became moral from the moderate perspective as an act of romantic love; sex did not automatically become pure through the marriage ceremony itself. This second direction was crucial and contained a potentially revolutionary idea. For if romantic love, not marriage, purified sexuality and affirmed its moral standing, then the morality of premarital sex had to be decided on the basis of emotional commitment and feeling. Applying the unvarnished logic of the heart to sexual expression before marriage provided one basis to challenge the double standard and the long-standing taboo against premarital intercourse. This is not to say that Victorian culture accepted this challenge or followed the rule of love without contradiction, resistance, or denial, but that it was already a powerful cultural ideal in private as well as public life whose full implications the twentieth century is yet struggling to comprehend.

The apparent disjunction between nineteenth-century sexual prescription and practice, ideology and everyday beliefs, was not so wide as one may have been led to believe. Many powerful similarities existed between public and private expression in nineteenth-century America. The medical and moral advice books did not portray women as asexual, though sometimes they were seen as less passionate than men, slower to be aroused, less knowledgeable, or shyer. A restrictionist might urge "cultivation" of less ardent womanly passion; a moderate might argue that coldness or lack of passion in a woman was a defect; and an enthusiast might insist that a woman's passion was naturally cultivated by love; but the basic sex drive of both sexes was rarely questioned.[143] This range of attitudes corresponded to the everyday beliefs expressed in the personal correspondence under review.

The emphasis on the mutuality of sexual expression in all three types of advice books also corresponded with the private views of men and women. Edward Foote, an enthusiast, urged sexual reciprocity in marital relations and saw it as the "all essential" element of marital felicity.[144] H. R. Storer, a moderate, stated flatly that "No true conjugal enjoyment can exist, unless it is mutual . . . ."[145] The restrictionist author of *Satan in Society* urged men: "Let nothing interfere with your determination to wait for and obtain entire reciprocity of thought and desire, and let this always be your guide . . . ."[146] Mutuality was double-edged, however. Through this device, restrictionists especially hoped to dampen male lust, men having to "wait upon" women's slower arousal or less intense sexual interest. Thus, mutuality was sometimes a Trojan horse for lowering the frequency of sexual relations. Many advisers simply assumed that the rule of mutuality, if followed faithfully, would decrease sexual activity. Whatever their motives, advisers from all three camps showed respect and concern for women's sexual feelings.

Restrictionists, moderates, and enthusiasts alike placed a high value and immense significance upon sexuality. For example, Mrs. Duffey claimed, "The strongest motive power in the world is the attraction between the sexes. It even exceeds the desire for gain, because in most cases this latter desire is born of a still stronger one to win or retain love or to gratify passion."[147] O. S. Fowler, an enthusiast, related everything to sexuality: moods, temper, moral tone and character, courage, pride, ambition, and talents. Fowler went so far as to insist that "No man can ever become extra great, or even good, without the aid of powerful sexuality."[148] Though not as excessive as Fowler, private love letters also reflected the weighty emotional values the middle class attached to sexuality.

The condemnation of sexual sins and the calls for reformation issued forth from some of the medical advisers to the people need not necessarily be interpreted as evidence of actual Victorian prudery. Historians now agree that Puritan ministerial denunciations of a backsliding people were *not* evidence of actual seventeenth-century social decline.[149] The jeremiad, in whatever guise, is a traditional American ritual device for coping with change; the rhetoric of condemnation is a symbolic stance, not a positivist mirror. Medical advice in the jeremiad mold should not be confused with sociological investigation, for Victorian prophets revealed the habits of society more clearly by the sins they denounced than by the restrictive reforms they proposed. From this angle of vision, the extreme restrictionists who bemoaned the sexual excesses they found at every turn and called for the reformation of American life were practicing "purgation by incantation."[150] In that sense, especially the most lurid restrictionist advice books were more than primers for prudery. They were formulaic laments for change, endemic to American cultural expression for three hundred years. No one, least of all these prophets, expected people to answer their cries for reformation. Certainly one of the mainstays of the prophetic role is that it is *not* representative of popular sentiments or opinion.

Moderates were less prophetic and more representative. They insisted that "sex is as divine as the soul" when love infused passion.[151] Reflecting the private values and behavior in male-female relationships, moderates approved of the erotic as an expression of love.

One of the deepest similarities between public and private expression in Victorian America is that ministers, doctors, reformers as well as middle-class lovers passed "everything having to do with sex through the endless mill of speech."[152] From nineteenth-century romantic novels, to anti-masturbation tracts, to advice for young men, to purity crusades, to love letters, sex was spoken of ad infinitum in nineteenth-century Victorian America. Because the stance of the authors of public discourse on sexuality was often cautionary, if not always condemnatory, it is easy to miss what is at least equally significant—that there was an immense public, as well as private, verbosity on the subject. Sometimes sex became a topic for public discussion under the rubric of cautionary, occasionally luridly fearful sexual advice, but it was incorporated in the public realm nonetheless. Ironically the attempts to constrain sexuality in medical advice books functioned as inducements to speak of sex. Even those advisers whose stance was most negative could not attack the evils of masturbation, birth control, and sexual excess without lengthy, often detailed discussions of sexuality. Even the most negative obsession with sex contributed to the opening up of sexual expression in the public sphere and the definition of sexual matters as public concerns. The evidence suggests that the history of American sexuality from the Victorian period to the late twentieth century might be more accurately conceived not as some Whig saga of progressive sexual liberation, but as a narrowing of the gap between public and private sexual discourse.

Victorian America has been called a climactic period of modernization. But those who believe that in the modernization process an industrializing society must inhibit and suppress sexuality for capitalistic expansion have seriously misunderstood nineteenth-century American culture.[153] It appears rather that Victorian America was modern in the sense that nineteenth-century Americans talked incessantly about sex—whether they condemned it, defended it, exalted it, or campaigned against it, the pleasure in speaking of sex predominated.

Any century which had a lively public debate over whether women should be allowed to ride bicycles, with opponents arguing against women cyclists because the seat might become a source of intense female sexual pleasure is, to say the least, erotically sensitized.[154] To focus upon this debate as an example of Victorian sexual repression and prudery misses once again the obsessive erotic imagination of so-called repressed Victorians as well as the symbolic weight and social importance attributed to sex.

In the interaction of public and private Victorian life worlds, sex was a symbol of both pleasure and danger, necessarily clandestine and yet imperiously exposed, an object of secrecy and also of intense investigation, a fearful expression of anarchy and individualism run amok but also the ultimate embodiment of romantic love and human spiritual communion.

5

Blurring Separate Spheres

Sex-Role Boundaries and Behavior

S EX-SPECIFIC CHARACTER TRAITS and gender roles were matters of serious
concern to nineteenth-century middle-class men and women. Nonetheless, male-female relationships had their lighter moments. William Nebeker held up a mirror, however unintentionally, to both the light and
dark sides of nineteenth-century sex roles. Writing to his mother-in-law
in 1880, Nebeker, still deeply anguished by his wife's death one year earlier, felt a compelling need to unburden himself, which, he openly acknowledged, drove him to a confessional mode: "The house was cleaned
shortly before I went to the city in March [the letter was written in July]
and though I have tried to be very careful, there is much dirt on the floor
nevertheless, which in appearance resembles dried mud, plastered on. I
strongly suspect that careless visitors have done a good share of it, supposing there was no need of being careful, where no woman was around.
Well after all it is what may be styled 'clean dirt,' being dry and apparently not particularly 'onhealthy.' When the weather was stormy and I
could do little else, I did a time or two wash up some plates used, but
being hurried now, and besides the remnants that adhere soon drying, I
do the best I can without washing, hoping that when the coating gets
thick enough it will crack and peel off, but if it don't do so soon, there
will be little or no room left in some of the plates for food."[1] Ostensibly
out of reverence for his deceased wife, Nebeker turned the foundation of
nineteenth-century sex roles—the separation of male and female spheres
of work and influence—into an absurdity.

Whether Nebeker was playing a deliberately wicked little joke on his mother-in-law, appealing to her sympathy in order to elicit immediate aid, or unselfconsciously reflecting the separate spheres ideology of sex-typed work, is not clear. Admittedly, the rest of his letter is a pious, sincere, and sober reflection of standard nineteenth-century sex-role ideology. Resonating to this gender ideology, Nebeker mused: "My ranch proper appears usually well—possibly better than has generally been the case at this time of year, the grass being very green and good, but the garden spot and the door yard show sad neglect, and in that respect present a sad contrast to their appearance one year ago. The rank weeds and the thrifty grass growing around the house but too plainly show, that the idol wife and the darling children, so full of life and animation around here one year ago, have long been absent, and alas the main one of them all never more to return in mortality. When I reflect upon how we have toiled, struggled and hoped here together this spot seems, as it were, hallowed ground."[2] His deceased wife's feminine sphere was so literally an embodiment of her reality that Nebeker was reluctant to disturb anything in her physical domain, which included not only the interior of the house, but also the garden and door yard. Thus, it appears he let house and garden "show sad neglect" as a sign of his respect and homage to her memory.

What is most noteworthy in this context is how powerfully he associated his wife in death with the commonplace sex-role prescriptions of the time. "In getting things to use, I frequently find them just as she last placed them," Nebeker confided, "and it is not without regret that I disturb them. Some things are tied up, and when I think it was *her* fingers that tied the bow knot I am loth to loosen it   Oh Sarah! Sarah! Why was it necessary for you to be taken from a husband who appreciated you so much and from little children, who your guiding hand and fostering care so much needed?" Emphasizing women's special abilities as nurturers and moral guides, Nebeker closed his letter by asking his wife's spirit to take care of him and his children.

Letters such as Nebeker's indicate deep belief in and self-conscious support for nineteenth-century sex-role prescriptions purveyed by the magazines, medical and moral advice books, and other popular culture of the time.[3] Examining such evidence, historians have developed three general labels for nineteenth-century feminine ideals. First, the doctrine of separate spheres encapsulated the belief that women should work in and exercise dominion over the home sphere and men should control the world outside the home. Second, and justifying this division of labor on the female side, was the cult of true womanhood—an idealized image of woman as pious, pure, domestic, and submissive that was also reflected (if understood correctly) in the evidence of men's and women's everyday beliefs about themselves.[4] Third, and closely related to the cult of true womanhood and the doctrine of separate spheres, was the ideology or canon of domesticity, which again reinforced woman's special role as nurturer in the home and as a domestic spiritual guide and moral authority to her husband and children. These beliefs, surrounding the home with a halo of sanctity, glorified woman's role as wife and childrearer.

The concept of masculinity, the ideals of manhood, and the private reality of the male sphere in American life have not been as exhaustively examined. Thus far historical research suggests the dual role of the nineteenth-century middle-class man as both provider and companion within a value complex that emphasized male self-control, economic aggression, Christian kindness, worldly authority, and emotional attachments to wife and family.[5]

The Christian Gentleman who "eschewed excess in all things" has been identified as a nineteenth-century masculine ideal.[6] Also accepted as a dominant ideal of masculinity is the self-made man who might rise to the top of the occupational and social ladder in American life through hard work, perseverance, and a competitive edge.[7] Another image of manhood is contained in the Daniel Boone myth of the western hero (and all its successors), leaving civilization to test himself against Nature, but also to renew himself in the sometimes violent struggle to overcome the natural elements.[8] The domineering, undemonstrative Victorian patriarch is yet another popular image of nineteenth-century masculinity which coexists alongside a more recent view of masculine role ideals that emphasizes the compassionate provider who was expected to comport himself respectfully and affectionately as a husband in a partnership of mutual cooperation.[9] In short, the Victorian ideal of the "true man" encompassed self-control in public, companionate marriage, occupational competence and ability to support a family, strength, courage, toughness, and a competitive spirit.

🎄 There is a tendency in studies of Victorian sex roles to characterize the ideals of male and female in static, absolute, and mutually exclusive terms. There is also a tendency to emphasize the difficulties of men and women finding common ground in a gender-dichotomized culture of separate spheres. The accent on the separation of masculine and feminine spheres has, however, obscured the actual relativity of Victorian conceptions of male-female differences. It has also masked the basis for intimacy, understanding, and mutual identification. Moreover, the focus on sex-role standards has diverted attention from the varied ways those standards were applied. Nineteenth-century men and women, for example, sometimes accepted the violation of sex-role boundaries with the help of protective rhetoric. Even in a fairly rigid gender system, Victorians skillfully used their sex roles rather than slavishly followed them.

Gender lines were more fluid in daily life than the static view of separate spheres has allowed. As men and women negotiated in private over their goals and behavior, even firm beliefs about the differences between the sexes and their proper spheres did not preclude "violation" of those beliefs in actual practice.

Ironically, the focus on female role limitations may have given the historiographical outline of Victorian gender a more rigid edge than it deserves. Not only were nineteenth-century gender dichotomies sometimes redefined in practice, but ideas about sex-role divisions were themselves less taken-for-granted and somewhat more ambiguous than in the previous century. Private exchanges indicate that the female role came under increasing scrutiny and woman's role boundaries were a more self-conscious issue between men and women after the Civil War. Furthermore, in the antebellum period, one crucial gender distinction was already somewhat blurred.

Within the long tradition in Western thought which extolled the rational capacities of mind above the irrational passions or emotions, men have been seen as the sex with the stronger intellectual capacities and women as the tenderer, more emotionally susceptible sex. However, in eighteenth-century America, under the influence of religious awakenings in both North and South, the heart took on a new significance. As historian Nancy Cott suggests: "Both men and women with religious aims wished to be 'tender-hearted.' "[10] Secular eighteenth-century culture also contributed to the general elevation of the heart.[11] But the higher valuation of emotions did not eliminate the distinctive identification of men with the qualities of superior intellect and women with the superior capacity to love or feel.[12] Historians of women and the family often agree that nineteenth-century women, presumably unable to find sufficient and satisfying sensibilities in men, looked to other women for their most intimate sharing and fullest emotional expression.[13]

This, however, is only a partial formulation. Private correspondence demonstrates that tender feelings and sympathetic emotions were an integral component of the male role. Victorian women were not left to "find answering sensibilities only among their own sex."[14] Repeatedly Victorian women also sought and found such sensibilities in their romantic relationships with men. Both sexes believed that men could and should open their hearts in a love relationship. If masculinity had actually been defined solely in terms of rationality and if men had identified only with the head in contrast to women's heartfelt feeling, then men would have been de-sexed by the romantic love which Victorian courtship required. Without a belief in men's capacity to feel, the romantic basis of mate selection is nonsensical. Moreover, there is hardly a trace of evidence that Victorians thought men as a gender lacked the emotional capacity for true love. In fact, men expressed themselves to women with an emotional fire that burns fiercely even at a distance of a hundred years or more.

This has been obscured by the powerful appeal of secrecy in Victorian culture. One female adviser described the male as accustomed "to keep the emotional part of his nature out of the sight of the associates of business-hours."[15] The masculine even more than the feminine ethos cultivated and encouraged a secret self. Victorian men were expected to hide their emotions in public and to loosen their expressive controls in private, communicating their hidden selves to the woman they loved. From a romantic perspective, public coolness and emotional control were not a disability so long as men were more emotionally expressive in private.

Charles Godwin explained the public-private polarity within the code of masculinity to a woman he loved deeply but who herself loved someone else: "You must not judge us men too harshly. If we are solemn or even sour sometimes, bear with us for we have much to try us of which you know not. Beside it is only our outer self hardened by contact with the rough world which you see. Could you look into the hearts inner chambers, and see the forms traced there, you would have more charity for us may be. Neath many a rough face a warm heart is beating . . . ."[16] Nineteenth-century middle-class men did not usually define their masculinity as antithetical to emotion in private. Culturally speaking, head and heart were split within the masculine role: "heart" was assigned to men's private world, while they expected controlled rationality of themselves in their public persona. Eldred Simkins expressed relief and delight at being able to "simply feel —not think" when he and his girlfriend were alone together.[17] Men were expected to respond in kind to women's feelings in private romantic relationships.

Meeting the performance demands of the opposite emotional poles in the male role sometimes proved difficult. That men experienced strain as well as failure to meet the bipolar emotional requirements of these contrasting masculine values is not surprising. In the midst of a budding romance, Samuel Francis Smith wondered if his letters were too sentimental: "There has been in them but little of that 'manly' energy which you might have anticipated. The head has been too much subjugated to the dominance of the heart."[18] Then, in some confusion, he added that he would not have less heart. Regardless of the difficulty of alternating between head and heart, nineteenth-century middle-class men were expected to express intense emotions in their romantic relationships. Tenderhearted feelings were not usually perceived as unmanly or as troublesome when confined to private relationships with women.

But the emphasis on rationality in the masculine sphere was still powerful. Victorian men were expected in public life to have much more expressive control than women. They were required to exercise their rational capacities to the fullest extent and to curb their feelings. Men continued to value emotional control in public. Eldred Simkins admired one man's undemonstrative approach to an upcoming duel: "I love to see such coolness—tis always characteristic of the perfect gentleman."[19] Charles Godwin said of men: "Eyes that the world says never weep, shed many a secret tear over what would seem but mere trifles."[20] As if to illustrate this point, Albert Janin described his emotional reaction to a play: "I felt at one moment as if I must leave the theatre and seek relief in the cold air outside, but I managed to control my feelings and to conceal my tears with my glasses—After all men are but children of a larger growth. I thought that I had done with weeping, but I find since your departure that tears start constantly unbidden from my eyes . . . ."[21] Men disguised tears in a public place, but acknowledged their emotional nature in private.

By comparison, women's role was not as internally polarized by the public-private split. In public, women were allowed a broader range of emotional expression and were not required to maintain the same level of rational control as men. Consequently it was a commonplace when John Marquis mused: "A womans heart is more easily wounded and more susceptible of sad impressions than ours and the same boldness cannot be expected from them in battling against difficulties."[22]

Though middle-class men might see the opposite sex as more emotionally vulnerable and even more irrational than men, they also recognized women's intellectual capacities. Depending on their educational achievements, men might value educational accomplishment and intellectual concerns as qualities of a future wife.[23] Samuel Francis Smith contrasted his fiancée Mary's "relish for knowledge" with that of another young lady whom he called upon three or four times, but stopped seeing because she had no interest in intellectual matters.[24] An educated man, Samuel recognized the importance of education as a criterion of compatibility and companionship in his ideal marriage.

In another example, Nathaniel Wheeler expressed approval of Clara Bradley's effort to earn a college degree. He told her—with no trace of irony—that he was glad she endeavored "to become something *more* than a mere girl."[25] He also remarked that he was happy she had done so well.

However, his approval of her intellectual endeavors was qualified: "My wife, I have always thought, will not need to ask to be excused at my study-door. I am sure of it, now. I do not want you to be a bluestocking, to *talk* wisely in public, or to be crammed with unassorted bits of knowledge but I do want you to be a cultured woman, who can share my mental as well as my social life."[26] Nathaniel would restrict her intellectual activities and his emotional feelings to the private sphere.

The companionate ideal of marriage was one of the chief foundations of nineteenth-century masculine approval of women's intellectual endeavors.[27] As Nathaniel's censorious attitude toward bluestockings and women who "talk wisely in public" underlined, the growing acceptance of women's intellect and intellectual training was restricted to the private sphere. The rationale of women's education was sex-directed: to make better wives and mothers. Another suitor, Will Harbert, strongly supported higher education for women, but he objected to the application of female education to the public sphere. Thus, he found his fiancée's public speaking on behalf of the woman's movement quite distasteful.[28]

Some men also expressed ambiguity about the extent to which women should engage their minds in private. They may not have been actively hostile to women's education and sometimes expressed approval of female educational goals, but they were concerned to keep intellectual parity between themselves and the women they loved. Thus James Bell queried Gusta Hallock after her announcement that she was returning to school: "Why: I wonder when you will think you have school enough I would advise *anyone* to go to school all they could. I can see now where *I* missed it Speaking of school reminds me of something I must tell you? I want you should note the difference in us. I've *often, often* thought of it; and I am afraid that at some time you will be disappointed in me. Will mind answere to mind?"[29] James, the sharecropper, was worried that he might be intellectually overshadowed by Gusta, the schoolteacher. The basis of his concern was one of male-female compatibility.

While middle-class men in the nineteenth century may have been reluctant to give up their long-standing claim to intellectual superiority, many willingly accepted more intellectual parity in their romantically defined involvements with women. What men found threatening was any hint of women's mental superiority. With regard to morality and religious sensibility, women were seen as superior to men according to standard nineteenth-century sex-role prescriptions; but in the area of mind, cultural expectations were strained, and Victorians still leaned toward masculine superiority. In intimate relationships there was an uneasy recognition of the intellectual accomplishments of women. Albert Janin, a lawyer, told Violet Blair before their marriage: "Most men I believe would be jealous of so learned and accomplished a wife, for fear of being dwarfed by her superiority, but you know very well that such is not my character."[30] While Albert may have been nonplussed, Eldred Simkins had absolutely no interest in a woman of intellectual intensity or stature: "when ever I get in the presence of great Genius or an old professor, for whom I have deep respect, my thought and words flow as sluggish as a stream of tar. Why then is it so, when I am with you? Not for your learning, thank Heavens, for you are not a *blue* . . . . "[31]

The success of private female academies, the inclusion of girls in publicly supported education from the early nineteenth century on, and the breakthrough in higher educational opportunities for women after the Civil War reflected a significant shift in public attitude, but individual men still found women's educational aspirations and achievements a source of apprehension and even distress.[32] Typically, however, men accepted women's intellectual capacities if they were contained within the feminine sphere.

Along with the doctrine of separate spheres, the head-heart division of gender characteristics remained a generally acceptable sexual identification throughout the nineteenth century; but for Victorians this rational-emotional split was a relative rather than an absolute reflection of men's and women's capabilities. Victorians may have thought women were more emotional *in toto* than men but manliness was not seen as antithetical to emotional expression. Victorians may have believed that women were less inclined to logical thought than men, but women's intellectual powers were not seen as absolutely at odds with their femininity. Thus, while nineteenth-century American culture is often characterized by the extreme separation of masculine and feminine spheres, it should also be described as a century in which mainstream conceptions of gender differences were narrowing.

🐝 This modification of the image of an elaborate Victorian gender gap is even more strongly supported by displays of gender ideology in everyday life, for sex roles are not only defined but also transacted. A role, therefore, can be seen not just as a reified essence contained in a binary structure of conformity or nonconformity, but rather as a dynamic and contextually fluid relationship between people. Based upon a learned set of values, beliefs, images, and rules, male and female roles may be understood, not as static essences to be possessed, but as a complex repertoire drawn upon in male-female interactions in a dynamic and creative fashion.

Thus, while the discussion of nineteenth-century sex-role ideals has been admirable, particularly for women, there is a need to examine more thoroughly the ways men and women used their cultural values and beliefs in their dialogues with each other. Individuals both demonstrate their culture's gender prescriptions and enact their sex roles in ways that are not captured by simplistic categories of conformity or nonconformity, belief or disbelief.[33]

Personal correspondence uniquely reveals the dynamics of nineteenth-century male-female interaction because it is in itself a communication between men and women. It allows glimpses, however fleeting, of the process by which ideals were translated into action. Letters between men and women suggest that nineteenth-century sex roles in actual interaction were less rigid than any description of their content might imply. Although most of the correspondents affirmed their belief in the ideology of domesticity, the cult of true womanhood, the doctrine of separate spheres, and the ideals of true manhood, they also demonstrated that sex roles did not necessarily remain one-dimensional and static in everyday life.

🦢 In the actual exchanges between men and women, certain gender expectations remained trouble free either because they were readily fulfilled or because they were irrelevant to day-to-day interaction. Other gender issues were sources of friction, discomfort, fear, pride, or elation. Negotiation between the sexes occurred both where sex-role conformity was maintained as well as where the standards of masculinity and femininity were violated. In fact conformity sometimes created the anxieties that seemed to encourage more flexible responses to gender expectations.

One of the central tensions of conformity to the male sex role in nineteenth-century America was created by the almost universally accepted claims of economic provider. Men identified manhood with work.[34] And while it is wrong to claim that men "alone" involved themselves—even "centrally" in the new economy, it was certainly true that nineteenth-century middle-class men as a rule identified "dollars and cents" as a masculine province. As one suitor, Charles Strong, told his fiancée: "Having been so busy today have not seen a soul except on business therefore cannot report to you what anybody has said excepting in the dollars and cents line which is not in a lady's province excepting when they *declare to go on their own acct.*—of course I don't mean this to exclude the dear creatures from exercising full and undisputed control over their own 'pin money' and such as buying stockings for the children, etc."[35] His perspective on women's role was in context: a lady's province changed, depending on her relationship to men. Charles allowed that women who disassociated their lives from men had a legitimate concern with the "dollars and cents line," but he insisted that women who married (or intended to) must give way to men in the economic sphere.

Women sometimes expressed a similar identification of masculinity and the economic world. Jane Burnett, a married woman, exclaimed, "Why should the cares and anxieties of business life intrude on my happiness, Everything is for the best I believe. And I shall try not to fret."[36] At least some nineteenth-century women embraced the ideal of economic independence and achieved a measure of it, or were forced by their circumstances to support themselves or their families. Still other women made what economic contribution they could to the family economy by taking in boarders, doing laundry, or selling produce. However, more privileged women probably rarely experienced the feelings of Dorothea Lummis when she rejoiced: "how vastly pleasant it is to get money, that is all one's own, and honestly earned."[37]

Men and women identified nineteenth-century middle-class manhood with gainful employment outside the home. They also identified masculinity with family, home, and love. Separated from his wife by half a continent, Lincoln Clark lamented: "I dread the return as much as I want to be installed in my own little Sanctum of domestic peace and enjoyment: —nothing but duty or gain would induce me at this season of the year to embark upon such an enterprise. I can truly say 'there is no place like home': and why?—only because you are there . . . ."[38] As Clark reflected, the juxtaposition of job, ambition, and money to wife, family, and love sometimes left men in an uneasy tension between two sets of values, both deeply embedded in the masculine role.

This tension is dramatically exemplified in the correspondence of James Blair and his wife Mary. As a naval officer, his work routinely took him away from home, intensifying the normal masculine role strain between work/ambition and home/love. While away from home James imagined being with his wife: "How I will enjoy your love and petting when that time comes it makes me joyous to think—indeed my pet, you know not how you have spoiled me with the lazy happy life of your house and how the height of my ambitions is to roll over on your bed and *quarrel* with you."[39] In this instance, James dreamed of collapsing the tension between ambition and love.

But James's ambition was not actually contained within his wife's bedroom, as he conceded while describing his professional duties two years later: "The only source of sorrow and unhappiness has been my profession which has from necessity and pride to be worthy of your love, separated us from each other more than the usual fate of man . . . but I feel and know that the life and fate of a Naval Officer is hard to endure, especially when he has, as I have, a happy home; made so by a loving and devoted

wife, and her sweet and charming little child . . . ." James warmed to
the tension he felt between home, family, wife and his work: "It sounds
very well to talk about distinction, reputation, and ambition but at the
end of a man's career in this world, he considers it all 'vanity and vexation
of spirit'—he does not look upon his struggles in the world with any
pleasure but turns to those hours which he has devoted to his wife and
children and looks upon them as the only really happy portion of his
life . . . ."[40] The antagonism James expressed between "distinction, reputation, and ambition" and a "happy home" was intense. While much
consideration has been given to the tensions that the division of men's and
women's work spheres caused women, there has been less appreciation of
the tension this division created for men. James wished to be with Mary
and his child, but in order to be worthy of the love he felt for them, he
was also obligated to fulfill the masculine demand "to provide for you
and your little cherub . . . ."[41]

Since his career required long absences from home, creating frustrations for him as well as his wife, he struggled to assure her and himself
that the sacrifices he made for his career were compatible with the duties
of the heart. "[A]nd my pleasure and pride," he said, "will be to see you
enjoy what I can afford to give which shall be all I have—I pledged my
soul in the presence of God and man to love honor and serve you my
Mary. I have done so with my whole heart thus far in life . . . ." But
almost immediately following his identification of economic gain with the
heart, he hastened to reassure his wife that the provider role he was just
extolling would not interfere with his love for her: "Fear not Mary no
ambitions for distinctions among men shall win your place in my heart.
. . ."[42] Nineteenth-century men had the dual and not always compatible
obligation to be in the world as economic providers, aggressively pursuing
money, status, and power, and also to be in the home, loving wife and
children and expressing tenderness, sympathy, and consideration toward
others.

The dualistic demands of the masculine role created tension in men
as well as women. This is apparent in the case of James Hague and Mary
Ward Foote, an upper-middle-class couple whose courtship began after
James was a well-established mining engineer. James was the first assistant
geologist on the U.S. Geological Survey of the fortieth parallel led by
Clarence King.[43] After 1870, he devoted himself to a consulting practice
in engineering and eventually made a fortune from mines he purchased
in Grass Valley, California. During his courtship, however, James was an
ambitious but not yet wealthy man in his early thirties.[44]

Throughout their courtship, James and Mary fought a tug-of-war between his career ambition and their romantic relationship, between money

and love. Separated by a continent—James in San Francisco and Mary at home in Connecticut—he justified the hardship: "Dear Heart, I've been thinking about you all day and longing to see your blessed face and take your hand and kiss you; and when I think how long it will be before I can do so, I get almost discouraged with impatience. But I know that our prospects for happiness by and by are all the better for my attending to business now—so if that takes me away from you I must make the best of it . . . ."[45] In this transaction, James deliberately chose the side of provider over lover, accepting the opposition of love and money. He reconciled the dominance of business over romance as a temporally limited one in exchange for a future and unlimited married life of love and family. Temporal sequencing was one strategic response to the opposition between economics and emotion in the male role.

James could, however, approach the identical tension differently: "Oh, you dear, darling, old Lovie, if I could only be where you are this minute and take you in my arms, I should make you reassure me by telling me you don't care about money but that you love me more than anything else, as I do you. Upon my word I don't care whether we are rich or poor, if you will only love me and I love you, all my life, as we do now. . . . You dear sweet girl, I want to howl with impatience to see you and kiss you and do some more courting!"[46] The tension between love and money was critical, but in this instance James sought reassurance from Mary that she valued his love above any future promise of wealth.

As was typical of younger swains, James worried over his ability to support his future wife adequately: "It grieves me to think that your welfare depends so much on mine—because that is so precarious."[47] James, who had placed the highest premium on his career by remaining a continent away "on business," now hoped that his future economic success was not love's dependent clause. Playing both sides of the polarity, at one time he insisted that business success was the necessary condition of their future love and happiness, but at this moment, he asked Mary to separate completely her love for him from his economic fortunes.

Immediately before their wedding, they were still in conflict. James, who had been absent nine months, repeatedly promised Mary a whirlwind courtship upon his return to Connecticut. When he left her "on business" a little more than two weeks after their Connecticut reunion, Mary was distraught. She accused him of indifference, suggesting that "there is a lack of 'honest reasons' for not spending more time at Guilford [her home]."[48] Significantly, in the face of her accusations, he denied any conflict between his economic pursuits and her emotional fulfillment: "How will it be by and by, darling, when calls for bread and butter take me off into the country on my horrid mining business—will my letters from 'home' cheer a fellow up—or will they complain of him for having a profession that takes him out of the house? . . . What can I do to make you happy? You know I love you dearly—that the great desire of my life is to have your love and to make your life a happy one—what shall I do that I have not done to prove it? give up my business? stop trying to get a living? of course not, you will say—but my present business and mode of getting a living will take me away from you now and then—perhaps often—and I shall be distressed if you fail to see that in pursuing business I have your welfare at 'heart.' "[49]

James voiced three different responses to the tension between business/ ambition/money and love/home/family. In the first, he admitted the dichotomy but used the strategy of temporal sequencing, claiming the necessity of business in the present and promising Mary that romance, love, and family would take priority in the future. In the second, he also admitted a polarity, but this time he elevated love above business success, seeking reassurance that she agreed. In the third, he denied the polarity altogether, insisting that in choosing business he was also choosing love. The tension was the same in all three cases, but the transaction was different. The point is not that James Hague was insincere, but that even rather rigid gender divisions were used creatively by people in actual interactions.

Nonetheless, despite variations, there were behavioral patterns. Courting males, most often assuming James Hague's last stance, presented their masculine obligation to be economic providers as congruent with their obligations as lovers, comforters, and intimate companions. Nineteenth-century middle-class men attempted to reduce the tension between economics and emotion before marriage. They phrased their identification of love and money in terms of their ambition to provide for women's comfort and happiness, often insisting that their economic concerns were motivated by the heart. But submerging the dualism of his role in romantic love added not only emotional intensity but also anxiety to a man's understanding of his economic functions.

Many men expressed deep anxieties over the economic responsibilities they had to assume for the women they loved. John Foster confided to Mary Appleton: "My *greatest fear* is that should I be the means of removing you from the station of peace and usefulness, which you now occupy;—should I be able to induce you to quit the safe prospect of ease and competence, which is now before you, to share with me the risks and labours of my own station in life—that I should be goaded by the reflection that I had diminished the happiness of the being whose happiness was more dear to me than any earthly good. Could I, my dear Miss Appleton, could I have felt half as certain you would be as safe in everything, as in my warm, constant, and undivided affection, I should not so long have restrained the expression of my wishes . . . ."[50] His economic fears were typical as was the contrasting security he displayed toward his affection.[51]

What is perhaps most telling is the responsibility the nineteenth-century middle-class male took upon himself for a woman's self-fulfillment by becoming her husband. Men strongly identified their economic success and women's happiness. From Nathaniel Hawthorne, the literary genius, to James Bell, the articulate sharecropper, men believed that their exertions would in large measure determine a woman's happiness. Hawthorne told Sophia that he "desires nothing but to be with thee, and to toil for thee, and to make thee a happy wife . . . ."[52] Associating "toil" with making Sophie happy, he also intended to write "beautifully, and make myself famous for your sake."[53] James Bell also associated male labor and female happiness: "We are pledged to each other, and the question with me is, where can I go? What can I do, and how am I to do? to make you happy and give you a good pleasant home." He added, "This is a subject that troubles me more than all else."[54] Underlining his sense of responsibility for his future wife's happiness, James Bell worried about his ability to fulfill this diffuse yet demanding masculine obligation: "I pray that you may always be as happy as your love makes *me*. 'I dont know but you will scold me for saying it' but I feal unworthy and inadequate to the task of making you happy, that is just as I *feal* . . . ."[55] James also articulated a powerful identification between his labor and his love for Gusta: "Well Gusta for a long time I have had an object to work for, And that was *you*. I have lived and labored in the hope that at some time I could have the happiness of calling you my bride . . . ."[56]

Eldred Simkins promised Eliza Trescot that when they were at last together, he would "think of neither Past or Future—but I will ever strive to make that present comprise your happiness and contentment."[57] Though Eldred was untroubled over his future economic prospects, most males were rarely as sanguine.[58] The last letter Lyman Hodge wrote Mary Granger before marriage reflected his anxiety: "I have long known that it would be best for me to [be] married, as it would for you. But I did not dare to marry, until I had a reasonable prospect of being able to take care of you.—You know that for a few years we must both avoid all expense . . . which I regret, as I know that you like nice things, and no one, more than I, likes to see your tastes gratified . . . ."[59]

John Marquis also worried that he could not provide Neeta "What every young man should be able to offer before offering himself," that is, a good home with all its comforts. "But I could not endure the idea of giving you up." He envisioned a future in which he would work hard to improve his material condition "especially if I could only have you Dearest Neeta to smile on my efforts, to cheer and assist."[60] And he pleaded, "I am not a 'man of wealth' but it would be my life effort to make you happy."[61] Men expressed an unambiguous responsibility for women's happiness through their economic achievement. The association of their labor with feminine happiness is one of the more intriguing aspects of nineteenth-century gender roles. "Making" a wife happy appears to be a burdensome agenda for a husband as well as a chancy task in the most favored of male-female relationships. That such a masculine role obligation may in some way have been functional for the dependent nineteenth-century wife in no way changes the potential emotional difficulties for both sexes.

The power of association between women's happiness and the economic side of the male role was so strong that even in a desperately unhappy marriage, a man such as Charles Watts, seeking his fortune in the Klondike, continued to believe or perhaps hope, "If I do right, am industrious and honest some of my investments will strike it and we three will be happy."[62] While an unremarkable statement on its surface, Charles was, after a number of years, married in name only. At his sudden death in Alaska, his wife was so alienated that she refused to pay for his burial costs with insurance money expressly designated for his funeral expenses, and rumor had it that he remained two years "above sod."[63] His wife's deep animosity raises serious questions about his steadfast faith in economics as the solvent for the problems of his marriage.

There was a powerful association of money and happiness in many aspects of American life, but particularly in men's sense of male-female relationships. The pressure for economic achievement in the masculine role weighed heavily upon nineteenth-century men who did not scale the economic heights. Charles Watts poignantly exposed the link between economic success and masculine self-esteem in his desire to make some money so he would "not be another of those worthless son-in-laws."[64]

In the now classic version of the Protestant ethic still prominent in interpretations of American culture, seventeenth- and eighteenth-century Protestant men believed that they were working for God, in the context of a theology that defined work as a religious duty. In these terms they closely identified religious commitment and spiritual character with individual economic success, seeing the economic rewards of this world as signs of the state of their relationship to God in the next world.[65] By the nineteenth century, the Protestant ethic was secularized, and the commonweal or public good filled the place once given to God.[66] Dwelling upon the need to be useful and the social duty to produce, nineteenth-century moralists shrunk in horror from the spectre of idleness and encouraged dreams of individual success.

One important element in the nineteenth-century work ethic, however, has gone unnoticed. While moralists urged a life of diligent toil in order to fulfill a social duty and to avoid scarcity and the evils of idleness, the nineteenth-century work ethic took a less abstract and more concrete form in individual men's lives. Many men expressed their desire for economic success in terms of their commitment to wife and family and phrased their own economic motivations in relationship to women's happiness. In fact, nineteenth-century men claimed they worked for women and children in a way analogous to an earlier generation of Americans who claimed they worked for God. By the mid-nineteenth century economic success was viewed from several perspectives. One neglected way of understanding economic achievement in nineteenth-century culture was as a male sex-role obligation tied to and reflecting a secular, romantic commitment to women, love, and family.

Striving to the fullest in the economic realm in nineteenth-century America was identified by men themselves with loving a woman. They turned to economic accomplishments not only as proof that they were doing their social duty, but also as a sign of their emotional commitment to a woman. For some men romantic love claimed a more powerful pull upon their economic performance than any abstract concern for usefulness, scarcity, or moral decay. By the early nineteenth century, the work ethic for middle-class men may also have been an ethic of love.[67]

🐟 It is widely accepted that the nineteenth-century middle-class canons of femininity emphasized religious piety, "a susceptible heart," and "the whole train of domestic virtues and capacities," exactly as Samuel Francis Smith defined his ideal wife.[68] There is, however, a difference of opinion over how much power this role definition gave women in practice. Sometimes this canon of domesticity is perceived as requiring unremitting physical and psychological labor from which women reaped little real power inside or outside family life.[69] Alternately, the demands of woman's role are seen as less objectionable, perhaps fostering some areas of personal fulfillment, even though marital relations were so limited by the law, economic barriers, restrictions of movement, and differentials of physical strength that the balance of power was heavily weighted on the male side.[70] Finally, it is claimed that the canons of domesticity, particularly elements that elevated motherhood and touted woman's superior moral and spiritual character, actually resulted in increased authority within the home and served as a stepping stone for augmented power and activities outside the home.[71]

The conceptualization of power itself is the conundrum of this debate. How is power to be measured in family life? Historians have used different yardsticks, such as various types of decision-making (for example, regarding children's lives or geographical relocation); economic responsibility and control of financial resources (for example, in naming executors of wills); family law and legal restraints; control of fertility decisions; the grounds of divorce petitions. Different conclusions about power relationships may well be the result of looking at conflicting factors.

In addition, most of the data on power in nineteenth-century family life is missing a key element: the actual interaction between men and women during the process of arriving at decisions. Critical questions remain: When did women assert themselves and when did they defer? What aspects of their definitions of masculinity and femininity were troublesome to them? What circumstances created regular tensions through conformity itself? When did they violate sex-role boundaries and how did those violations work? As was true for Victorian men, women's responses were contextual and demonstrate that even in what many perceive to be a fairly rigid gender system, there was room for some creative adaptations.

Nineteenth-century women took control of some situations with assurance and cut a wide authoritative swath through certain thickets of their social experience with men. They asserted themselves—often with confidence—in areas within the female sphere. Many men agreed that the exercise of female authority was "permissible" within woman's domain by gender right. But when women ventured self-assertion or leadership outside their sphere, they were usually more circumspect and were often greeted less enthusiastically.

Jane Burnett provides a good example of the "permissible" assertion of female authority when she told her husband of eleven years: "Go to church—and do take care of Ikie [their son]—Keep yourself warm—I fear when you have no house to go to you will not take good care of yourself. Buy a weter proof overcoat—They are all the rage and keep out the rain."[72] Jane Burnett commanded her husband with assurance in all the areas which were woman's special province in nineteenth-century male-female sex-role ideology. Women were expected to take responsibility for their family's moral development and religious training; for children's care and proper upbringing; and for men's health and physical well-being. What it meant to be a woman in America at the time was defined by these tasks. Thus, Jane Burnett freely expressed her sexual identity in this transaction by pressing the significant features of her feminine role. Such women unambiguously asserted themselves with men in areas where their sex role gave them mutually agreed-upon authority.

The authority that the nineteenth-century gender-specific division of labor gave to women can also be seen in another incident. James Bell, having learned to do many domestic tasks for himself during the war, asked Gusta Hallock in 1863 whether she would like a husband who was adept at housework. She responded: *"Of course I would.* I think that it would be very nice and *always* thought so. Tho I should calculate to do all the housework *myself* Yet it would be very handy sometimes."[73] Gusta "calculates" to do all the housework, asserting her "territorial right" to this work in spite of the fact that James appeared open to the possibility of sharing at least a few traditionally female domestic chores.

A southern woman also insisted upon control over a task she obviously considered crucial to her feminine identity. Neeta Haile adamantly laid claim to milking the cows after marriage, in spite of her midwestern fiancé's assumption that it was his responsibility. John Marquis, sensing the import of her vehemence on milking, responded: "Though if you really insist that it is 'womans work' why of course I shall give up to you just 'like a little man.' "[74] Given the chance to have John assume this task and lighten her everyday work load, Neeta insisted upon milking those cows because that was woman's work. Routine tasks were symbols of sexual identity, transforming work into a ritual which publicly announced the significant differences between men and women. Even a spoilt southern belle, after a year of marriage, decided to set up housekeeping in order to assert herself in the feminine sphere.[75] Though the impulse quickly faded in her case, nineteenth-century middle-class women were understandably reluctant to abandon authority from which they derived power and identity. Typically, women asserted themselves and men acquiesced in concerns of childrearing, health and sickness, domestic purchases and household management, food preparation, and standards of morality and behavior within the home, extending to certain non-economic features of community life such as religion and mixed-sex social activities. In these areas nineteenth-century women exercised what both men and women respected as an authoritative voice.

One must ignore nineteenth-century women's perceptions, feelings, and behavior to dismiss the power women exercised within the feminine sphere as unimportant or trivial. It would be a serious oversight, however, to neglect the tensions that conformity to the feminine role ideal created. Nineteenth-century men and women felt strained by a similar contradiction in their roles but were pressed in opposite directions; they were tugging on the same rope but pulling at opposite ends. What might be characterized as men's domain by gender right was defined as money, economic success, and worldly achievement, but men were also expected to identify with love, home, and family *in relation to women.* While woman's domain by gender right was defined as love, home, and family, women were also supposed to identify with money, economic success, and achievement *in relation to men.* Men felt the tension between economics and love from the worldly side and women from the family side but both were strained by a similar polarity.

In interactions with her husband, Alice Baldwin illustrates the tensions between economics and love within the female role. "A woman," Alice said, "is a forlorn creature indeed without a husband (a good one like mine) unless she is blessed with riches."[76] She accepted masculine territorial possession of the larger economic world and women's subsequent dependency on this male "domain by right." Though she was ambitious, her economic and social aspirations revolved around her husband's job and promotion. She recognized a tension between the feminine domain—love/home/family—and masculine role demands: "I hope affairs will turn out as you hope for my own dear boy your interests are mine if it warent for the faster promotion and increase of pay I should feel very bad indeed to have you in the Cavalry for you know unless I could be with you all the time it would be anything but a happy life for me." After recognizing a potential conflict, Alice emphasized their mutuality of interests: "but after all these objections I think of the idea of your always remaining nothing but a Lieut—and then I feel as if I would endure all sorts of torments for your promotion for you know how it galls me to be beneath any body . . . ."[77] Alice saw herself as working *for* her husband's promotion. Though she stated the potential conflict between her love and his job, she submerged this tension in her own identification with her husband's economic success, status, and professional recognition.

Only a little more than two weeks later, however, she wished she could earn something herself, "But pride and a fear that you would be greatly displeased if I should seek employment kept me from even *trying* to do something. I know you could get along nicely if it wasnt for me I feel like an encumbrance on you and I am I know it."[78] Earlier she pictured them working together for his promotion, now she saw herself as a burden.

She soon shifted her ground again: "I love you so well that I shall always *wish* you had *wealth* and *rank* and *power* and I will confess it has made me very unhappy that you dident have more *rank* but it never made my *heart* any the less warm and full of love for you." In this transaction Allie reassured her uncertain and disheartened husband that their love outweighed his worldly success. She hedged her affirmation of love over wealth and rank, however, by adding a temporal qualifier: "But you are a young man yet—and much to hope for."[79] Like James Hague, Alice Baldwin also used temporal sequencing to mediate the opposition of economics and emotion in nineteenth-century sex roles.

Three years later Alice again changed her stance. She initially identified with her husband's achievement and self-consciously affirmed that his job and promotion should take priority over their private life. However, after her husband, Frank, became a hero in the Indian campaigns, she wrote: "Frank the more I think of it the more I realize that your position is most undeniably established in your Regiment and that is the most important and necessary thing for an officer to attain. Now if *I* could only be somebody and be a helper to you . . . ."[80] Alice now distinguished between his achievements and her own, expressing a strong desire to "be somebody" herself. She feminized her dream of ambition and success by adding, "but beauty is the chief thing in woman and that I have not got in the slightest degree neither in face nor figure . . . ."[81]

Alice continued to yearn for her own distinctive success in the face of her husband's public praise: "I said to myself It is *my* husband that all these fine things are said about and I could not be thankful enough. Palmer and his wife I know both think highly of us both—particularly *you*—and I am willing they should—but I being a woman have had no opportunity to distinguish myself as you have had and perhaps if I had I couldn't."[82] Clearly jealous but not rebellious, Alice was restive in her domestic role, but accepted the standard definitions of her proper sphere because she conceived of no alternative. Very few nineteenth-century middle-class women escaped the identity-defining power of the doctrine of separate spheres and the ideology of domesticity. Even the least conventional and most independent woman, such as Dorothea Lummis, could write to her husband before her graduation from medical school that she would "try my deepest and best to be to you, not a physician wife, but a true wife."[83]

Alice Baldwin again lamented her insignificance, measuring her self-worth by both masculine and feminine sex-role standards: "I do try so hard to get along better than I do and to make the best of what I have and succeed so—wretchedly and feel so discontented all the time I can't endure it. . . . my miserable ambitious proud nature and disposition is the cause of my unhappiness *principally*—but not *all*. I would be someone if I had a chance—but with neither money influence or *beauty*—which is *every thing* in a woman what hope have I?"[84] Alice wished to "be someone," but she herself saw most achievement as antithetical to the female role. Claiming that beauty was practically "everything," Alice defined women's success as more a "given" of nature than a result of personal effort.

Alice Baldwin's discontent is an indication that the values of economic achievement and independence were already tugging at some women. It is also a reminder that sex-role transactions varied across time in male-female interactions. In the first instance, Alice identified herself with her husband's ambition, and elevated his job over home and family. In the next, when her husband confessed that he was depressed and disappointed in his work, Alice reassured him that love was paramount. In later inter-actions, after her husband had become relatively successful, she expressed a yearning for her own public recognition and success, no longer suffi-ciently identifying with his achievements, but complaining of her own insignificance.

Deeply committed as she was to the prescribed female sex-role ideol-ogy, Alice was also insecure about her ability to perform her own role. The more her husband achieved, the less adequate she felt to be his wife. While he was struggling, she remained unthreatened. But his worldly success made her feel extremely inadequate in her own sphere: "You are too good for me and *I know it to be.* I wonder what makes you love me, theres nothing to attract a mans love like you. I feel sometimes my hold on you is a precarious one when I compare myself with other women, for I always loose in the comparison."[85] Anxiety and discontent with her own role performance partially fueled Alice's plaintive cries of "I want to be somebody."

Though she yearned for control in the world, economic independence, and a sense of power and excitement, she herself identified these possi-bilities with the masculine domain. Physical beauty, inherited wealth, and influence were the only avenues to power she conceived for women like herself. Alice literally could imagine no alternative to the standard nineteenth-century female role prescriptions that she so strained against but so often accepted despairingly.

Little wonder some women were more content than others within nineteenth-century female gender boundaries. Women's discontent (as well as men's), however, was not necessarily a constant in their lives. During times of crisis or special personal need women might wish to assume male prerogatives. Dorothea Lummis, frustrated by what she took to be the constraints of her sex, wished she were a man by day but a woman by night.[86] Significantly, her discontent did not extend to the "after-dark" expression of her femininity.

She was not the only woman to suggest a male-female crossover while maintaining a strong feminine identification. Eliza Trescot, a much more conventional woman, also wished she could be a man. Her sex-role frus-trations were linked to the intensified feeling of vulnerability that the Civil War created, especially for southerners near the fields of battle. "If I could I would join the Army," she wrote; "why didn't nature make me a man?"[87] After President Abraham Lincoln had proposed to grant am-nesty to all who participated in the war on the Confederate side, except high government and military officials, Eliza fumed: "I suppose the only course to follow would be quiet submission but I *wouldn't* take the oath of allegiance—*never*! I can fancy I see you smiling at my fancied bravery . . . ."[88] Later in the same year, after Atlanta fell, Eliza expressed her vulnerability: "the truth is now I would not feel safe anywhere. I tell you Eldred I have never wished more earnestly that I was a man to be in the army instead of being a helpless dependent creature to whom a Yankee raid would be a far worse thing than camp life and everything else the soldiers have to endure."[89] Eliza was not always at odds with the constraints of her feminine sphere, but like many women, she manifested discomfort with her role when she was involved in situations that caused her to fear her dependency on men.

In response to the frustration and acute sense of powerlessness that overwhelmed her in the midst of the Civil War, Emily Lovell also wished she were a man: "Where then can you find food for *such a civil* war, in a *democratic* country, it has *literally eaten* itself up—the huge monster has crushed itself in its dieing struggle—*Emancipation proclamation* say I—I never did wish I was a man until now—seems to me *I* could do something . . . ."[90] These women tended to be most overtly discontented with their feminine role when their dependency on men was most physically threatening, as in war, or immediately challenging, as in a troubled marriage. Dependency was a key issue, but it would be a mistake to see it as a constant concern. Though an unceasing reality, it became problematic in response to the many unpredictable elements in any nineteenth-century life history.

For most individuals there was one phase of the life cycle, courtship, in which the tension between economics and love was a predictable source of anxiety. Both men and women felt the weight of the choice of a marriage partner, but the scales were tipped in different directions—men worried over assuming economic responsibility for the woman they loved, and women feared giving over that responsibility to the man they loved. This moment was especially difficult for women; they thus required more intense and elaborate emotional reassurance than their partners.[91] While most Victorian women accepted their domain by gender right and derived some satisfying authority from their role, their very conformity to the separation between men's and women's spheres created tensions in their interactions with men.

🐝 Inherent tensions between home/family/love and worldly achievement/economic success/career pulled upon both men and women. These tensions created discontent that could sometimes be translated into outright violations of the boundaries of the male and female spheres. These boundary violations involved shifting in and out of standard sex-role expectations. Such sex-role shifts are acts of reframing.[92] When reframing occurred, a sex-role prescription was violated, followed by the redefinition of the behavior to fit the socially sanctioned standards of masculinity and femininity. This usually reestablished sex-role equilibrium. Paradoxically, reframing was an interaction that both violated and respected nineteenth-century sex roles, both transcended and held them in place. Though nineteenth-century sex-role ideology was rigid, through reframing it became a more flexible system on the level of actual behavior between men and women.

Nineteenth-century middle-class men reframed their interactions with the opposite sex under very different circumstances than women. When Clara Bradley appeared to have secured teaching positions for both Nathaniel Wheeler and herself, Nathaniel responded by aggressively claiming priority for his role as economic provider: "He [the principal] may have your services next year—tell him—if I secure the appt. At least I consent now and till a certain indefinable day next Summer must refer him to you. After that—we'll see who's who."[93] Threatened by Clara's initial success in the job market, Nathaniel asserted his masculine claim to future economic control. He even predicted his future economic conquests: "Never mind—I'll be even with you some day."[94] In this exchange with Clara, Nathaniel illustrates the pattern of masculine reframing. Clara encroached upon the central male domain of economic provider and took the lead in finding her fiancé a job. He responded by shifting the scene to the future at which time he would decide whether any school obtained her services; it was a future in which he would find jobs for them both. What is significant is that Nathaniel reframed the incident to make it appear to be an instance of masculine dominance when in fact it was not.

Men most often reframed a transaction with women when they renounced the power given by their sex role. The opposite was true for nineteenth-century women, who reframed a transaction most often when they assumed power not given in their sex role. Men demonstrated a need to compensate for their self-renunciation, women for their self-assertion.[95] Nathaniel Wheeler, reflecting on the meaning of his manhood, teased Clara: "Well, we've had quite a chat. I must go now. What! would you keep me as of old? No, I'm a *man* now, and must assert my prerogative. Goodbye."[96] For Nathaniel, self-assertion was the most significant emblem of his newly acquired manhood. This self-assertion appears symmetrical with the self-abnegation demanded at times in the feminine role. Of course, for a variety of reasons sex-role imperatives were subverted in actual interactions. Reframing was a way to honor them in the breach.

Will Harbert and his wife, Elizabeth Boynton Harbert, illustrate how sex-role flexibility could be gained through reframing. Elizabeth was active in a variety of nineteenth-century women's groups, and while attending a Women's Christian Temperance Union meeting, she wrote to her husband and children that she would return home next week, unless she received a telegram indicating that they required her immediate presence. "Take *good care* of each other," she exhorted, "and let me know *if* for any reason I should hurry home—."[97] Elizabeth regularly left her home and family for woman's suffrage conventions, and W.C.T.U. conferences, but by offering to return home at any moment she translated her "violation" of feminine sex-role obligations into a long-distance "fulfillment" of them. By confirming to both herself and her family that they were her priority, she offered self-abnegation in theory but self-assertion in practice.

In another example, Emily Lovell asserted her will, and then asked her husband Mansfield to decide what she already told him she intended to do. All the other generals' wives had left town, having been sent for by their husbands, and Emily waited forlornly: "I am the only *she*, General left . . . ." Emily's patience was exhausted: "Now darling, let us come to plain terms—tell me . . . about your movements—so that I may know if I am doomed to be so far separated all winter—I have decided so far myself dear,—if you can say that you are to remain—two months longer where you are, or near where you are. I am going to move *Heaven* and *earth* to get to you, that far I say, now I want to know if such a movement will be in accordance with your [word unclear]."[98] Nineteenth-century women rarely asserted themselves outside their defined sphere without attempting to restore at least the façade of male dominance. They were not simply engaged in face-saving gestures to assuage masculine pride, but appear to be responding to a deeply felt need on the feminine side as well.

Another example of reframing involved Lincoln and Julia Clark. In the late 1840s they planned to move out of the slaveholding South. Raising a family in Alabama, they worried over their children's involvement in slavery and the corruption of character they associated with it.[99] After much discussion, and Julia's subsequent departure with the children for New England in 1847, Lincoln was still unable to decide where they should resettle. He solicited Julia's advice and opinions on where to move, carefully maintaining the distinction between consultation and decision: "I do not mean to cast upon you the responsibility of this movement—for when you were here, I know you would have cheerfully remained had I said so—you may be less willing to return now, altho' in one of your letters you expressed yourself ready to do so if I desired it: This I know as a good and faithful wife you would cheerfully do: but I can not bid you do that in which your *heart* is not, for it is that which I wish to command in all things, not the will."[100]

It was important for Lincoln to reassure himself that Julia would have followed him anywhere, even though he did not intend to require such rigid wifely role performance. Fretting over their final destination, Lincoln even made an argument for staying in Alabama. Then he symbolically surrendered and gave the decision to her: "I do not wish you to wait for further advise from me—if you think we are right then take measures to proceed to Chicago . . . . Should you think it best to return here you had best take the western route . . . . I shall wait an answer to this with much solicitude."

Lincoln lobbied against the move, even as he left it in her hands. "I perceive on looking over your letter again," he continued, "that you write as tho our home were to be in the West—if you advise me to encounter all the difficulties and dangers that look us in the face there I shall cheerfully do it, and think *perhaps* it will be best, tho I admit I shrink from it." He phrased their move to the West in such negative terms as to almost preclude it, but not quite. "And now I close with this remark, if there is any feeling in your heart which seems to say that you can not be as happy here as in some other place then we will go to that place; for wherever you are happy, there the sun will shine."[101] Lincoln in the same letter moved from shouldering ultimate responsibility to recognizing their joint responsibility for the move, to finally making it a matter of his wife's singular emotional preference. He was so anxious that he added a long postscript the next morning in which he again expressed pessimism about his ability to succeed in the West. Though charging her with the final choice, he put all his weight behind staying in the South.

Julia's response was strong, clear-headed, and unwavering: "You leave the decision of our future home with me. If you do, I think I must say on the children's account but on no other I would prefer a residence at the East or West . . . . Had we not better go to Chicago, and stay there through the summer, it will be no more expense than to board in Alabama, and then if we find they will not *let us live* at the West return to the South. That was the plan we first framed, we gave ourselves till the fall courts [Lincoln was a circuit judge] to determine about the future. Perhaps you will reply, I have the responsibilities, but if you are not successful, the *trial* will perhaps be mine as fully as any . . . ."[102] First, and perhaps most important, Julia openly acknowledged that Lincoln had left the decision to her. She calmly gave her reasons for steadfastly preferring to move out of the South, and then she asserted her own share of the responsibility, refusing to be intimidated by Lincoln's misgivings. Continuing to lobby against the move, Lincoln lamented several weeks later that fifteen years of work to gain his reputation in the South would be lost. At forty-seven such a relocation was difficult, and he feared that his career would suffer irreparable harm. Reluctantly and with some foreboding he told her: "But I shall go to meet you at Chicago . . . ."[103] Hardly a ringing endorsement of the choice she had made for them both.

While nineteenth-century women might become adept at reframing, so also might the men. As Lincoln Clark and Nathaniel Wheeler demonstrate, men reframed episodes to assert symbolically their authority where it had wavered. On May 21, 1847, Lincoln, with Julia's help, began to reframe their move: "You say you think now that you shall go to Chicago if I so decide—I have so decided because that seemed to be your wish and desire . . . ."[104] After trying in vain to change Julia's mind, Lincoln reclaimed the decision-making power, noting that it was based on her preference. This is a clear example of a sex-role transaction that had multiple layers of conformity and nonconformity, belief and disbelief. In the end, Lincoln greatly feared the move he contemplated for so long. Renouncing his prerogative to control the final decision, he manifested his affection for Julia, as well as his emotional ambiguity over the move, by asking her to decide their geographic fate. Shifting sex-role responsibility in this case may have eased his psychological burden as economic provider for her and the children in a new and uncertain environment. Julia willingly shouldered the added responsibility, but she used their children to justify her choice in terms that conformed closely to part of the standard female sex-role prescription. Apparently at Julia's prodding, Lincoln then reasserted a masculine frame upon the experience, claiming the decision as his own based upon the values of a companionate ideal—her wishes. Thus nineteenth-century sex-role ideology was both violated and respected in a complex interaction that reveals how malleable even rigid sex-role prescriptions became in actual relationships.

In another interaction between the Clarks, recorded some twenty years later, Julia was living in Kentucky, working for the Women's Sanitary Commission. She responded negatively to her husband's request that she return home to care for him, justifying herself by the importance of her work and her own success. She was proud of what she had accomplished, turning an old run-down building into an efficient Union hospital kitchen. "It is said here . . . ," she told her husband, "that I am the only one who could have done it. . . . and I do not think it would work to my credit or the good of the Institution to have me leave till this plan is carried thro'—."[105] Then she requested permission to remain. Although she had already told him, in so many words, that she intended to continue her work in Kentucky, his authority was preserved by asking for his consent.

Women's ability to disguise their power in relation to men is a widely held bit of folk wisdom. But the dynamics of female reframing differ from the folk view of feminine guile. Julia did not beg, cajole, act hurt, or otherwise use her weakness to get her way, nor did she disguise her intentions. She stated her case in terms of her competence and strength—she asserted herself—then she reframed the transaction. Reframing on the female side involved genuine and open self-assertion outside the feminine domain, followed by some gesture of outward conformity to the standard sex-role precepts. While Julia's work for the Sanitary Commission can easily be seen as an extension of woman's sphere, she felt enough legitimacy in her husband's request to return home that she tempered her decision to remain.

Reframing illustrates the complexity of male-female relationships and the need to consider processes of interaction in interpretations of power. Women were encouraged to be assertive and men passive in the feminine sphere of influence. Moreover, women might take control in areas that were ambiguous or clearly outside of woman's sphere, reframing their interactions with men to make their self-assertion more acceptable just as men might reframe when they let their "rightful" authority waver. The pressure on men to be assertive and women to be submissive was expressed through the reinterpretation of behavior that did not conform to their images of masculinity and femininity. Thus some sex-role flexibility was achieved without abandoning allegiance to the more rigid Victorian ideals of masculinity and femininity.

Although men's and women's outward role behavior may not have changed greatly, evidence indicates there was more tension over sex roles from the Civil War to the end of the century than in the antebellum period. Certainly the Civil War itself may be a factor in what appears to be a new level of sex-role insecurity in postbellum male-female relationships. War is one situation which functions simultaneously to cast women's dependency and submission in a negative light, forcing them out of the feminine sphere—throwing women on their own resources, encouraging them to learn new skills, and compelling them to manage in unaccustomed ways. War also affects the masculine role, but less attention has been paid to this important possibility.[106]

Sex-role anxiety after the Civil War was concentrated upon the question of male-female contrast. Gender contrast was a particularly vexing problem for nineteenth-century men, whose sense of masculinity was linked to their ability to counterpose male to female behavior. The necessity to establish a contrast was one of the most pressing claims of nineteenth-century masculinity.

In that respect, and in several others, the nineteenth-century woman's movement appears to have played a significant role in increasing sex-role insecurity in post-Civil War America. Though the majority of nineteenth-century women may have had nothing to do with feminism or suffrage, the post-war male-female exchanges indicate that the first woman's movement contributed, perhaps centrally, to a new level of self-conscious sex-role awareness. The cultural scope and significance of the first woman's rights movement has been underestimated.[107] Though admittedly only several thousand women can be counted as woman's rights activists, a much larger portion of the middle class was probably touched by the movement.

By initiating a highly visible cultural debate, the nineteenth-century woman's movement drew opponents as well as friends into a dialogue over woman's nature and thus her proper social role. One of the most unappreciated effects of the first woman's movement is not measured in women's direct participation or ideological adherence to woman's rights, but rather in a growing uncertainty over female commitment to standard sex-role boundaries. Although most middle-class women conformed to the doctrine of separate spheres, Victorian men signaled an uneasiness and insecurity over women's commitment to the standard feminine role.

Significantly, there were few analogous expressions of uncertainty and doubt by either sex over definitions of the masculine role. Furthermore, there was almost no parallel social discourse over alternative conceptions of masculinity. This disparity translated into a noticeable contrast in sex-role awareness. Both nineteenth-century middle-class women and men were considerably more attuned to issues surrounding woman's role.

Men displayed more uncertainty over women's commitment to dominant feminine role prescriptions than women showed over men's commitment to the dominant prescriptions of masculinity. Women rarely questioned the standard masculine role model. For example, Dorothea Lummis, soon to be graduated from medical school and possessing at the time a more promising set of educational credentials for economic success than her husband, told him that they should have a child "as soon as you can support him."[108] On the whole, women admired men's physical strength and self-assurance and expected male protection. When James Bell wrote a letter complaining of camp life soon after he joined the Union army, Gusta responded unsympathetically that he should stop being an old maid.[109]

Women sometimes reassured men when they expressed role-performance anxiety, but they rarely discussed the abstract nature of masculinity and the male domain as such. Perhaps a sizable portion of nineteenth-century middle-class women were afraid of analyzing masculinity because they were dependent in so many ways on traditional male-role performance. Also, of course, the self-assertion inherent in women commenting upon the masculine sphere in interactions with men may have silenced many who were sensitive to the limits of their prescribed feminine role.[110]

Middle-class men, by contrast, rather freely dissected the female role in exchanges with women, almost as if they saw a masculine prerogative in such debate. Moreover, when they discussed gender issues, men made much of the differences between themselves and women, especially when those differences were blurred. Clara taught alongside her husband one academic year at Lawrence University, and the similarity in their work lives bothered Nathaniel. "You were formed for home," he told her, "and to be *its* mistress, not to be my helper in breadwinning and my rival in my special work but to be my intelligent and sympathetic home guide."[111] Men insisted most strongly upon the contrast between themselves and women when confronted with the possibility of female crossover into their domain by gender right.

The gender boundary most hotly defended by men centered on issues of independence and dependence, often but not always economic. Eliza Trescot confessed to Eldred Simkins that "we all ought to try to improve and be all that we ought to be, don't you feel a contempt for the idle way in which most young ladies pass their time, leaving their hearts and minds neglected and uncultivated? I do and Eldred it is not only for my own sake, but for yours also that I am determined to follow a different course and one of these days I will be worth something."[112] In a small way, Eliza minimized an important male-female sex-role distinction when she asserted: "We all ought to try to . . . be all that we ought to be." Eldred confirmed that she crossed a significant gender boundary when he responded directly to her sentiments by saying: "You are not dependent upon your exertions for a livelihood but are engaged in a work of supererogation . . . ."[113] Although he dismissed his own remark as a humorous aside, the contrast he drew was stark and staccato. He compared men's work, which was real because others depended upon their labor, with Eliza's exertions, which he called superfluous—"a work of supererogation." He scorned her program of self-improvement because it was not connected to economic responsibility for herself or others. While softening the blow by quickly adding that he admired her desire for self-improvement, the transaction revealed a deeper tension.

Masculine anxiety over gender boundaries is even clearer in the next exchange. Feeling threatened by what he perceived to be the possibilities of women's social crossovers into the male sphere, John Marquis told a woman he met in 1862 and who eventually became his wife, "The use of tobacco, drinking, and swearing are the worst vices men are addicted to and in woman they are positive crimes. I don't know whether you are a very strong upholder of 'woman's rights' but I am sure you will agree with me on this. Man is a courser animal and too much must not be expected from him. But woman I consider a higher class of beings. . . . Guides, sent to lead us poor sinners up to higher and better things."[114] Embedded in this standard nineteenth-century sex-role rhetoric was an uncertainty over this woman's role commitments which suggests one effect of the woman's movement on private relationships.

The nineteenth-century woman's movement, from the Civil War on, increased middle-class men's and women's awareness of the permeability of women's sex-role boundaries. John Marquis, for example, referred to woman's role not as a taken-for-granted part of the culture both he and his correspondent shared but as an area of potential conflict. He attacked nineteenth-century feminists for what he portrayed as almost sacred boundary violations. "I have a very high opinion of the female sex occasioned I suppose by the associations of a good mother and sisters but while an insult to a lady will hurt me quicker than almost anything else, nothing will disgust me so soon as unladylike conduct. Especially do I dislike these 'Amazon Lucy Stone' women. Women who not content with showing how little respect they have for their own sex, try to disgrace ours by putting on masculinity."[115] What most disturbed John Marquis was female behavior that crossed over the sex-role boundary into male territory.

But John himself imagined another kind of crossover: "I used to often wish I was a woman and not obliged to *fight* my way." He continued: "If I *was* a woman I would be an old *maid* certain. It was always a mystery to me how a delicate tender woman could love a great rough wiskered man. Why I should as soon think of loving a rhinoceros."[116] John worked hard to be a man—to fight his way because he had no choice—and it infuriated him that women, with no need to meet masculine role obligations, wanted to "put on masculinity" and were sometimes successful. He found army life difficult, and the threat to his masculinity from those "Amazon Lucy Stone women" appears to be a reflection of his struggle to be a "man." His effort to reject the siren song of passivity and dependence, which he associated with the female role, may have made women's crossovers into masculine territory more intolerable to his psychic economy.

Whatever the case, John was not alone among nineteenth-century middle-class men in being unsettled by the first woman's movement. James Hague had a cooler and more sophisticated approach but he was also threatened. Traveling west in the summer of 1871, he met Elizabeth Cady Stanton and Susan B. Anthony on the train across country. The two women were campaigning for woman's suffrage in the West, and Hague introduced himself to Stanton with whom he chatted pleasantly until Anthony joined them in Laramie. "Arrived at the station," Hague jabbed, "we found Susan on the platform with an attendance of strong minded women and (weakminded?) men awaiting the arrival of the distinguished Stanton and the subsequent departure of both apostles of your injured sex."[117]

Hague teasingly expressed an amused and distantly satirical response to the scene, but the opposition of strong-minded women to weak-minded men revealed a certain uneasiness. "I was amused to see the two ladies kiss each other on meeting," he added; "Fancy two politicians, off on a campaign, kissing each other on meeting after a brief separation. Do you think the influence of woman in Politics and the Legislative halls will [be] making kissing fashionable? I can believe it of the opposite sexes—and can readily understand how some perfidious male representative in Congress should be very glad to have a pretty girl for a colleague—though the country is safe enough from any objections of that nature while the likes of Susan are selected as representatives." Though the humor was good-natured, through it Hague trivialized the nineteenth-century feminist goal of political equality by concentrating on women as sex objects.

While warmly disposed to Stanton, Hague found Anthony less appealing. Nonetheless, he granted that even Anthony was "good in her way; but is more severe in expression and inclined toward violence without being really any more vigorous or effective than her co-laborer." Spending an entire day on the train in conversation with them, he reported on a spirited discussion of a soon-to-be published book by Catharine Beecher opposing Anthony's and Stanton's views on women. "[I] could only say that the few pages I had seen sustained the view of woman's *subordinate* position, which enraged those two females intensely. Susan said 'Twaddle' and 'Trash' and various other uncomplimentary ejaculations while Mrs. S. said that nothing provoked her so much as such talk from a woman."[118] Hague responded by defending Beecher; "not only because she is a near relative of the girl whom I adore . . . but because her views entitle her to great respect; so I said that I thought Miss Beecher had been very useful to her generation and if not agreeing with her vituperators on all points she had at least advanced their objects by promoting the greater education of women; but Susan would hear nothing in her favour . . . ."

Catharine Beecher argued and supported, throughout her distinguished career as author, lecturer, and educator, a division of power and influence based upon gender. She urged each sex to exercise power in its proper sphere of influence: women in the home and men in the world. James Hague accurately assessed the greater educational opportunity for women justified by this doctrine of separate spheres, which gave great weight to women's influence over children and over the moral character of the entire community. But he failed to recognize, as a contemporary observer, what others have later seen, that while the doctrine had its virtues and opened "certain avenues to women because of their sex, it barricaded all others."[119]

Hague, like some other men, was envious of the privileges of feminine weakness, but also threatened by the potential blurring of separate spheres inherent in the first woman's movement. He teased his fiancée, "Well darling, be a good girl and get strong and well and so be prepared to share some of these disagreeable errands which your Reformers claim as your rights. When you come into full possession of your rights perhaps you will go down the mines and study them in the mud and water and candle grease and let me write the report in comfort."[120] While Hague had pride in his own "masculine" struggle with existence, he also yearned for a feminine existence "undefiled by any groveling search for useful or practical qualities."[121] But his yearning existed alongside contempt for woman's weakness, in an ambiguous tension between pride in the masculine mastery of the world and envy of the "privileges" of the feminine sphere.

Under other circumstances, Hague wished he could have supported and protected his fiancée on her journey to the Peekskills for medical treatment. "Poor child! If I had been there you should have had my shoulder, even at the risk of exhibiting some sign of 'sheer disgraceful affection.'" But his subsequent quip, "if you think it consistent with your views on Womans Rights and the Equality of the sexes" carried an implied threat that his protection and support depended upon the inequality of the sexes.[122] Again, Hague appears to have been ambiguous about the privileges of the female role but also proud of his contrasting and thus masculine-defining strength and power.

William Harbert was another nineteenth-century male concerned with the contrast between men and women in light of the first woman's movement. He commented extensively on the woman's movement and his fiancée's role in it. Elizabeth Boynton, an active participant in the feminist cause, lectured for woman's suffrage in the Midwest. At one time, she apparently accused him of being unsympathetic to the movement. He responded by defending his position: "The avowed object of the movement the amelioration of the condition and the amendment of, and addition to, the object and aims of women, is certainly good—and should be favored by all who regard man or fear God—The movement comes none too soon and by proper means cannot be too zealously supported—This much is granted—Now what next—This! I cannot consider with favor that class of professed reformers who suggest (evils as they call them) *inconveniences,* that attach to womanhood which are of the same nature as other inconveniences with which the bearded portion of the human family is compelled to contend—inconveniences grounded in the nature of things—for which there is no remedy—."[123] What he meant by male and female "inconveniences" is unclear. Whatever his specific meaning, he expressed some covert sex-role jealousy when he objected to any changes which might result in a more favorable situation for women. It bothered him that women's role might become less burdensome than men's role.

Will Harbert also defended men against those women who continually berated them "as if they were more naturally cruel, selfish and unfeeling—This is not only impolitic but unjust . . . ." Hoping that men and women might work together for woman's suffrage, he declared that men could be counted on for gallantry and justice if the enthusiasts were quieted and more reasonable women were recruited to the cause. As was true of many female supporters, he believed the movement should aim to expand woman's domain, rather than change any basic sex-role ideals. "I believe," he insisted, "that I respect womenkind as much as anyone, I would have womens sphere so enlarged (and it will be) and herself so educated and prepared by being the welcome guest at all seats of varied and polite learning, that crowned with a wealth of learning and accomplishments she shall be the heart of home of church and society."[124] Arguing not for the equality of the sexes but rather for the ideology of separate spheres, he believed women's role inherently justified greater opportunities for their sex. However, Harbert's support for an enlarged economic and social role for women rested upon their special feminine qualities.

Doubtless, too, the majority of women also upheld the division between women's and men's spheres in order to enhance their status. By stressing the differences between men and women they avoided the issue of female inferiority.[125] This was the "answer" that most nineteenth-century middle-class men and women gave to the sex-role question. A minority of both sexes argued for equalitarian feminism based upon the common humanity and equal claim of both men and women to the entire range of economic, cultural, and social experience. A majority of women and men strongly supported a definition of gender that limited women's access to certain social worlds and did not blur sex-role differences.

Will Harbert argued that colleges and universities should open to women and that if women were no longer excluded from higher education, "I think all those avenues of labor for which she by nature is fitted would open . . . to her accomplishments education and feminine tact—There are some classes of employment for which she is especially fitted and some for which man is especially fitted and they are about equal—away with the idle questioning as to which is superior, man or woman—each is dependent upon the other, each necessary to the full development of one other, each the complement of the other."[126] Harbert was in favor of women's access to higher education as well as greater employment opportunities for women, but he was at the same time against women giving public lectures because they were better suited "by nature" for quiet literary employments.[127] Both arguments were grounded in sex-specific characteristics elaborated in the two-spheres ideology. As he illustrated, this ideology was a two-edged sword—it could be used to justify both opening and restricting women's access to a broader range of social experiences and economic opportunities.

While behavior may actually have changed very little for the majority from the Civil War to the end of the century, during this period the volume, intensity, and uneasiness of sex-role comments between the sexes increased dramatically. The nineteenth-century woman's movement contributed to greater self-consciousness concerning women's role and insecurity over sex-role boundaries in middle-class men and women. There was a noticeable absence of dialogue over men's role in private correspondence. This was most likely an indication of a taken-for-granted attitude toward masculinity on the part of both sexes. Men's role, however, did not exist in a vacuum. The public discussion surrounding the possibility of female crossover into the male sphere elicited comments from men on women's role that also betrayed some anxiety over nineteenth-century masculinity.

One picture that has been drawn of Victorian America emphasizes the gulf that separated men and women. Victorian marriage, in this view, originated from pragmatic motives disguised by a romantic rhetoric which merely gilded the inevitable conflict, misunderstanding, or simple indifference of two sexes deeply divided by socialization, personality, and the division of labor. The widely accepted conclusion is that the deepest interpersonal relationships in nineteenth-century America were same-sex, thus underlining the "separate" in separate spheres.

Though not completely untrue, this picture of Victorian domestic relations misses the actual cultural and characterological ideals that provided a firm basis for some measure of male-female compatibility and behavioral intimacy. In fact there was a meeting ground in actual behavior, undergirded by a less divergent conception of masculinity and femininity than has been recognized. By examining the interactions of men and women, it is possible to show how divergent roles met and how role prescriptions were translated into relationships.

Even at the level of cultural expectations, men's and women's natures were not defined as antithetical. Though men continued to be identified with superior intellect and women with tenderer emotions, the heart took on a new significance for both sexes. The cultural differentiation between the sexes was diluted by religious ideals and the secular romanticism of mate selection and marriage. Victorian men and women found a common ground in the culture of separate spheres through their very conceptions of masculinity and femininity.

Expectations of mutual sharing and understanding between the sexes were neither culturally illogical nor exceptional. Masculinity was not defined solely in terms of rationality, and women's intellectual powers were not always seen as inherently at odds with their femininity. Nonetheless both sexes confronted limits in the cultural standards used to measure their behavior. Victorian men were expected to hide their emotions in public but to loosen their expressive controls in private, communicating their hidden selves to the women they loved. Women were allowed a broader range of public emotional expression, but were expected to channel their ambitions through men, circumscribing their public activities and thus their economic opportunities.

Conformity to gender ideals clearly produced strain in the experiences of both sexes. This strain originated in tensions between the economic world of career, achievement, competition linked to the realm of rationality and emotional control and the other world of companionship, children, domesticity linked to the tender-hearted feelings of love and sympathy. Men lived with one foot in each camp, actively moving between the two. They were expected to perform differently in each area. Women's vulnerability was created by actual limits on their opportunities in the public sphere. Their reliance upon the bonds of love for economic support and protection from men created painful role strain for them in mate selection, the troubled marriage, and in arenas of great physical threat. The same man or woman, however, might sometimes be content with his or her role in one context and dissatisfied in another: sex-role terms had some elasticity in response to the needs of time, place, and condition.

The woman's movement brought certain gender issues to cultural consciousness in the postbellum period. Women's incursions into public spheres raised the spectre of gender renegotiation in private relationships. Men became increasingly uneasy over women's role commitments and were troubled by potential female crossovers into their domain. Even for those not actively involved, the woman's movement precipitated a new dialogue in intimate relationships that led to greater sex-role self-consciousness in both men and women.

Both sexes found room for maneuverability in this gender system through a method of softening gender-boundary violations. Men and women violated sex-role expectations and then reframed them to restore equilibrium. The analysis of the process of decision-making is necessary because couples reinterpreted their own history as they interacted. Summary reports of a completed decision may be misleading. Women sometimes asserted themselves and men sometimes abnegated their power in transactions that allowed the violations of sex-role boundaries without a radical revision of sex-role ideology.

The relativity of the head-heart gender distinction, the growing consciousness of female role boundaries spurred by the woman's rights movement and masculine uneasiness over potential female role crossovers, the tension within both the male and female roles produced by conformity itself, the resulting fluidity of role behavior, including violations of masculine and feminine ideals which were sometimes protected by the system itself, all challenge a static view of separate spheres. Thus, while nineteenth-century American culture has been characterized by the extreme separation of masculine and feminine spheres, it should also be described as a century in which sex-role distinctions were beginning to blur. A fully drawn history of Victorian sex roles demands paradoxical recognition of both the rigidity and fluidity, the division and overlap, of gender lines.

6

Testing Romantic Love

Victorian Courtship Rituals and the Dramas of Private Life

U NDER THE ENTRY ON courtship in a nineteenth-century *Dictionary of Love,* a pseudonymous Theocritus, Jr., observed the predictable contrast between courtship, "the most tremulously delightful season of the tender passion," and marriage, which he described as "sober, philosophical, meditative, and sets itself, with good intent to the stern practical duties of life." He continued, "But, in courtship, it is all romance, excitement, hope, desire, expectation, sweet dreams etc. . . . ."[1] While not without some basis in fact, Theocritus, Jr., missed a serious dimension of nineteenth-century courtship. Caught between the lower expectations for marital happiness in the seventeenth and eighteenth centuries and the increasing acceptability of divorce in the twentieth, nineteenth-century middle-class suitors were forced to cope with the soaring emotional expectations of marriage and the still socially tainted prospect of divorce. Though full of what he described as excitement, desire, expectation, and sweet dreams, nineteenth-century courtship was also wracked by anxiety, pain, doubt, and suffering.

Middle-class courtship usually featured at least one dramatic emotional crisis, precipitated by the nineteenth-century woman as a test of her potential husband's professions of love. Women threw large and small obstacles in the path of the courting male to measure the depth and intensity of his romantic love. Men tested, too, but often less intensely and less resolutely than women. Nonetheless both men and women relied upon nineteenth-century courtship testing to gauge their partner's emotional commitment. This testing involved the creation of major or minor crises in a courting relationship. Often the participants used the life materials at hand to set obstacles in the pathway of love, which one or the other partner had to overcome, usually through either actions or words of re-assurance.[2]

This ritualized courtship testing filled several needs created by the privatized and highly individualized Victorian mate-selection process. First the rituals of courtship testing helped to develop a couple's romantic at-tachment. This was beneficial to the entire social order for in strengthen-ing the participants' romantic bonds, courtship testing contributed to the stability of companionate marriage.

Furthermore, since women often set more vigorous, dramatic, and difficult courtship tests than men, Victorian courtship functioned espe-cially to intensify men's romantic attachment to women. This was advan-tageous to women who, often without other means of support, relied upon the sentimental ties they forged with men. Courtship testing increased the value of women in men's eyes and perhaps strengthened a woman's own sense of self-worth.

Finally, Victorian courtship testing was a private mode of checks and balances in a system that had all but lost any outside supervision. Parents no longer had control, even in the form of veto power, over their chil-dren's choice of a marriage partner. A courtship structure that had inter-nal mechanisms to ensure a strong emotional identification between men and women was vital to the privatized, autonomous, and sentimental choice of a lifetime partner.[3]

The ideal of companionate marriage challenged the very ground upon which parents exercised control over the choice of a husband or wife. They could objectively measure duty and judge domestic, business, and other performance skills, but they could not evaluate the privatized expe-rience of emotional openness and personal satisfaction with another. Par-ents bowed out, not just because the family became less of an economically productive unit in an industrializing economy, but because acceptance of the ideas and values of love and the self gave them no basis to act upon—except as advisers and manipulators of the pool of eligibles.

A drastic change in world view occurred that made individual percep-tion a law unto itself in mate selection. Love became defined as the shar-ing of an essential self which was autonomous, private, and beyond the social conventions of everyday roles. Consequently, parental estimations of the prospective mate's ability to perform the duties and responsibilities of a spouse were largely irrelevant to couples who saw marriage as, initially, the union of romantic selves.

As parents were removed from the fundamental cultural approach to choosing a marriage partner, middle-class courtship developed an internal system of checks and balances. External checks, such as chaperonage, were out of place in this increasingly dominant ethos of romantic love. Emotional intimacy, now anchored in the private ritualistic testing of emotional commitment, was the basis of the Victorian mate-selection system. This emotional testing sharpened individual self-definition and intensified the intimate sharing known as romantic love. It also helped the participants to measure the depth and strength of their emotional bonds before marriage. And, closing the circle, setting and passing courtship tests deepened the commitment of both partners entering marriage.

Subtle power undoubtedly still remained in parental hands, but by the 1830s at least, men and women were engaging in courtship, agreeing upon marriage, and only then seeking parental blessings. Middle- to upper-middle-class children were making their own choice of mates, whatever formalities were still observed in the matter of familial consent. In typical fashion, one set of New England parents gave their prospective son-in-law, Samuel Smith, permission to marry their daughter Mary after the couple was already engaged. Mary wrote to Samuel before receiving her parents' approval: "the solemn deed is done. Promises have been mutually exchanged and I trust are already sealed in Heaven."[4] Clearly Samuel's request for permission to marry their daughter was a formality, the husk of an older parentally controlled courtship system. Mary's parents admitted as much in their letter of consent: "And as your mutual happiness very much depends upon our sanctioning an affection which it is certain makes you already one in heart, we will only add we give our full and hearty consent . . . ."[5]

Another mother, responding to a formal request for permission to marry her daughter, wrote, "If in many things I am governed by selfish feelings—in this case I would forget myself to secure her happiness, it would be barberous to make her feel 'How sharper than a serpant sting is disappointed love.' "[6] Though this widowed mother had counted on her daughter Julia to be "the stay of my declining age," she would not put familial needs above romantic love.[7] By the early nineteenth century the claim of her daughter's affection for Lincoln Clark was so culturally paramount that she described any action she might take to thwart her daughter's choice as "barberous," literally uncivilized.

Though observed as a formality throughout the nineteenth century, the ritual of obtaining consent from the woman's parents was even more attenuated after 1860. In 1863, James Bell expressed uncertainty over whether he should write to Gusta Hallock's father on the subject of marriage.[8]

Charles Strong and Harriet Russell were, Charles wrote, "cemented . . . in the ties of Betrothal" just before 1862 took "its leave forever."[9] Early in the new year Harriet asked Charles to see her father: "So I must 'ask Papa' must I?"[10] He was surprised but willing: "All right it shall be done first opportunity. I must confess that I never had that job to do before and it never occurred to me but that it was all fixed."[11]

Many courting couples throughout the nineteenth century continued to seek consent from the young woman's parents. Scant attention, however, was paid to the young man's family.[12] Nineteenth-century men enthusiastically supported the value of familial independence. Eldred Simkins boldly declared to Eliza Trescot, his fiancée: "But I am waiting till all the family know it and then I shall await the result but dearest, I must confess that to me personally tis a matter of little importance for who has a right to decide on our destiny but God and ourselves? And you are willing to entrust your happiness to me and what difference will it make to others. So you must pardon me if I am not suitably impressed with the importance of their opinion."[13] Eliza was not as sanguine over the spectre of parental displeasure and insisted that her mother must approve the match with Eldred. But Eliza's mother at first refused to give her consent. She was unhappy over her daughter's secret engagement and miffed by the tardy request for her blessing. However formalistic the ritual of parental consent had become, Eliza too obviously denigrated her mother's sense of importance in the choice of her mate.[14]

It is noteworthy, however, that while Eliza insisted upon her mother's consent, she maintained her own right of relatively unfettered choice in mate selection. "I told her that I had agitated the question a long time in my own mind," she reported, "but that a positive engagement was very recent, that as soon as I made up my mind on *that* point I also decided that she should know all about it, and, to tell you the truth one reason I kept you so long uncertain, was that I was reluctant to enter into any engagement without her knowledge and then it would have been foolish in me to consult others when my own mind was in a state of uncertainty."[15] Both Eliza and Eldred anticipated her mother's approval, which was given after a brief delay.[16] Though Eliza's mother was irritated with the secrecy of her daughter's engagement, she did not question Eliza's exercise of autonomous choice.

While some men may have thought it necessary to go through the formality of receiving permission to marry from their own parents, Nathaniel Wheeler was more typical of the nineteenth-century swain in that he acted independently of his family in affairs of the heart.[17] He became engaged without his parents' knowledge and confided in a close friend before he informed his parents. It was years before Nathaniel Hawthorne told his mother and sisters that he was engaged, and then only because Sophia, his fiancée, patience exhausted, wrote his family herself.[18] Male independence of even minimal formalities of parental control over their marital choice was common; men seeking the approval of their own parents was rare.

From 1830, and perhaps earlier, men acted on their own behalf. This was certainly true in a case where a prospective mother-in-law met her son's fiancée for the first time. This mother was completely surprised by her son's engagement. The prospective daughter-in-law, Eliza Shepard, described the first encounter with her mother-in-law to her sister: "Then she held me a little way off and *looked* right *through* me, as it seemed and then she said '*she's well-grown*'—I love her already. Then she said 'But you should have told me before Raphael, I cried all night when I got your letter.' Then I talked to her and told her that he wasn't to blame, because it was very sudden all round—Then she asked me if I were not afraid to take him so soon?"[19] Raphael Pumpelly's mother was obviously chagrined by the short courtship and impetuous engagement of her son, but she had little recourse. Feeling helpless and also somewhat threatened, Eliza's future mother-in-law asked her: " 'Do you know what a treasure you've got—O you *ought* to be happy, What a happy woman you must be!— Did you know he is the finest man in the U.S.? O I don't believe you know how good he is—' I was fairly overpowered—," Eliza confided to her sister, "but I believe I convinced her after awhile of my appreciation of her darling. I was sitting beside her in the car and bye and bye she said 'I want to see you, move—sit—where I can look at you!' Did you ever hear anything like it Nelly?"[20] Eliza's rhetorical question demonstrated both her amazement and exasperation at this display of motherly jealousy. Raphael was also displeased by his mother's emotional outburst. "It made him feel terrible to have her say so much and she told me afterward she had never spoken so before, but she was very much excited—."[21]

The nineteenth-century middle- to upper-middle-class mother was expected to accept her son's choice of a wife more impassively. But marriages that crossed class, race, religious, and other sensitive social boundaries may have been exceptions to that rule. If Eliza's marriage improved her class position (as she hinted), this might account for some of her prospective mother-in-law's perplexity. Eliza was initially intimidated by the wealth and distinction of Raphael's social circle. She expressed an almost awed reverence for the Pumpellys' material possessions, including original etchings by Albrecht Dürer and Rembrandt.[22] She also wrote to one of her sisters that her mother-in-law "says I *must dress* elegantly for my position and she has quite terrified me—Raphaels friends are mostly people

of wealth and distinction and she says she wants to feel proud of me.
. . ."[23] Whatever their private objections, parents such as the Pumpellys
were incapable of controlling the romantic love that had become the main
criterion of Victorian mate selection. Consequently, they often accepted
their children's choices of marital partners with more or less grace.[24]

Courting couples, on their side, insisted on the priority of their feelings over all social barriers or familial restraints. In response to the active
opposition of his fiancée's family, John Marquis replied defiantly: "I trust
we may be able to bring about an agreeable state of feeling but if we fail
Neeta then let us be all in all to each other."[25] The self of the loved one
loomed very large in the emotional current of romantic love, as it swept
family, relatives, and friends aside before its powerful ego-centered force.
Many couples reflected the belief that their romantic relationships took
emotional precedence over all other social ties. Thus, John asked: "Why
should your brother make such objections? He professes to have your
happiness at heart. Does he expect to secure it by making you miserable
first? I too, desire to make you happy and wish it far more than he *can*
do. I also feel that I have now a claim upon you which will precede *all
others* and one which I never *can* and never *will* yield unless I feel that it
is *your own wish* and formed after seeing me and knowing me better and
not formed from the opinions of others."[26]

In social structures that allowed it to be unleashed, romantic love might
be an antidote to loneliness, but it might also act as a socially disruptive
and alienating force. Not surprisingly, lovers usually ignored the anti-
social potential of their romantic feelings. Like John Marquis, they sometimes pictured a hostile world of social relations, which was only made
tolerable and livable by a saving romantic love. Such was the case when
he wrote, "Do not weep all the lustre from those black eyes of mine for
I both love them and am proud of them. All will come right by and by.
If your friends truly love *you* they cannot cherish and nurse their wrath
against me especially if they see you happy If they do Neeta we must *kill
them* yes, I mean it *Kill them* but do it with *kindness* you know."[27] John's
metaphorical aggression against those who would stand in the way of
their relationship bespoke volumes about the isolating and socially disruptive potential of romantic love.[28] On March 5, 1866, John Marquis and
Neeta Haile were married. Whether her family grudgingly gave in or
not, romantic love moved against the grain of familial bonds, and parental
objections proved not to be insurmountable obstacles.

Lyman Hodge boldly expressed the emotional power of romantic love
vis-à-vis familial ties when he declared: "I love my father, mother and
sisters. . .[but I] love you so much more."[29] Dorothea Lummis admitted
the appealing consistency and balm of mother love, but she nonetheless

believed that "husbands exercise over us women a more potent sway and charm, than mothers or any other woman can . . . ."[30] Mary Smith echoed such sentiments when she said: "As well as I love my parents—as well as I love my connections to friends—Yet all *all* could I resign most willingly—most happily for *your* sake. My affection for them dwindles into comparative insignificance when I think of what I bear towards you."[31] Eight months after married life commenced, Mary returned to her parents' charge to await the arrival of her first child. This circumstance did not weaken her sense of her husband's emotional priority: "I think I may safely say that not 5 minutes have passed without my thinking of him who above *all others* so fully possesses my affection—and who above *all others,* so fully deserves them. I am indeed at home, my love, but oh, I feel *alone* . . . ."[32]

Though romantic partners might take emotional precedence, parents nonetheless had some influence with their children, even at a distance. One daughter asked her father's opinion about a romantic prospect. Though he was in the far-off Klondike, Charles Watts's daughter solicited her father's marital advice. He responded: "Husbands are like to painted fruit; of this much take my counsel and advice; *marry not in haste* for should you get the best of husbands you put on a golden fetter."[33]

Bereft of direct power, parents still expressed concern about their children's choice of a mate. One astute mother, Mary Hague, wrote her husband from Europe that it was imperative she and her children return home. She expressed a self-conscious desire to control the associations of her children and a clear-sighted recognition that this was the only available avenue for parental action in the mate-selection process. "Marian is at an age now, when she likes to know young people," Mary reported. "I do not think much of the society one meets over here, for people of her age. The Americans who live here are either students, people whose domestic life is unsatisfactory, invalids, or people with limited incomes, whose sole aim is to be as comfortable as possible on what they have."[34] Dissatisfied with the group of young people her daughter had met in Europe, Mary Hague preferred the provincial possibilities: "Marian at home, would meet young men who have passed through their student life and are starting out for themselves. . . . There is a class at home, the upper Middle— children of professional men—who are better than any similar class here, and I should like Marian to see something of that class at home, in a comfortable home with her family around her. One very good reason why . . . [the Howland's daughter] married an officer in the army, was because her mother and brother never took the trouble to have a suitable home for her, and bring into it, the class of young men, whom after all, they would have liked her to marry. I should not like to make that mistake with our Marian."[35] Mary Hague neatly summarized the indirect manipulation of associational networks which was all that remained of parental control in the mate-selection process. The choice of a mate was highly personalized by the acceptance of romantic love as the dominant criterion of marriage. Still, the attempt by astute parents to manipulate their children's circle of friends reduced the chances of an inappropriate match from the standpoint of economic and social status.[36]

🐝 Since the direct exercise of parental control was rare in nineteenth-century middle-class mate selection, little wonder that parental supervision of courting behavior, contrary to popular belief, was lax. Courting couples expected to be together without a chaperone.[37] In fact, they were commonly left alone and reported on evenings that sometimes continued until two, three, or even four in the morning. These lovers, however, were not the post-World War II adolescents who made the transition to marriage at the youngest ages in American history. The average marriage age of men in the nineteenth century was mid- to late twenties, and women went to the altar only a few years younger.[38]

Nineteenth-century couples were given many opportunities to speak, walk, and meet together alone. While the South may present some regional exception to this rule, courtship privacy most certainly was a dominant cultural expectation in the North, Midwest, and West. Mary Granger, a young Buffalo woman, observed, "I am thankful that the customs of the South do not prevail here, especially those in regard to courtship and marriage, if never allowed to meet alone, I fear we would not now be engaged, with such restrictions . . . upon their intercourses, what a poor chance do they have to become familiarly acquainted and how much they *lose, we know.*"[39] This northern woman appreciated the value of privacy in developing romantic love.

One swain enthusiastically described his late night hours alone with Harriet Russell: "I wonder if she is sick today after being kept up so late by one that ought not to have done it, but the few dear and sacred moments which thus far we have been permitted to enjoy together to me seems like violence to end them even at 2 o'clock in the morning."[40] Late hours in courtship were common. Nathaniel Wheeler teased his girlfriend: "Certainly, if that person knew you as well as I do, she wouldn't accuse you of having kept *late* hours. For they were the *earliest* hours I ever knew a young lady to keep habitually,—from one to four A.M. But 'we won't do so any more.' "[41] Nathaniel and Clara's "one to four A.M." was not habitually early, but late hours were also not unusual for courting couples.

Four years later Nathaniel wrote that he ached for the time when "you and I were cuddled up on the dear old lounge in the corner—with Mother fast asleep in her astonishing confidence in her wicked little daughter, and the city clock trying to arouse my conscience by striking *three* in that peculiarly solemn way that clocks have at such times, ah the good old times!"[42] In Victorian culture, it was not uncommon to find mother fast asleep while daughter Clara and boyfriend Nathaniel petted in the parlor. Parents seemed to accept the need, even the right of courting couples to be alone.

James Hague reported on the cooperation, almost conspiracy, of Mary Ward Foote's family to assure their courting daughter's privacy. "Isn't it funny as well as very nice to note the carefully brought about accidents by which the rest of the family are somehow wanted elsewhere so that the two spoonies may be left by themselves? With what caution any thoughtful member of the family, having occasion to come into the room thus occupied, gives due notice of his or her approach by slamming of doors, or calling 'Mary' before appearing, pretending to be uncertain whether you are there or not, but really intending to afford us an opportunity to assume positions of indifference or of the most unquestionable propriety."[43] A Confederate army officer, Eldred Simkins, referred to routine late-night meetings with his girlfriend: "You know I always kept you up till 3 o'clock at night . . . ."[44] He also dwelt upon a special night when "daylight parted us."[45] The Civil War may have loosened southern courtship customs or perhaps they were more similar to northern and western patterns than the earlier northern correspondent recognized.

Though serious courtship in the nineteenth century was always a pairing-off process in which privacy was an accepted norm, peer group activities with male-female pairing also existed. Gusta Hallock reported on a picnic where ladies invited gentlemen and nine couples attended.[46] Two years later Gusta attended a party with her two sisters and a platonic male friend from which they returned home at 3 a.m.[47] When James was in town, Gusta followed the usual late night, private, paired-off courtship behavior. Much to her chagrin, Gusta's sister guessed exactly how late James left her the night before his move to Illinois: 4 a.m.[48] Whatever other socializing occurred, between the beginning of a courting relationship and marriage, a vast middle ground of shorter or longer duration existed in which nineteenth-century couples interacted in private. They were left alone to create, if they could, the mutual self-disclosure, self-definition, and empathy which resulted in psychological intimacy. Alongside this was also the opportunity, in being alone together, to experience an active physical intimacy.

The loss of parental control over marriage and the recognition of limited familial power in the face of a cultural commitment to individual preference set the stage for the central experience of nineteenth-century courtship: a cycle of emotional testing which required ever more convincing demonstrations of romantic love. The warp and woof of Victorian courtship was a series of crises, created by the participants themselves, which led to the altar if successfully resolved. Nineteenth-century courtship was usually intensely exciting because its basic motif was inherently dramatic.[49]

Doubt about the choice of a mate was channeled in nineteenth-century courtship into elaborate emotional tests. These dramas of crisis and resolution took different forms but were more than random troubles in love's path. Women particularly took the materials of life at hand and shaped them into obstacles that men had to overcome to prove the emotional depth and sincerity of their love. Illness, debt, family, other men, religious differences, character flaws, and personal inadequacy all formed the plot material of these private cultural dramas. While the individuals involved seemed only partially aware of their own complicity, the tests of courtship formed such a consistent pattern that they cannot be explained as mere coincidence.

The highly emotional courtship testing ritual met the need for assurance created by the falling away of parental controls and also contributed to strengthening the emotional foundations of marriages based upon romantic love. If marriage was to be entered for life on the strength of romantic bonds, the need to intensify those emotional ties was palpable, especially for women who most often faced a limited range of economic opportunities outside of marriage. Generally, though not exclusively, women were more anxious about their choice of mates than men and exhibited greater intensity and duration of doubt before marriage.[50]

Emotional testing usually began early in nineteenth-century relationships, and often reached a crescendo after an apparent commitment (such as an engagement) and emotional understanding had been reached. In the spring of 1862, John Marquis met seventeen-year-old Neeta Haile in Columbus, Kentucky. A Union soldier, John left Kentucky but began a correspondence with Neeta that lasted the length of the war. Early in their relationship, Neeta engaged John in a minor testing cycle. Before they had exchanged two letters, John asked her: "Did you *really, honestly,* and *solemnly,* and *devoutly* believe that you occupied but a 'small place in my memory.' If you did, *I* really, honestly, solemnly and devoutly believe that you are mistaken."[51] Neeta's direct expression of doubt elicited a reassuring response from John. The same letter recorded a similar transaction: "I am afraid your hope that I would 'fall in love' with some pretty Memphian is a vain hope for I have neither the opportunity nor inclination to make acquaintances here."[52] Neeta tested John by suggesting that he might fancy other women. Of course, she hoped he favored her, and John again reassured her. Two years later John at last openly avowed his love, and the pair exchanged pledges of their mutual commitment. But Neeta's testing continued in typical fashion. After an apparent understanding and commitment had been reached between them, she administered the climactic test of love: she broke off the relationship.

John's identification with the Union cause was deep and passionate. Neeta had Union sympathies, but her brothers were Confederate soldiers. Outraged by Lincoln's assassination, John forgot himself and announced that if the Confederates had sanctioned or directed the murder, they should be "exterminated."[53] With the war at an end, John returned home to Illinois and waited for a response from Neeta, fearing that his "assassination" letter had offended her.[54] Though John wrote her several times following the end of the war, none of his letters reached her. Interpreting his four-month silence as indifference or worse, Neeta wrote to break off the relationship, tersely asking that he return all her letters. Shaken and struggling to win her back, John explained: "I wrote you a letter just after the assasination in which I used language I should not have done, but it was in the heat of the moment. *That* letter I never rec'd an answer for. I then wrote again, supposing I had given offence, apologizing for my rudeness and begging you to overlook it but again failed to receive any reply. You say in your letter to sister that you immediately answered my letter of April 19 and that my letters then suddenly stopped the cause for which you never knew."[55] Neither John nor Neeta received the letters each wrote to the other after the end of the war.

This appears to be more than a mere coincidence but John remained noncommittal: "How this happened I cannot conjecture, whether by accident or design. If by design who could it possibly be. I know of no one to even suspect." John's sister was less disingenuous and more suspicious of Neeta's immediate family. She suggested that John send his next letter to Neeta in an envelope addressed in her hand. "I will however, as Dappie suggests, enclose this under her hand to make assurance doubly sure." John then pleaded "to be taken back to your heart again." He urged: "Dear dear Neeta do not let this one sad occurrence obliterate all the many pleasant memories of our friendship and although your feelings may have been deeply wounded, do not do me injustice and pray be very, *very* careful you do not trifle with your happiness and mine."[56]

John labored to explain himself: "I know I have given you reason to doubt the sincerity of my intentions by not being more explicit in my letters long ago, but I am sure you will believe me when I tell you I deferred making a formal request for your hand, because I wished to wait until I was free from the army and could come to you and plead my case in person. Had I not thus waited this estrangement had probably never occurred. We would have had a better understanding and more confidence in each other."[57] After the war ended, with no word from him and no visit as he promised, it was not unreasonable for Neeta to conclude that he was indifferent at best. Once engaged in courting, what appeared to be deliberate silence was translated as indifference, anger, or antipathy in the language of the heart.

But externally justifiable action in romantic love may still be symbolically ambiguous. Neeta was simultaneously breaking off their relationship and testing the meaning of it. Though she wrote John a terse letter ending their relationship, she wrote his sister, who lived in the same house, a full explanation of her action.[58] Fighting to save his romance, John responded by passionately affirming his love. Pleading for reconciliation, John asked if "[I might] come to you again and come expecting to bring you away. . . ."[59] The letter addressed in his sister's hand reached Neeta, and she not only agreed to resume their correspondence but also accepted his tentative proposal.[60] After three suspense-filled weeks, Neeta reaffirmed her unambiguous commitment to him. He wrote jubilantly, "The relationship in which we now stand to each other gives you sacred rights only second to those rights which will be yours when a few short words of ceremony make us one before the world as I believe we are now one in heart."[61] Neeta's and John's actions fit the pattern of the private drama of courtship testing.

In some relationships there was evidence of severe courtship testing. In one extreme case, Violet Blair tested Albert Janin repeatedly. Albert, in a moment of insight, invited her if she doubted him to "put me to the test," which she did with great skill.[62] Violet used other male admirers to measure Albert's commitment to their relationship. For example, soon after she left for an extended European tour, Violet encouraged Albert to think she was being pursued by an arch-rival.[63] Albert reacted with frenzied jealousy in the following weeks, tracking down his rival in order to gauge the extent of Violet's involvement. However dimly, he recognized the import of Violet's actions when he pleaded: "Violet, *never doubt me* under any circumstances. There never was a more loyal heart than mine; it is not given to man to respect anything more than I respect you, my very ideal of all that is noble and highminded and pure."[64]

The seesaw of doubt and resolution in their relationship was so extreme that it took on a rather comical air. At one point, she warned him against marrying her, employing the most common romantic testing device of self-criticism: "But my dear Albert I am such a true friend to you that I advise you to let me off, I hate to see any one I am so fond of trying to make himself unhappy for life—You have no idea how miserable a life would be spent with such a discontented unamiable vain woman as I am—."[65] Albert responded by saying if she should drop him, he would commit suicide, were it not "for my religious scruples."[66]

This transaction must be understood more symbolically than literally. Violet did not think that a marriage to her would destroy Albert's happiness and Albert did not intend to kill himself if Violet refused to marry him. Violet confessed: "[I] made up my mind thoroughly to bother you for the rest of your life," while Albert was quick to point out that his "religious scruples" made suicide impossible. Nonetheless, both were serious. They were engaged in an extreme but not uncommon courtship ritual of doubt, testing, and reassurance.

Doubts about the choice of a mate had to be expressed to the potential partner in some external form in order to qualify as a courtship test. Secret feelings of doubt and fear might erode a relationship, but only when they were visible to the partner could they function as a test of love. In a nineteenth-century courtship test, doubt was dramatized in an event outside the self or as a problem within the self, but in either case, some life experience was used to create an awareness in the romantic partner of an obstacle in the path of love. Participants in this ritual courtship testing were not necessarily aware of the purposes of their acts or of the common pattern of their emotional drama. They sometimes, though not always, perceived "through a glass darkly."

Though not necessarily conscious of their own propulsions, nineteenth-century men and women staged courtship tests using whatever materials were at hand. These courtship tests usually involved identifying an obstacle that the partner in love was challenged to overcome. For example, Violet presented her "personality" as one obstacle to marriage with Albert. She told him he would be miserable married to her because she was a vain, unamiable, and discontented woman. The reassurance offered by Albert was contained in his dramatic response that he would rather be dead than without her.

Fearful of marriage, Violet continued to test Albert, seemingly never satisfied by his stream of reassurance. Albert asked her at one point why "torment yourself and me by conjuring up all sorts of imaginary dangers for the future? You know how deeply and exclusively I love you; you know how similar our tastes are, how admirably we suit one another in disposition, how agreeable our intercourse during the last four years of our great intimacy has been. What further guarantee, then, could you desire, what stronger assurance of our future happiness."[67] Apparently, it would have helped if Albert had walked six miles in the rain to see her.

Albert was incredulous but never flagged in his efforts to reassure her. "How can you, Vivie, seriously ask yourself whether you have not perilled your chances of happiness by selecting me instead of D.K.? You must feel in your heart that it is impossible that he, or anybody else, should be more devoted to you than I am. I can hardly believe that you were serious in asking whether I would walk six miles in the rain to see you. I would do that and a great deal more to do any friend the simplest service. You make a mistake, Vivie, in not fully believing what I say."[68] Albert, after several years of courtship testing, appeared to become self-conscious about his involvement in a patterned ritual. "If I had leisure for literary pursuits," he commented, "I should like to compose a work entitled: 'Violet's Reasons' or 'The Sophistries of Celibacy' accompanied by a *pendant* to be called 'Albert's Answers' or 'The Triumph of Logic and Love.' "[69]

"Violet's Reasons" for not getting married and "Albert's Answers" continued literally up to the altar. As their planned marriage approached in May 1874, Violet stepped up her campaign of doubt. She cited religious differences, ill health, different dispositions, and her bad character as reasons why she should not marry Albert.[70] In one last dramatic act of courtship testing, Violet wrote: "Uncle Montgomery has been talking to Mama and has worried her very much—He told her that you have absolutely *nothing* and that it is very imprudent for me to marry you—She knows that I am not fitted by either inclination or bringing up to struggle with poverty, and that I have only enough of my own to keep us alive and clothed—Of course while I am in Washington I will live at home, but it is when I am away that she fears discomfort for me . . . If you had a name that had ever been distinguished like Viry's it would not be so hard, but oh I can't help trembling—."[71] Violet noted slyly that "I will not mind marrying you as much as I would anyone else," but then delivered the ultimate taunt at the close of a two-and-a-half-year engagement: "I was wrong to accept you when I did—I should have told you to prove that you were worthy to be chosen by me before I said 'Yes' . . . as it is I know that all who know me well think I am very foolish and imprudent . . . I will never be satisfied until you have won a case in the Supreme Court—Tell me honestly will your father let you take charge of one—It provokes me so that people are able to say that I am going to marry a pleasant talker and that you have never proved yourself to be anything else. . . ."[72]

Albert responded with dignified resignation. He agreed to postpone the marriage, telling her that he had no desire to humiliate her. As a result of hard-won confidence, insight, or perhaps simply wounded pride, Albert met Violet's last and most insulting objection to their marriage by stepping aside. For the first time he did not respond with elaborate or frantic reassurance but accepted a postponement until he was elected to Congress or distinguished himself in some other way.[73] By thus agreeing to her objections, Albert forced her to take the part of reassuring him or actually delaying the marriage. He turned her test of him into a test of his own. The sources are silent except to tell us that Violet and Albert were married as planned on May 14, 1874.[74]

The intensity of Violet Blair's courtship testing was unusual; the function her testing performed was ordinary. As Albert, after three months of marriage, explained: "If I had succeeded with you at the outset and been spared all the trouble and suffering which I underwent in my pursuit of you, I fear that I would not have appreciated the great value of my conquest and might have proved a less grateful and self-sacrificing husband than I intend to show myself."[75] In a system of multi-leveled female dependency, increasing the value of a wife in her husband's eyes was desirable from women's perspective, and beneficial to the stability of the marriage system itself.

The testing of love in nineteenth-century courtship took many guises and was coded in many different symbols. Testing could be done in a minor key, or to provoke an upheaval. When Sophia Peabody asked Nathaniel Hawthorne if she wrote too often, in a very small way she was testing his love. He reassured her with considerable emotion, obviously in tune with the import of her question, "My dear Sophie, your letters are no small portion of my spiritual food, and help to keep my soul alive, when otherwise it might languish unto death, or else become hardened and earth-encrusted, as seems to be the case with almost all the souls with whom I am in daily intercourse."[76] Using the code of the letters, he told her: "I keep them to be the treasure of my still and secret hours , , , ."[77] Again using correspondence as her testing medium, Sophia repeated the question four months later: should she write him so many letters at the custom house, since his "brother measurers" teased him?[78] His long involved answer can be summarized in his own words: "My beloved, you must write whenever you will, in all confidence that I can never be otherwise than joyful to receive your letters."[79]

Letters continued to be the symbol of love's measure and intensity in Hawthorne's relationship with Sophia. In late July, he asked under the guise of humor: "How does my Dove contrive to live and thrive, and keep her heart in cheerful trim, through a whole fortnight, with only one letter from me? It cannot be indifference; so it must be heroism—and how heroic! It does seem to me that my spirit would droop and wither like a plant that lacked rain and dew, if it were not for the frequent

shower of your gentle and holy thoughts."[80] Sophia immediately understood his innuendo that she loved him less than he loved her because she
needed fewer letters.[81] She exhibited the typical sensitivity to the nuances
of doubt of those involved in nineteenth-century courtship. Apparently
reassured by her pained response, Hawthorne denied that the question
was serious: "And why was my dearest wounded by that silly sentence of
mine about 'indifference'? It was not well that she should do anything
but smile at it."[82] But humor in this case probably disguised at least a
modicum of anxiety. The trifling nature of the whole incident and others
like it makes it difficult for those outside the circle of romantic love to
appreciate the emotional significance of such encounters to the participants. Both minor and major testing in nineteenth-century courtship,
however, helped to reinforce a relationship and contributed to the emotional identification of the romantic couple. Whether testing was minor
or provoked a major crisis, it was a response to uncertainty over feelings
and doubt about the correctness of a major life decision.

Men, when they were not reassuring women in the courtship drama,
sometimes reversed roles and tested in return. But men were often (though
not always) more direct than women when they expressed their doubts.
For example, Hawthorne asked Sophia point blank, "You love me dearly—
don't you?" or again a few months later, "Do not you long to see me?"[83]
Hawthorne even asked and answered his own question as doubt welled
up and then subsided, "Doest thou love him? Yes, he knoweth it."[84] Men
could be aggressive about the reassurance they needed in a romantic relationship, as when James Hague remarked that he "should make" his
fiancée reassure him if they were together. "Oh, you dear, darling, old
Lovie, if I could only be where you are this minute and take you in my
arms, I should make you reassure me by telling me you don't care about
money but that you love me more than anything else, as I do you."[85]
Though men played the role of reassurer more often than women—passing tests of love rather than setting them—the need for reassurance was
mutual. After several years of courtship experience, Hawthorne still questioned Sophia, "Dost thou love me at all? It is a great while since thou
hast told me so."[86]

Women could also be direct in their testing, as when Mary Ward
Foote asked James Hague, who had recently embarked on a long business
trip to the West, whether he loved her. He responded, "Do I love you?
That's the question you ask in your letter of June 30th just received—
Why, old Blessing, you *know* there's no need to ask that, Sweetie, I love
you with all my heart and soul."[87] But flat declaratives, however absolute,
were usually not enough to settle the question in nineteenth-century wom-
en's minds. James declaimed his love for Mary over and over again. For

example: "Dear girl, you keep asking, 'do I love you?' when you know well enough whether I do or not. You know I love you better than anything else in the world—and that I long for nothing so much as to get home and prove it to you during the rest of our lives."[88] Still, Mary's sense of uncertainty remained intense, perhaps owing to a prolonged series of separations from James during their courtship.

Absence might enhance the state of doubt, but the degree of aggravation appeared to depend on the emotional uncertainty with which a couple parted. Not at all reassured by James's protestations of love, Mary employed various devices in a constant stream of ritual courtship testing.[89] One of the ways she put James's love to the test was to withhold letters from him. He reponded: "The idea of your not writing often because you thought I have other things to think of and wouldn't care about hearing from you! Sweetie, you know better. There's nothing I like so much here as your letters. They do me more good and make me happier than anything else. . . ."[90] The hidden agenda was articulated by James in this same letter when he wrote: "and be assured, Darling, of my sincere and earnest love. . . ."

Mary, who suffered intermittently from a nervous illness, also used it as a touchstone of her fiancé's love. When her physical condition worsened, James encouraged, yet at the same time pressured, her to get well. He urged: "Do, Polly, take something or do something to cure up that blessed old nerve of yours, so that when I get back we may have a beautiful time."[91] News of her recovery sent him into raptures: "and now when I get back I shall find you, I hope, not stretched on a sofa or bed but ready for a tramp or a scramble or, if the Lord prospers us,—a Wedding Journey. Won't it be gay! Darling, I could 'holler'—yell—with delight to think of your getting strong and hearty again."[92] She reacted expectantly, affirming her readiness for his proposed "Wedding Journey." But in response, he backed away from their future intimacy. "At present I do not think of coming home until I have seen something of the probable result of explorations at *our* mine—and that will take some weeks yet."[93] Following his unexpected withdrawal, her letters returned abruptly to the theme of her ill-health, as she sought treatment in the mountains.[94]

Following James's withdrawal, Mary's testing became deeper and more intense. Using the medium of her illness, she created a notably dramatic scene of courtship testing. She announced in a letter written September 15, 1871, that she would probably be an invalid for life. Under those conditions she was releasing him from any obligations he might feel to marry her. Suddenly, the letter ended in mid-sentence. James had no idea whether a sheet was missing inadvertently or she was overcome by illness and could not continue writing: "you couldn't have chosen a more dreadful place to leave off—for you hold out the prospect that you are to be a hopeless invalid for life, requiring somebody's constant care and that you will never ask such a sacrifice of me. And at that point I am left to wonder what happened next—and no prospect of another letter for several days."[95] James was given the ultimate courtship test: would he marry her, even if she was to be an invalid for life? The melodramatic quality of the question illustrates again the dramas that nineteenth-century women often staged in testing men during courtship.

James passed this test by symbolically agreeing to take care of her for life, if necessary. "I do not know myself well enough to answer for years to come. I am too sensible of the changes that time works to be certain that the feelings and motives and impulses of the present moment will endure unaltered through life. But this much I do know—I know my *present* mind and purpose—and that is if I were called upon tonight to 'sacrifice' (as you say) my future to your happiness and welfare I should not hestitate a moment. I have no more cherished purpose in my 'future' than to devote myself and my life to you—no more desired object than to enjoy and to be worthy of your earnest, hearty love and intimate affection."[96] James's sincerity on a literal plane is at least questionable, but beside the point. What he was communicating was not his willingness to be her nurse, but the level of his commitment, the intensity of his love, and her significance in his life plan. James's affirmation of love for her was in the medium she had chosen to test him in: sickness. He thus assured her that even prolonged illness was not a barrier to their marriage.

His closing lines, however, indicate that he may have had some measure of understanding of the courtship testing pattern of doubt and reasurance, expressed in her case in the language of disease. "So, darling, cheer up and don't 'go back on' your boy because you get a few pains and aches—but do him the credit to believe that he wants to stick to you while you've got a nerve left in your body to bother you."[97] Before his letter could possibly have arrived, however, she wrote him that she could be cured.[98] Whether from insecurity or fear or mere coincidence, she did not sustain the test. Her improvement in the weeks following the arrival of his "passing" letter, however, was dramatic. He wrote on October 14, "I cannot tell you how glad I am that you are in such good spirits for that is half the battle . . ." and on October 17, "Lovie, I am glad you love me more and more. . . ."[99] On November 16, she told him that she was well enough to marry him and take a journey across the continent.[100] After more delays on his side, they were married the following spring.

The structure of Victorian courtship was a highly dramatic series of small private exchanges, often punctuated by emotional uncertainty on one end and greater self-definition and relational commitment on the other. Women generally precipitated the most intense crisis in the courtship. In fact, they often waited until after men made an initial commitment, either an engagement or something less formal, to test them most severely. It is almost as if nineteenth-century women were "rechecking"—in the language of the heart—the soundness of their choice and the strength of the partner's love.

Though both Clara Baker and Robert Burdette had deceased spouses and were much older than the average courting couple, they nonetheless engaged in a predictable pattern of courtship interaction. Just as the relationship deepened, Clara raised a series of objections to their union, including her own strong, fixed habits, friends, and her ambitions, plus the memory of his first wife.[101] Robert answered her objections, particularly the last, and then with great firmness insisted that he would never stop loving her. He returned to the problem in a postscript: "I open this letter once more—this is the second irresolute time, and even now I do not know what I want to say. It is as though I held you in my arms Sweet My Own, saying good bye. At every movement of yours to disengage yourself, I hold you the more closely. And every time my arms relax themselves, of their own motion they tighten their grasp upon you. It is a lover's good night, only. Not good bye. Come what may dear, dear Clara, you are My Sweetheart, My own dear Love, My darling—my darling— my darling. And always and always, fond, and loyal and true, I am your Lover."[102] Four days later he confidently announced: "I am willing that my love for a saucy Little Girl in Pasadena may be tested."[103]

While Robert reassured her, Clara's doubts increased his own anxiety. Thus, he asked in an excessive, but more direct masculine style: "Clara dear, Clara, my darling, do you love me? Do you? DO you? Do you, Clara? Say, do you? Do you love me? Do you? Say, you do, don't you? Do you, Clara? CLARA, DO you? Do-you-Love-me? Yes dear, I know you do; but I want to hear you say so."[104] Soon after this exchange, Clara invited him to tell her about himself. In his response, he described himself as an orthodox Baptist, a Republican, a protectionist, and in personal habits a total abstainer. But it had not always been so. "Five or six years ago, a long smothered appetite for alcoholic stimulant bore out, mastered me, wrote sorrow and shame across my life, and made me trouble, trouble. . . . It is all over and I am my own master."[105] In confessing his alcoholism, Robert was following the standard conventions of romantic love that required self-revelations beyond anything expected in other social roles. Within the cultural assumptions of romantic love, personal unmasking was supposed to be at least partial proof of genuine love. Robert offered his dramatic revelation of personal weakness in this spirit. But in the intricate romantic spiral of doubt and reassurance, he anxiously awaited her reply. So anxiously, that as he expressed it, he had "a mere passing nervous breakdown."[106] After a week of rest, Robert returned to the lecture circuit and was overjoyed by reassurances from her that his dark secret had not alienated her affections. He believed that every barrier was now down between them.[107] Little did he know that he had yet to pass his most serious test of love.

Guilt-ridden and worried that his confession might spoil their relationship, Robert continued his own testing in a characteristically direct masculine style. "Oh my darling," he wrote, "if you could banish your Robin, maybe it would be better for you. . . . And yet—and its joy through my tears to think so,—nay, dear, to know it—you won't. You will be forgiving, and gentle, loving and tender. You will be? You are. Dearest and Best, dear faithful, loving loyal heart, my joy is part of my punishment."[108] Clara responded with even more intense expressions of love, as the ritual of doubt, anxiety, and reassurance was enacted between them.

At this point, Robert and Clara were still far from a placid resolution of their courtship drama. Clara heard rumors that Robert's past transgressions might have involved women as well as alcohol. In his defense, Robert declared that since his wife died fourteen years earlier he had "loved no woman." He told her he had passions which "blaze in my blood as fiercely as volcano fires" but they "break forth only where I love. I will look into your eyes with a clear, unchallenged soul, Clara, when I meet you."[109] Clara believed him, and their relationship grew progressively more intense.

Robert reflected upon the upsurge of romantic love between them: "I try sometimes, to measure your love for me. I sometimes try to set limitations to it. I say, 'Because of this, her love must falter. Because of that, her affection must change to toleration. . . .' And when I have reasoned it all out, and set metes and bounds for your love that it may not pass, lo, a letter from Clara, and in one sweet, ardent, pure, Edenic page, her love overrides my boundaries as the sea sweeps over rocks and sands alike; crushes my barriers into dust out of which they were builded, over whelms me with its beauty, bewilders me with its sweetness, charms me with its purity, and loses me in its great shoreless immensity."[110]

Robert and Clara spent over a month together at her home in Pasadena in late June and July 1898. After their rather idyllic interlude, Robert wondered: "Why do I plead so earnestly for your love when I know already that I possess it? When I know that you do love me, tenderly, dearly, faithfully?"[111] This question was relevant to many nineteenth-century courting relationships. When love was mutually established, and clearly articulated, why was it that this was often the point at which extreme doubts and greater anxiety erupted? Robert's answer was widely applicable: "as my love for you grows stronger, my fears awaken."[112] The spiral of doubt, anxiety, reassurance, and more anxiety fed on the very reassurance that nurtured deeper love.

The final "crisis of doubt" had yet to be enacted between Clara and Robert. True to form, she precipitated it by identifying a serious obstacle in the path of their union. After they had both sworn a deep and abiding love for each other, she queried: "Are the things which I need—freedom from responsibilities, and tender care and love—in your power to give, is the question to be answered." She concluded that he could not meet one of her most pressing needs—freedom from responsibilities. "Not many years hence I should have to be the entire burden bearer. It might mean the giving up of our nice home—an inability to live as I do now—in my present state of health, you can understand how great—perhaps unduly so—seems to be the need to be relieved of burdens. It is my second greatest need—if not the first. If the first is love, it ought to be a love, if possible, that does not bring a sum total of added burdens. Isn't that true?"[113] Robert was older than Clara by eleven years and had suffered serious financial set-backs through some disastrous investments in the past. Thus, Clara suggested the possibility that her love for him might cost her a financially secure and even happy future.

During the Victorian courtship ritual, courting pairs responded to obstacles set in the path of love in three general ways. One was to attack the obstacles directly, attempting to remove them immediately. This was the most typical response to tests of love, illustrated by most couples. Another response was to propose ways to surmount the obstacles to love in the immediate future, as James Hague did in response to Mary's illness. The third was to step aside, accepting their existence and admitting that they blocked the relationship, as Albert Janin did in response to Violet's objection that he was not distinguished or famous enough to marry her. The last course was particularly effective when an obstacle and a lover were seriously mismatched. In that case, the relationship could continue only if the testing partner withdrew the insurmountable barrier.

Robert responded to Clara's impossible objections—he could make himself neither younger nor much wealthier—by stepping aside for an imaginery lover who would satisfy her needs. "God give you perfect happiness, and strength, and health as you now are, dear! God send you a Love that can be a 'burden bearer' for you! And my love for you will not let my heart feel one throb of bitterness against you for that. My love shall never stand between you dear, and your peace and happiness. Never— Never—Never! Dear Violet!"[114] Robert continued for pages in this vein: "as then and now your happiness was and is my greatest and first desire—

so now, dear love, I give you back all the freedom I have taken from you. Be free, dear, dear Violet, to greet love if he should come with clearer mercantile brain and stronger arms—with deeper, truer, tenderer love he can never come. Be free, Sweet Santa Clara, to live as you now live, in the sweet and happy independence with which 'God has blessed you.' . . . I charge you, let not my love darken your life with any shadow, or weigh upon your thought like an 'added burden.' "

Given the cultural presumption that women were the more burdensome sex whom men carried on stronger financial and psychological backs, Clara's suggestion that Robert might add to her life's burdens was both insulting and wounding. And Robert responded accordingly. He scorned his love as a "helpless, useless love. It cannot even lighten your burdens. May God send you a love that can! And when this perfect rest and strength and happiness comes into your dear life, truly will I be content, even tho I sorrow over my own helplessness to give it to you—my own pitiful failure in trying."[115]

Recognizing that she had to withdraw her objections or lose Robert, Clara apologized for the cruelty of her last letter and reassured him that she loved him still.[116] But this was not conciliatory enough. He responded that her letter was not cruel, simply plain and clear.[117] Apparently now fearful herself, Clara telegraphed: "Violet does not want renunciation. . . ."[118] But Robert remained intransigent. With the pride of the almost vanquished lover, he needed to be coaxed back with a more emphatic apology and extended incantation of love than a telegram could contain. Her test of him had turned into his test of her. He refused to concede, insisting she was correct: "I was never meant to be a part of your active, material life, dear. I see it now, as you have yourself revealed it to me. I was only sent into your 'dream-life.' On the 'gray days' I will come to you. . . . It would have been a dearer, completer happiness if I could have come into your life fully and completely; if your clear-seeing analysis of *me* had not laid bare certain disappointing limitations, that, when I was weighed in the balances found me wanting."[119]

Two pleading letters from Clara resolved this crisis. Placated and reassured, Robert responded with an outpouring of sentiment: "Out of your pain, dear Love, pain in head and heart, you have written with sweet assurance to me. Because you knew that I was lonely and troubled; because the ache in your own loving heart told you of the pangs in my own; because in the sobbings of its own grief, your dear heart knew that mine too, was longing for comfort and Violet, you sent me the caress that I love, the kisses that I hunger for. . . . Dear, dear Heart of mine! Don't cry, Sweet. It's all over, Little Girl. Everything is understood. You are back in your old place, my precious Little One. . . . My arms enfold you—never before did I hold you to my heart with such strength, my darling."[120]

Doubt and anxiety resolved, these lovers confirmed a common pattern: the dramas of nineteenth-century courtship testing strengthened love. "How much more we love each other," Clara observed five weeks later, "and how much we have entered into each other['s] lives since those early June days. We can never again be the same as then—to each other—to our-selves—[to] the world even."[121] After several lovers' trysts on the East Coast, Clara agreed to marry Robert.[122] Robert rhapsodized: "My darling, I turn your dear life and your dear love over and over and over in my thought, and more and more deeply do I love you in every new and changing light that I turn upon you—your life—your love for me—your tenderness—your perfect trust—your beautiful 'child-affection' for me, and your woman's love, deep as the sea."[123]

Robert's response to courtship testing was typical, for generally, as lovers resolved their crises of doubt, they reported that their love grew in intensity and depth. Nathaniel Wheeler, in one of the more eloquent tes-timonies of this kind, assured Clara Bradley that he loved her more than ever "though I can not write love poems for her now, as I did then. 'I love her to the level of every day's most quiet needs.' And that is higher than sonnets and vows, isn't it?"[124] Gusta Hallock told James Bell, "I always knew I *loved* you as I never could another, yet I never began to realize *how much* till within a year . . . ."[125] Almost a year later, she was astonished by the magnitude of her love for James and "that you love me in return—that you love me *so much* . . . ."[126] James wondered, "does Gusta know, or believe *how much* I do love her. I have told you many times that I loved you, that you was *dear, very* dear to me, yet when I write, it seems that I had never told you of it."[127]

Before marriage, men and women in romantic love were constantly measuring their love and found that it increased dramatically during their courtship. Victorian courtship testing contributed to the increase of love for a number of reasons. It spurred self-disclosure and mutual introspec-tion. Furthermore courtship-testing rituals demanded action, compelling belief through participation in the minor exchanges of denial and affir-mation as well as the major melodramas. Doing was believing in love as well as in other areas of cultural life.

The most paradoxical fact of courtship testing was that romantic love fed upon and thus gained strength from the anxiety cultivated by the tests themselves. As a nineteenth-century American *Dictionary of Love* astutely observed, doubt was "a great sharpener and intensifier of the tender pas-sions."[128] When mutual interest existed, doubt and anxiety actually inten-sified romantic passion, perhaps because contemplating the end of a relationship through an effective courtship test could force one to imagine life without the lover, thus pointing to his or her value and meaning.

The power of courtship testing and especially the role of anxiety in strengthening romantic bonds should not be underestimated. Eliza Trescot, after agreeing to become engaged to Eldred Simkins, asked him to release her until after the Civil War. Eliza gave Eldred no reason, he said, "except those I know. But asking it of me as a matter of 'happiness' and 'confidence' I yield: and *now release you of any promise you may have made to me* and now Lize you are free—free as air—our engagement which hung so heavy on your soul is taken off, and you breathe freer. As for me, of course my dream is over. I knew it could not last long—but to tell you the truth, dear cousin, I have no wish to survive this war." [129] Again, the personal interaction was melodramatic. Eldred's wish not to survive the war was a metaphor of the extent of her importance to him. This metaphor, of course, posed its own kind of challenge to Eliza. She responded with reassurance: "Why do you say Eldred that you do not wish to survive the war? I know of no sad changes that could happen to make you care so little for life. . . . You don't know how much unhappiness that one expression of yours has given me. I have read it over and over and wondered why you wrote it, but I have wondered over all the whole tenor of your letter. There was a tone of reproach through it all as if you thought I cared nothing for you and it would give me pleasure that we are simply cousins now and nothing more—but do you think if I had followed the dictates of my heart and not my reason I would have written that letter?" [130] She emphasized the possibilities of the future. "Don't write as if you hadn't one hope for the future. I have many." [131]

Predictably, Eldred was much cheered by her response as he "read" hope back into his relationship with her. He explained the gloom of his last letter: "I know of no objection you can urge now that [you] may not urge after the war, if you choose—and though I asked none of your reasons, and followed your request to the letter, yet how could I know that reasons so weighty in nature that you could not even let me know them could ever alter their nature and turn in my favor." [132] The spiral of doubt and reassurance continued in this relationship as Eldred staged his own test, employing the intrinsically dramatic real-life situation of war: "My own cousin—I may never see you again but if I am killed remember I died *loving* you to the last—God bless you—and if we meet again I shall certainly claim you—so look out—so don't you wish we may never meet. Write to me at battery Wagner—every day. I can at last claim that now—for each letter may be my last." [133] Eldred's plaintive "don't you wish we may never meet" was an obvious invitation to her to contradict him, particularly as death was the most likely reason for any permanent separation. Having been moved to the front, Eldred used this opportunity to test Eliza as well as demonstrate his love for her in the medium of fighting and dying. He told her: "when I am fighting I will think I am protecting you and you may be certain I will do my best."[134]

The ritual courtship testing proceeded so intensely in the two months after Eliza broke their engagement that Eldred felt sufficiently emboldened to propose again: "You said in your last speaking of *my* home, 'I have no right to it *yet*.' Did you really mean it: for the sentence implies a future certainty—or did you write it without thinking?"[135] Eldred approached her about a re-engagement in the characteristically direct male style of nineteenth-century courtship. In their case, as in many others, the courtship-testing ritual of doubt and reassurance was quite effective. For in his next letter, he remarked that Eliza's positive response was not surprising.[136]

Eliza was prompted to break off her engagement to Eldred because of fears that her love was not strong or deep enough. Eliza's "moments of indifference" worried her and created doubts in her own mind about whether she loved Eldred.[137] These doubts precipitated her most severe test of Eldred's love: calling off the first engagement.

Trying to reassure Eliza, Eldred held himself up as an example of how to handle doubt. "[A]nd I further state that I never now resolve the question in my mind, whether I love or not. Were we continually inquiring into the feeling which prompts us to worship the Creator we would in time become infidels. . . . So I let the matter rest, always remembering, whenever I think of the matter—'I love'—without questioning or perplexing myself with doubt: can you not follow my example?"[138] Eldred was less troubled and more secure than Eliza, until she precipitated their crisis of doubt. Women questioned and perplexed themselves more vigorously than men with regard to love, requiring deeper levels of reassurance and rarely letting the matter rest at first resolution. Nineteenth-century women were also often more indirect in expressing their feelings and desires than the men they loved. As Eliza self-consciously explained to Eldred: "All my life I have been afraid of saying more than I felt, but dear Eldred, you must always take for granted that I mean more than I say when I write to you."[139]

Both men and women searched each other's words and actions for signs of true feelings and both needed reassurance. Eldred confessed: "I scan your letters to find out your feelings, when you were writing. Need I tell you how much I prize the slightest word of love and affection?"[140] He described the process of "reading between the lines" that characterized almost all nineteenth-century courtship exchange: "You tell me or rather imply you scrutinize my letters to find out this 'hidden spirit.' Do yours escape? But with this difference, when I would draw a conclusion that I wish, I am afraid of being too sanguine."[141] The participants themselves recognized the "hidden spirit" in their letters and the metaphorical way they were speaking to each other of love and commitment. Eldred continued: "Dear Lize, why don't you write exactly as you think? If you continue to write so, don't be surprised if I turn out a regular Oliver Twist. . . . when I ask for 'bread' I don't want a 'stone.' Often I ask for some certainty. How can you tantalize me with vague probabilities."[142] Eldred conveyed the perplexity of deciphering the language of the heart.

Eliza also complained of the difficulties of interpreting his emotional intentions. She wrote "of being mystified" by Eldred, while claiming her own "perspicuity and plainness in words, style and meaning."[143] Eliza engaged in a process of reading between the lines that was very similar to Eldred's, as she confessed: "you talk of reading over my letters to discover my feelings, it is my greatest happiness to read your last letter over and over again and think and wonder over what you were feeling and thinking when you wrote."[144]

Eldred passed his major courtship test—a broken engagement—and Eliza wrote in January of her hard-won assurance. Continually re-reading his old letters while waiting for the next to arrive, she observed: "They give me even more pleasure now than when they were received for then I was often tormented with doubt but now all bears the peaceful calm of certainty and Eldred every expression of affection gives me such exquisite pleasure, more than I could express."[145] Eliza knew that she had many more doubts than Eldred: "In reading over your letters I find them all alike. You never changed but upon that point mine are so different. I often wonder that you did not think me trifling and silly but although I wavered and took a long time to arrive at certainty it is now a certainty forever my future is fixed."[146]

Eliza's hard-won certainty did not change her mind about waiting until the end of the war to get married. Women exercised more *formal* control than men in only one area of nineteenth-century courtship outside the wedding ceremony itself.[147] They were expected to decide when a marriage would take place. Men accepted women's control in the matter of timing, even if it meant considerable delay.[148] Don Graham urged Margaret Collier to "please settle as soon as you can the day of our marriage. . . ."[149] John North responded to Ann Loomis's reasons for delaying their marriage, "they have great weight with me; besides I recognize the right for *you* to decide that question."[150] Eldred Simkins also agreed that the timing of marriage was a woman's prerogative. Though pressing for an immediate marriage, he conceded: "I think it a matter in which the Lady has the 'largest say.' "[151]

In waging a dogged campaign for a more immediate marriage, Eldred observed a paradox of romantic love: "but Lize were you mine I believe the few moments of pleasure we could enjoy with each other would far overbalance the pains of anxiety we might feel and another reason [to get married] anxiety tends to increase love . . . ."[152] The truth of this statement is not invalidated by its use as a rationalization of Eldred's matrimonial desires. In the limited area of ritual courtship testing, anxiety did indeed intensify romantic love. Thus, as nineteenth-century men and women employed the life materials at hand to test their prospective mates, they both created and allayed anxiety and doubt. The spirals of doubt and reassurance illustrated in so much courtship correspondence included anxiety as a silent concomitant. Women, whose anxiety was greater over the strength of the romantic bond, kindled anxiety in men through various courtship-testing devices. Though not self-conscious about the effects of their delays, women more often than men held off the marriage day. Their postponements were part of the testing ritual itself, cultivating the anxiety in men that drew the couple together in ever tighter emotional circles of love.

In raising men's level of anxiety, women contributed to an intensification of the couple's emotional bond in relationships based upon a preexisting commitment. This qualification may seem superfluous yet it is essential to a full understanding of nineteenth-century courtship ritual. Emotional testing worked to deepen love only if there was a prior commitment on the part of both partners to developing a romantic relationship. The testing rituals that were so effective in the context of mutual attraction turned into self-pity and worse without it.

One unsuccessful suitor, Charles Godwin, claimed only to intend a "brotherly" correspondence with Harriet Russell, yet he clearly hoped to nurture a romantic spark in her. "If I write you a foolish word or two," he began, "you won't be offended will you. Perhaps if I were not so foolish or sensitive I might not write these lines at all. I have proposed a correspondence between us. Now this corresponding sometimes becomes a serious matter."[153] Godwin confessed his attraction to Harriet, but he claimed to "have conquered every feeling toward you save that which a brother might give a sister. Will you write to me something on such an understanding?"[154] Hoping to circumvent her potential resistence or even refusal to correspond on romantic terms, he asked Harriet to write him as a sister. She agreed.

Godwin gave every indication that he had not completely lost romantic hope in Harriet's regard. Soon he began testing her, hinting that he would like to see her on his next trip to town, and wishing aloud that she would write more often and at greater length. Harriet did not respond promptly to his letters, and feeling sorry for himself, he pleaded petulantly: "please write as often as your beaus will permit you to share the time."[155]

Much to his dismay, Charles Godwin soon discovered that Harriet had just become engaged to another man. "Excuse my writing so often to you. I send this tonight to tell you, you must not write to me any more. I was not concious what a selfish undercurrent was running in my soul, and which prompted me to encourage this correspondence, until this weeks unheavings revealed it to me. No you must not write any more. I must forget that ever, any one fair and young, said kind words to me. I must banish every thing bright from my memory or the darkness before me will surely engulf me."[156]

Harriet responded kindly to such self-pity, reassuring him that he was still her friend. Godwin, whose style always bordered on Mark Twain's parody of Victorian lugubriousness, soared, "Like cadences of inexpressibly sweet music, your kind words came to me: causing every nerve to vibrate as though electrified by some far off strain of heavenly harmony; and thrilling my inmost soul with gratitude that though I have met with some trifling hopes, I still have the principle treasure of such a friendship as yours left."[157] He then confessed what had been clear from the beginning of their correspondence: "True I had permitted myself to entertain some bright hopes which (like the swan that tis said sings sweetest as she floats away to die) seemed more exquisitely beautiful as they went out forever in that hour of darkness."[158]

In his next letter, Godwin clutched desperately at their relationship. Having paid her a visit only to find she was not at home, he sarcastically suggested that she purposely avoided him. Almost groveling, he told her that he would return next Monday or Tuesday, so "that if your guardian angel forgets you can be on your guard and leave home . . . ."[159] Harriet was obviously trying to ease him gently out of her life, and thus proposed to end their correspondence: "You spoke of 'not troubling me with any more of your poor letters' . . . ."[160] But Godwin would not make the break an easy one. "My dear friend your letters have never failed to give me pleasure and comfort. I prize them more highly than you think. I would prize them more than any thing else if I had a right to."[161]

Harriet continued to correspond with him. But longing for intimacy and unable to control himself, he growled, "It is very kind of you to forget better company long enough to write to a poor wreck like myself. You must be a charitable creature."[162] His self-pity was now uncontainable and he closed this letter with an unctuous benediction: "Here is another stupid, long letter for you to read I am truly sorry for you."

Godwin's self-deprecation was an example of Victorian courtship testing gone awry. Under conditions of mutual interest, Harriet would have vehemently objected to his self-criticism. Romantic involvement, however, had to be mutual for a courtship test to evoke the requisite reassurance. Therefore, with no romantic commitment on Harriet's side, Godwin's tests of love escalated into ever more desperate expressions of self-pity and self-flagellation. In his next letter, still engaging in useless testing, he asked Harriet if it was "getting to be a task for you to write to me. If it is do not write."[163] Harriet complied and ten days later, he was reduced to pleading: "If you ever get lonely, you may (if you can do no better) write to your lonely friend and brother."[164]

Charles Godwin's last hurrah was written on April 30. He had called on Harriet and received a decidedly cool reception. This may have prompted him to write a letter whose express purpose was to make Harriet feel guilty: "Your coolness—I may as well call it coldness speaks louder than words how much you detest me! But like a dying man my thoughts like to linger around the only being which can save me; my life is a blank without you and when you are gone my days will glide along on the worn tracks of an accustommed routine, cheerlessly . . . . One thought will console me though; you will be happier away from me, the rejected one, the despised one."[165] He was not yet finished with Harriet, saving his most self-absorbed line for last: "May you know a happy journey and may the omnipotent Being convey on you all the blessings that He and You withold from Your friend."[166] To Godwin, God and Harriet—on some equivalent plane of meaning—were in a conspiracy to punish him. Unfortunately for him, Harriet was not emotionally identified with his suffering. Thus in testing her, he failed to generate the sympathy he hoped would lead to greater intimacy.

The Victorian courtship rituals of doubt and reassurance functioned effectively only when a couple could already lay claim to a mutual commitment to one another. The difficulty was that the preliminary commitment to romantic involvement usually had to be based upon the *hope,* not the reality, of emotional identification. The willingness to engage in courtship activities that nurtured romantic love had to precede the full experience of love itself. In some cases it would be appropriate to call this initial impulse to love another an act of faith. In others it was based upon long acquaintance. But in either case courtship initiation was a sensitive undertaking.

The achievement of romantic love in courtship was, of course, the chief motivation of Victorian marriage but does not adequately account for the initial selection of a prospective mate. Though lovers believed their choice of a mate was triggered by a mysterious and irresistible attraction and though love might have irrational emotional consequences, it usually required time to build intensity and was often tamer at its inception than at its fulfillment. In the very beginning, people made what were probably fairly willful choices among their courtship prospects.[167] The initial decision to cultivate a romantic bond with one individual over others was, however, shrouded in the ideology of love as a mysterious attraction.

Selecting a romantic partner was crucial and somewhat uncertain in a mating system that by 1830 primarily relied on personal preference and the bonds of sentiment. While courtship was the dominant cultural incubator of romantic love, the initiation of a courting relationship demanded the prior identification of emotional potential. This judgment included a complex evaluation of moral character, personality, physical appearance, emotional and sexual attraction, and intellectual capacity.[168]

One young man, Nathaniel Wheeler, attempted to dissect the causes of his interest in initiating a romantic relationship with Clara Bradley. Following encouragement from his friends, and her non-verbal flattery, Nathaniel listed the qualities that attracted him to Clara: "For I saw . . . that you were of a happy sunny temper, given to establish intimate relations with those you liked, full of sympathy, impulsive, warmhearted, inclined to 'snuggling' (if this is entering dangerous ground, remember I'm a *man*) not by any means an unpleasant armful.—'a creature not *too* bright or good for human nature's daily food!['']—and was attracted by all this."[169] Based upon his positive assessment of her individual characteristics, and confident of her interest in him, he was willing to make romantic overtures to Clara.

The period preceding a nineteenth-century courtship is best designated as the pre-commitment stage. In this stage, men and women exhibited a searching mode of behavior toward each other. False starts and stops, confusion, and even pain awaited the unlucky or unwary. Left to their own devices, young men and women ran the risk of losing face in this pre-commitment phase. But the risks, especially for men, had to be counterbalanced by the desire to move a relationship along the path toward greater intimacy and mutual commitment. Men were generally expected to take the overt lead in nineteenth-century courtship initiation, which may have given them more power but also increased their anxiety.

Women, however, had to encourage the man of their choice, but by nineteenth-century standards they could not appear too aggressive. Gusta Hallock wrote: "It makes me *smile* in reading your letter when I come to the place where you call me a *naughty* girl for not telling you that I loved you, and keeping you 'in suspense for a *long long* time.' . . . Twould'nt have looked well for me to have told you so before you asked me, would it?"[170] Men's leading role in courtship initiation was confirmed again by Eliza Trescot: "I suppose you meant me to take that scold about writing to myself too but I didn't for I certainly expected you to write first as it is not customary for a lady to begin a correspondence."[171] Gusta Hallock, however, reported on what was perhaps a bold counterexample of female courtship initiation: "The early part of last Summer—somehow Rebecca Elliott took a tremendous shine to him. The consequences was, she invited him to take her around to meetings and other places—which resulted in his taking a terrible shine to *her*. and this winter he has engaged board there—doing the chores night and morning—which is declared by all to be the handy'st sparking know[n] lately."[172]

Circumstances differed widely between couples, but men and women had to establish, on the part of their opposite number, a willingness or unwillingness to enter into a romantic relationship. Given the subtleties of friendship and familial relations, sorting out the various meanings of male-female interaction required considerable sensitivity to emotional nuance. Reading the signs of romantic interest was a complex cultural act that was not always easily mastered.

Since men were expected to initiate courtship behavior, they were usually more exposed to overt embarrassment than women. Both were vulnerable, however, to the ambiguities of identifying romantic preferences in another. J. H. Meteer wrote to Elizabeth Boynton, suggesting that they "cultivate the acquaintance which was just beginning to be formed before my departure." He respected her "as a lady of mind and culture and more, *a warm Christian heart* . . . ."[173] Elizabeth conveyed her lack of interest bluntly, but apparently without malice, for Meteer responded, "You have no notion of marrying me and I have not the remotest prospect of marrying you at any time or any one else very soon. So, having enough in common to make us friends I am sure we *can* be without acting the dunce."[174]

Another young man still yearned for a cousin who had no romantic interest in him. Perhaps hoping to make this cousin jealous, he spoke glowingly of a potential rival for his affections: "most charming moonlight walks with a very pretty girl—*all alone* would not you have liked to have been along—she became very sentimental—*very*—I am myself becoming very much so lately . . . ."[175] This testing behavior was useless because his inclination to court her was not reciprocated.

Another pair of cousins were more successful. Eliza Trescot and Eldred Simkins had lived under the same roof for long stretches. Eldred confessed that he deliberately stayed out late to weigh her interest in him: "I remember I used to stay out on purpose some evenings till very late and when I returned I always met you still there, but you never guessed how delighted I was to observe it. Again when I came home long after supper from lodge or elsewhere who used to wait in the parlor for me and sometimes all alone?"[176] Eliza eventually understood the meaning of Eldred's behavior: "when you stayed out late I thought it was from indifference and never dreamed that it was to test me. But I stood the test and you could trust me."[177]

Eldred carefully scrutinized Eliza's inclinations before he openly revealed his own affinity. But men sometimes were less cautious in deciphering feminine intentions during the pre-commitment phase. Women such as Maude Hulbert might take advantage of this masculine vulnerability. Maude cultivated male admirers, toying with the less scrupulous and gently rebuffing more innocent hopefuls. She told her fiancé, Jack Horn, that one potential suitor had called twice since Jack left town but she had been "under the weather—so he could not get to talk with me alone. But Monday night he was about to leave for Placerville and wanted to make doubley sure there was no show for him—He began by asking for my picture—Told him they were all taken—Then he had the cheek to try and put his arm around me—I very firmly told him I'd have nothing of that kind and he said 'Well, you're not married yet and he's a long way off'—I said 'It don't make any difference whether I'm married or whether I ever intend to be—' He didn't gain much satisfaction from what I said—and left almost immediately."[178]

While Maude was abrupt and blunt with her first admirer, she reported a much more sympathetic encounter with another suitor. Though she stopped answering his letters, this hopeful pressed to see her. "I let him come—for I thought the sooner we came to an understanding the better—Poor boy I really do feel sorry for him. He was so pale and nervous—he could hardly talk—up to then he has always felt that 'where there's a will there's a way' and has thought that though I now entertain but a sisterly feeling for him—it might change. So long as there was no one else in the way. He had heard the rumors about us but did not believe them—So I thought I'd be frank with the boy—I told him there was some one that I cared a great deal for—and that some one cared for me— and that I would always like him—as a friend. I could *never* even think of anything else. He did take it so hard—just broke down entirely—I spoke so kindly to him as a sister would . . . ."[179] Maude revealed some sense of her own part in this shower of unrequited love when she confessed to Jack, "Oh dear—I'm going to be so careful hereafter—not to let anybody care for me more than they ought to. . . ."[180] Men were more vulnerable to rejection at an early stage of the courtship ritual than women, due to their more overtly dominant roles. Flirtatious women might take advantage of this masculine vulnerability in order to flatter themselves and enhance their public image.

In fact, while men exercised formal control in the courtship initiation process, women sometimes saw themselves as the aggressors. One young woman, Sophia Child, wrote to a friend, Patty Newcomb, about a rivalry over a man: "I most *heartily* give you joy of the interest you have gained in the *affections* of the *youthful Adonis*. I hope you will take advantage of it—but I think it was hardly fare in you to take advantage of my absence to ingratiate yourself in his favor—I did not suppose it possible you would turn *traitor*—well I suppose I must submit, as I was without doubt born under an evil star—but depend on it *I will* not go unresigned." Sophia warned her that when she returned to the fray: "I will play off all my *artillery of charms* and endeavor to captivate him and hope I shall be as successful as you have been."[181] Women could conceptualize their role in courtship initiation as an active one in spite of the restrictions on their formal expressions of control. Though men were expected to launch the courting relationship as well as issue the marriage proposal, the origination of courtship was decidedly reciprocal.

Women participated actively in courtship initiation by giving men signs of their own interest and encouraging them to feel that their advances were welcomed. This was probably often accomplished non-verbally. Nathaniel Wheeler believed that Clara Bradley's expression of interest through eye contact was his primary reason for courting her. "—Ultimate Causes. —A. Those affecting the party of the second part—1) Primary—I felt an interest in you because I knew you were watching me. Have you yet to learn that that is the most effective flattery possible?"[182] James Bell revealed the effective use of this non-verbal romantic code by Gusta Hallock: "Dident you know that you told me more of your love with your eyes than any other way? There was an expression there when you met me that never was there when you met others."[183] James returned the compliment in watching her: "I used to look at you when you didnt know it . . . ."[184]

The emphasis on eyes and watching each other was noted by James Hague, recollecting his first meeting with the woman he was courting: "tried every now and then to make you look towards me that I might enjoy the light of your eyes—all through the croquet in the rain . . . the tea drinking and all—to the time I bade you goodbye—wondering whether I should ever see you again—not dreaming that you were any less indifferent to me than your manner implied . . . ."[185] While giving no definite reason for assuming her indifference, he hinted that her lack of eye contact was crucial. Though the evidence for the initiation of courtship was not extensive, non-verbal cues from women were probably a covert indication of female interest and a way to reassure men that their advances would not be rejected. Therefore who was taking the initiative with whom was sometimes difficult to determine.

🦋 Nineteenth-century middle- to upper-middle-class courtship assumed a distinctive character in response to the diminution of parental control over marital choice. A private, unchaperoned, mutually agreed upon relationship, Victorian courtship was an incubator of romantic love. It encouraged the growth of both physical and emotional intimacy. The latter, in the middle classes at least, was expected to reach a certain level of intensity before marriage. Men took the overt lead in courtship initiation but the process was decidedly reciprocal. Couples began the courtship ritual based upon the unstated agreement that the potential for romantic love existed between them. They founded this pre-commitment on personal preferences often molded by masculine or feminine sex-role ideals, an initial assessment of the individual attributes of eligibles, some peer group encouragement, and signals that their interest was returned in kind.

The central structure of the middle-class American courtship was a ritual of testing emotional loyalty and commitment, which made it an intensely dramatic and exciting period of interaction. The testing rituals of nineteenth-century courtship encouraged the development of romantic love which, by the 1830s, had become the most desired state of male-female interaction. Courtship spirals of doubt and reassurance intensified the emotions and solidified the identification of couples going into marriage. This was beneficial to women whose power in the nineteenth-century family stemmed from the companionate ideal of marriage often undergirded in reality by the emotional bonding developed during courtship.

What was most striking about nineteenth-century courtship was its dramatic structure. While historians have documented and explored the public dramas of American life from the Salem witchcraft trials to Watergate, the cultural performances of private life have been less systematically studied, except in individual biographies. Nineteenth-century courtship was cultural ritual with a cast of characters and an audience which involved families, not nation-states. Nonetheless, courtship was a private drama that performed functions similar to its more global relatives. It was a definitional ceremony that developed a couple's collective identity, strengthening agreement on values and reinforcing shared interpretations of the world. The couple served as its own audience in staging these emotional scenes, each dramatizing doubt, frustration, and fear to the other, and generating conviction in their future together through active demonstrations of reassurance and praise. The courtship testing that preceded marriage was a ceremony that helped to clarify and strengthen the couple's feelings for each other.

"Doing was believing" in nineteenth-century courtship testing as romantic love was not merely stated but enacted in the form of obstacles to overcome and crises to resolve.[186] The redress that followed a courtship crisis moved the nineteenth-century couple to a new and more intense level of commitment and understanding of their interdependence and shared identity. In a culture which by 1830 was relying more than ever on sentiment to bring couples together and then maintain their relationship, the rituals of courtship testing were essential preparation for companionate marriage. One suitor captured the preparatory significance of Victorian courtship in a musical metaphor: "To get married without it is like the opera without any overture."[187]

7

Husbands and Wives

Duty-Bound Roles and Unaccountable Love

I F THE PURPOSE OF nineteenth-century courtship was to nurture the growth and development of romantic love, strengthening the affective ties between men and women entering a companionate marriage, the challenge of nineteenth-century marital relationships was to adjust an essentially voluntary and volatile romantic bond to the involuntary obligations and duty-bound roles of husband and wife.[1] Victorians conceived of marriage in terms of love and personal choice. Yet spousal roles were largely defined as compulsory social obligations. An act of self-determined choice, Victorian marriage nonetheless imposed a set of mandatory sex-role specific duties upon husband and wife. This contradiction was at the heart of the nineteenth-century middle- to upper-middle-class conception of marriage.

The essential difficulty was in the clash of opposing ideas about love and spousal duty fostered by Victorian culture. Though the participants often described and sometimes analyzed aspects of their experience of love, they did not conceive of their affection as especially malleable. Romantic love was accepted as an essentially uncontrollable consequence of inexplicable forces of attraction. In the Victorian lexicon of emotions love could not be commanded by the will; it was understood to be beyond individual control and was not treated as a social obligation. Consequently, it proved difficult if not impossible for Victorians to hold their spouses accountable for the love they both agreed was intractable.

While the previous chapter suggests that nineteenth-century couples in fact exercised some deliberate control in choosing a love partner, the pervasive attitude toward love was, as one advice-book writer described it, "a mystery . . . the effect of a mysterious agency acting on two minds naturally suited to each other, but which had never before come within each other's influence."[2] Another adviser, Timothy Arthur, insisted in one breath that true love was *not* "a wild, strong fiery impetuous passion," but rather "calm, deep, and clear seeing," yet in the next breath he described it as "that mysterious attraction of heart for heart which comes from above . . . ."[3] Another Victorian marriage adviser, though chagrined by the fact, admitted, "There is a theory, too generally accepted, that love can not be evaded—that there is *destiny* in it—in a word, that you can not help yourself."[4] He advocated phrenologically controlled mate selection guided by reason and moral sentiments.[5] More than one marital adviser urged rational calculations upon readers, and a goodly number of rational matches were probably made in the nineteenth-century marriage market, but Victorians generally accepted the *theory* that love was an irresistible force and that "you can not help yourself." This view of love as will-less had profound consequences when linked to a system of lifelong marriage.

Nineteenth-century companionate marriage was based upon an atomistic ideal of two individuals mysteriously but permanently bonded by romantic love; yet Victorian marriage was also an institution steeped in the traditional obligations of husband and wife. Provider, cook, nurse, or protector, the sex-role specific duties required by the marriage estate were tied to emotions that remained mysteriously unfettered to the Victorian, above duty and outside conscious control. An intellectual contradiction, certainly, but did this clash of ideas translate directly into behavioral or emotional strain? We must discover whether the contradiction between the requirements of spousal roles and the latitudinarian attitude toward love actually troubled nineteenth century marriages.

The emotional history of middle-class marriage is more difficult to chronicle than courtship. For while a couple was naturally separated in almost every courtship, living in different houses if not neighborhoods or even cities, separation was an unnatural marital condition. For this reason as well as others, marital letters were often more intermittent and less revealing than courtship epistles. Uncomplaining and perhaps silent, it is possible that Americans were content to let love fade after tying the knot, tolerating a friendly rapprochement after the first few years of romantic passion had dimmed. But the extant evidence challenges this interpretation in two ways.

Victorians clearly distinguished between excitement and love. They expected the novelty, drama, and all the bracing uncertainty of courtship to fade with marriage. Both unmarried men and women predicted with relative equanimity that wedlock would prove less exhilarating and more routine than courtship. They did not, however, accept the absence of love in their future marital relations nor did they seem easily reconciled to the actual experience of a loveless union.

Victorian marriages were obviously characterized by a variety of life circumstances and complaints. But running through the "troubled" marriage was a half-expressed, sometimes shy, but also occasionally anguished and articulated lament for lost love.[6] The Victorian marriage was significantly strained by the conviction that one spouse did not love the other. Sometimes happy and sometimes sick at heart, Victorians continued to support the idea of marriage until death.[7] Nonetheless neither husbands nor wives appeared easily reconciled to a loveless denouement.

The Victorian conception of marriage was not a happily-ever-after haze of courtship excitement and passionate intensity. This reflected the fact that Victorians plainly distinguished between excitement and love—two emotional experiences which intertwined but were not identical. Obviously the sources of marital dissatisfaction were complex and sometimes difficult for the participants, not to mention an outsider, to decipher fully. Significantly, however, Victorians had been schooled to distinguish between a dull and a loveless marriage. When they complained about coldness, indifference, and the absence of love, there is reason to believe they did not easily confuse them with the muted allure of familiarity and routine.

The meaning of marriage in nineteenth-century America was strongly tied to calm, quiet, settled expectations. Nathaniel Hawthorne, worried over Sophia's reaction to his "nonappearance" on Christmas Eve and anticipating her anxiety and depression, urged her to "have the feelings of a wife towards me, dearest—that is, you must feel that my whole life is yours, a life-time of long days and nights, and therefore it is no irreparable nor very grievous loss, though sometimes a few of those days and nights are wasted away from you. A wife should be calm and quiet, in the settled certainty of possessing her husband."[8] Hawthorne reflected the clear dichotomy in nineteenth-century American mainstream thought between the perplexities of courtship and the repose of marriage.

Similarly Albert Janin looked forward to marriage as a "haven of rest." He conceived of marriage in one of the central dichotomies of Victorian culture: "No matter what turbulence or contentions may attend my out-of-door and public life, I shall have one blessed harbor of refuge where I recuperate my strength and prepare for a renewal of the struggle."[9]

Typical of the courting man, Albert contrasted the tumult and uncertainty of public life to the peace, calm, and soothing balm of marriage. For him this idea was one of the chief attractions of the marital state.

Men usually compared marriage favorably with courtship. Hawthorne and Janin reflected the attitudes of most men who happily anticipated the tranquility and predictability of married life: "Oh, I want thee with me forever and ever!—at least, I would always have the feeling, amid the tumult and unsuitable associations of the day, that the night would bring me to my home of peace and rest—to thee, my fore-ordained wife." [10] Generally, men's conceptions of marriage centered on the steady companionship and constancy of daily living together. This vision of a marital future was appealing and comforting to James Hague: "Duckie, I'm tired to death of living away from you. I want to see you—and that isn't all—I want to begin life with you—to know that we are one—and to live together day by day, having common interests and adapting our daily plans and occupations to each other— How nice it would be if we had a pleasant little home of our own out here and could live here comfortably and happily together." [11]

Images of marriage among nineteenth-century men in this study were dominated by visions of domesticity. Certainly domesticity was on the mind of Eldred Simkins when he rhapsodized about marriage: "visions of future years fraught with happiness passed before me when we should sit together by ourselves in 'our home' on the winter evenings by our bright fireside and talk about these years of separation and trial! And when I would come in tired and worn out my fairy would meet [me] at the door and charm away fatigue and care?" [12]

Though women shared some of the premarital idealism of men, they were often less prone to flights of fanciful imagination where domesticity was concerned. Gusta Hallock told the story of her married sister who, in the midst of a sublime natural landscape, "had been intently gazing at a rank young milk weed and wondering how it would taste cooked all the time that I had been enraptured by the beauties that surrounded me." [13] Though Gusta laughed at her sister's misplaced domesticity, she nonetheless accepted the role of a wife as someone who would tend to her husband's physical and emotional needs. When James asked her what she thought about marriage, she responded: "I'd advise you to get someone who will take good care of you—who has the power to scare off the glooms and can make good coffee (I believe you are a lover of said article)." [14]

Eliza Trescot was another woman who tempered her domestic idealism with a dose of skeptical realism: "Oh Eldred you don't know how I love to think of it, to write of it, that there is such a thing and at sometime our plans will surely be realized. No more separations and waiting and longing a whole week for the letters but rather a danger of seeing too much of each other. Did you never reflect upon the chance of getting tired of seeing the same face day after day, especially if the scene should not be varied and enlivened by some others? It would be unfortunate to tire of it for you couldn't get rid of it . . . ."[15]

Another young woman, most likely married for only a few years, discussed matrimony with her unmarried girlfriend Gusta Hallock: "Now my advice is to get married, that is if you can find any one that you like well enough, and that is worthy of you, not that I would advise you to throw your self away, for the sake of getting married. To be sure the romance soon wears off but speaking from my own experience, there is enjoyment after that, I have a kind husband, and we have got to have our first quarrel yet . . . ." This young married woman expressed no disillusionment over the fact that "romance soon wears off"; clearly, she expected her premarital ardor to wane, and instructed her unmarried girl-friend accordingly.[16]

Nineteenth-century men and women both anticipated that the fervor, exhilaration, and drama of courtship would recede after marriage. They were prepared for a diminution of emotional intensity in their marital relationship. The prescriptions for change varied. Dorothea Lummis wrote her husband Charles: "Someone says 'Marriage should be one year of joy, two of comfort and all the rest of content'." Dorothea, struggling with an unhappy marriage, hoped that she and Charles could find the content-ment of a happy marriage "*all* the time and bliss once in a while."[17] She reflected the dominant image of a happy marriage among the middle-class native-born: peace and contentment.

Mary Smith also accepted this image before her impending marriage: "We my dear one, are probably anticipating much that is happy although I trust neither of us are expecting perfect bliss. We know full well how often disappointment treads on the heels of pleasure or meets us in her flowing path—we know too much of life, the reality of life to suppose for one moment that the cup we have given us is *full* of sweet—with no bitter ingredients at all . . . ."[18] Mary warned that they should not set their expectations of marriage too high. Typically, she emphasized the values of comfort and calm in her future relationship with Samuel. Her hope was that they might be kept "calm and composed" through the "short pilgrimage of married life."[19]

James Hague illustrated a similar unruffled expectation that courtship romance and excitement would dissipate: "Well, dearie, I don't suppose we shall be spooney always—I dare say we shall become very matter-of-fact by and by—but don't let us give up what pleasure we can find in the present just because we can't foretell the future. I only know that if I were absolutely certain that we should love each other always as we do now, I couldn't love you any more this minute than I do."[20] James predicted that the charm and intensity of courtship would be dispelled. Almost no one approached marriage expecting that it would duplicate their courtship experience. Generally, they anticipated that routine and familiarity would create a more matter-of-fact atmosphere between them.

Not everyone was sanguine about such a prospect, however. One young man wrote to his aunt of not wanting to give up the excitement, novelty, and most especially the drama of the unknown for the settled predictability of marriage. "[B]esides were I married I would have no taste for one of the greatest enjoyments of life that which above all others contributes to make us happy I mean the pleasures of anticipation. I can now allow my fancy to paint entire constellations of stars, but did I attempt to pluck one I might find it no star at all but an ugly mortal like myself—no, liberty is a State too noble to be lightly exchanged for one of bondage, at least so I think."[21] Nonetheless, few were as gloomy about the prospects of matrimony as Violet Blair, who lamented, "What an unlucky letter 'M' is, to begin medicine, martyrdom, murder and matrimony—."[22]

Generally, marriage was not so foreboding to either men or women. James Hague teased Mary, his fiancée, about the probable contrast in their relationship before and after marriage, "Duckie, you must get strong and hearty to stand all the hugging you'll get when I am once more at home— otherwise your feeble constitution will fail altogether, and I shall hug you to death before we get married. Probably you will be safe *after* that, for I am told people get over those habits after a time of married life—but we will see—I only wish we were ready to begin the trial."[23] Middle-class expectations of marriage had much more to do with settled comfort than unmatched ecstasy and bliss. Though courting couples anticipated loving marital relations, they also predicted the routinization and habitual flavor of married life.

Clearly, nineteenth-century courting couples expected and accepted a contrast between courtship and marriage. They were not indifferent to the prospect of a bad marriage, however, and were not reconciled to a loveless future. They hoped their love would survive on the other side of the matrimonial divide. For example, James Bell remarked to Gusta that he would never think of getting married if there was a chance his married life might resemble that of his boss's eldest daughter: "They lead a Cat and Dog life all of the time." He reported that this woman told her husband of six months that she regretted ever marrying him.[24]

Three years later, James was again concerned with the bad marriage: "For my part, as year after year has pased away my love for you, has grown brighter and stronger. And if we are ever married I want it should be just so then. I believe it will to. Somehow, people after they are married dont love as they had ought to. Two thirds of them are unhappy; which you know is wrong (funny world)."[25]

James was not alone in his estimate that unhappy marriages vastly outnumbered happy ones. Gusta also saw an unbroken chain of miserable marriages surrounding her. As part of her compensation as a rural school teacher, she was given room and board in the homes of her pupils. On this basis she wrote back to James: "You say 'somehow people after they are married dont love as they had ought to.' Well that is something *I* have noticed too: and oh! how much I have seen of it. You know a *school teacher* has a great chance for observation and I have *wondered* and *wondered* why it is so, Tho I've thought there were a great many reasons why, that *I* could see, In the first place there is'nt one couple to 50, who are well matched, that is who agree in principal, opinion and taste, and that alone is sufficient to produce discord, for instance to unite a liberal benevolent soul, with one of a selfish, penurious, disposition—a pious spirit trying to reach the skies; with a worldly nature, pulling directly the other way . . . ."[26] Gusta saw other reasons why couples were so unhappy. They allowed the small irritations of life to have "a powerful effect upon the temper." She confessed that she also lost her temper over trifles, but predicted: "Yet I think we can be happy together if we each make up our minds to bear of the other."[27] In spite of the gloomy pictures of marital discord and unhappiness they drew for each other, both believed their marriage would be different.

In the face of such negative assessments of the institution of marriage, James's and Gusta's optimism about their own marital prospects might be dismissed as just one more romantic illusion. Or their reports on the epidemic of bad marriages might simply be taken as hyperbole. Both interpretations, however, miss the point. The question of sociological accuracy in these bad-marriage testimonies is not at the center of their significance. There were good marriages and bad marriages in the nineteenth century, and the complexity of discerning one from the other (leaving aside the difficulties of defining good and bad) should be enough to make one cautious about quantitative guesses such as James's and Gusta's.

Certainly some examples of marital success were available for comment. Yet almost no courting couple was interested in the good marriage. Why? It appears that courting couples used the bad-marriage testimonial as a reminder of what they could become if they did not exercise vigilance in their own marital relationship. Paradoxically, courting couples also used bad-marriage testimonies to reassure themselves about their uncertain future together. Bad-marriage testimonials allowed them to feel some certainty, if only of never reaching the level of unhappiness or marital discord they put before each other in these negative examples.

Bad-marriage testimonials might also be exchanged between young unmarried women. While discussing marriage and marital prospects, a young, unmarried girlfriend of Anna Dane recounted the marital woes of a neighbor woman: "When but sixteen she run away from home and married a boy of 19 yrs. Her parents did not want her to have him because he drank, but he told her he did not and she believed him[;] her mother would not forgive her until just before her little girl was born. In the mean time her father died and left her $2000. it was put in the bank in a way that her husband could not get it but he forged her name and drew $500 and drank himself most to death. he abused her and would not support his wife and child. so she left him and went to her mother, and just at that time she got a letter from her Aunt in Silver City Nevada telling her she could get $1.00 a lesson for teaching music, so she left her baby with her mother and went to Nev. and taught   well the first man she met was Mr. H—a widower with one child, a boy. he wanted her and her uncle and aunt told her she was a fool to teach when she could marry such a rich man, so finaly she applied for a divorce, and the day it was granted she got the news that her husband was dead   he had been dead some time, but his folks were so angry at her for leaving him that they kept it a secret. She married Mr. H—and he got her $2000 worth of jewelry, and every thing else to match, but he invested in stocks and lost all, so here they are. She has crossed the continent 5 times. She says her happiest days were before she married at all."[28] The moral of this bad-marriage testimonial is that it is better to wait than marry in haste, and by implication better to be single than unhappily married.

There is no indication that bad-marriage testimonials were presented or received on the level of factual evidence against marriage. They were tales told in moments of anxiety about marriage to reassure and comfort. The girlfriend of Anna Dane joked about their being old maids together, but the clear inference was that neither girl intended to remain single for long. Men and women who anticipated marriage, and were sometimes even on the verge of it, told these tales of bad marriages not because they rejected marriage or wanted to discourage it, but rather as part of their effort to cope with their premarital anxiety. Ironically, the bad-marriage testimonial was reassuring rather than defeating.

Mary Smith, sixteen days before her wedding, recounted an incident to her fiancé which began with a loud, urgent knocking at her door. Looking out, she saw a very sick man in a carriage. Vaguely recognizing this man, she expressed disbelief and chagrin that his wife was not by his side. She exclaimed: "Why Francis, if I believed we should ever fall into such cold apparent indifference, I would never, no *never* consent to enter the marriage relation. Yes, I would rather much rather remain as I now am, and indeed nothing would induce me to change my situation did I not believe that in doing so—I should possess a friend, a devoted friend—one who *would remain* unchanging and constant both in 'sickness and health.'"[29] This bad-marriage testimonial was a warning against certain marital traps. Mary's message, just days before her wedding, was that she and Samuel should guard against such insidious indifference and waning commitment. Almost six months earlier, Mary called Samuel's attention to a happily married couple. Nonetheless, the weight of the illustration was still on the contrasting majority of married couples who were unhappy. "And I cannot conceive why such cases of continued and unabated affection are so very rare. It need not be thus, I am sure. But I do suspect nay believe that much almost everything depends upon the cautious or incautious manner in which persons commence such a life though of course *much* depends upon the degree of congeniality of kindred tastes—of refined feeling—which characterizes the persons concerned."[30]

Not all bad-marriage testimonials occurred before marriage. One was written after four years of marriage by a man still deeply in love with his wife: "I *do* not understand why some men are so afraid of letting their wive's know they love them or rather are so afraid of telling them so. They seem to think it looks *soft* or undignified. I know men who seldom speak kindly not to say lovingly to their wives. I hope to never see that time when I cannot with all fondness put my arms around my wife or take her upon my knee and taking her face between my hands kiss her and tell her I love her better than anything earthly."[31] Again, one must exercise extreme caution in using bad-marriage testimonials as sociological description. John Marquis could not "know" what other men said to their wives in private. There is no doubt that some husbands treated their spouses unkindly. But bad-marriage testimonials were not so much factual descriptions as ritualistic incantations, which functioned to reassure the anxious and compliment the "good" relationship. They also indicate that though courting Victorians expected a more matter-of-fact marital environment, they feared the loveless marriage and did not confuse familiarity with deeper feelings of affection.

🦋 Levels of anxiety before marriage obviously varied, but not surprisingly reports from newlyweds were very positive. Less than two months after his marriage to Violet Blair, Albert Janin soared exuberantly, "It is a delightful satisfaction to me to feel that my old theory of life, vis.—that the realities of our existence can be made to eclipse in attraction all the beauties of romance, is verified in respect to our marriage. Since the 14th of May I have not known a day that has not been full of happiness to me when in your presence. And what is more, Vivie, I sincerely believe that this state of things will continue . . . ."[32] One year later, Albert was still enthralled by marriage and his wife: "I love my wife with more real tenderness than I ever felt for the girl I was courting, because the love for one's wife is a nobler and better love than any other. It is accompanied by peace and quiet happiness, cemented by mutual interests and sanctified by Heaven."[33] Violet, who rarely allowed herself any open or unguarded avowals of love for Albert Janin before marriage, wrote three months after their union, "I feel so terribly sad tonight—Oh, if you were only here to take me in your arms—I do long for you so—My heart aches to be with you again, my own beloved."[34]

After less than a year of marriage, Mary Smith rhapsodized about loving Samuel above all others; she called him her best "earthly treasure" and intoned, "it is my highest happiness to feel that I am yours and you are mine . . . ."[35] Three days later, she sighed: "Every orange I eat— every favorite dish of yours I see on the table, every cup of tea or coffee or tumbler of egg, milk and wine which I drink seems only to make your loved image more fresh in my memory."[36] There was no surprise when she declared, "We *are* happy in our marriage relation, my love, as happy I think as possible—there may, there doubtless are, some barriers, but perhaps as few as in the situation of any on earth."[37]

After two years of marriage, Nathaniel Hawthorne exclaimed: "I love thee, I love thee! Thou lovest me, thou lovest me! Oh I shiver again to think how much I love thee—how much we love, and that thou art soon, coming back to thine own house—to thine ownest husband; and with our beloved baby in thine arms."[38] Eight years of married life did not dim Hawthorne's romantic enthusiasm, for he told his wife he had nothing much to say except the "unimportant fact that I love thee better than ever before, and that I cannot be at peace away from thee." He closed this letter by claiming that away from her "there is only emptiness."[39] After fourteen years of marriage, Hawthorne was still rhapsodic: "Oh, dearest, dearest, interminably and infinitely dearest—I don't know how to end that ejaculation. The use of kisses and caresses is, that they supersede language, and express what there are no words for. I need them at this moment—need to give them and to receive them."[40]

As the years wore on, expressions of love between spouses varied considerably. Troubled marriages were often dominated by yearnings for love, and a sense of loss. But in less agitated relationships, enthusiastic affirmations of romantic love were not that unusual.[41] These affirmations of love, however, were intermittent markers of marital relations. It is possible that daily marital interactions belied these summary statements. Nonetheless, after nine years of marriage, Elizabeth Harbert wrote her husband Will, "My heart is full of love for you tonight and my earnest prayer ascends for your safety and happiness. When a few days separation is so hard to bear what would life be worth if we were separated by death. *it* makes me tremble to think of the possibility of anguish in the human heart—. How rapidly the years are whirling by . . . Let us crowd as much happiness as is possible into every day—."[42] She closed this letter "Goodnight, my precious husband." After thirteen years of marriage, Emily Lovell pleaded: "Promise me that the end of this war (if we both live to see its end) that you will never ask me more to be separated, I am determined to follow you where ever you see fit to go—I have had enough of this truly—is it not strange dear I am no longer 'sweet sixteen,' yet I am just as deeply in love—as though I was."[43] Though she felt strange and unusual to have maintained such strong romantic feelings, Emily was not alone.

John North assured his wife of thirteen years, "One thing rest assured of Dearest—distance does not weaken my affection for you, and I trust it will not. And yet I do not want to learn to live without you. I know you can not be as happy even at your old home as you would be with me, and I am anxious to have the time made as short as possible; until we meet. And what a happy meeting it will be to me, when I can greet you in my new home—Much love to you all; [meaning his children and other family members] but most of all to you *my good* wife."[44]

After nearly eight years of marriage, Margaret Graham wrote her husband: "that love passes out of expression, I wondered if you felt it. I think constantly of you and the more I see the more I know what a jewel you are in mind and soul, if indeed I had needed to see more to know it."[45] Her husband reciprocated her sentiments in the same year: "This will reach you on tomorrow your thirty first birthday. . . . I send a little handkerchief and thirty one warm kisses. My blessed, blessed darling. You are so good and so smart. I worship your very image. I like your photograph very much, it is so precisely like you, except that your hair is not at all at its best. It does not flatter you in the least. The little quirk in the mouth—is my own little quirk. I would not have the picture otherwise."[46]

Lois Dane wrote to her mother on her fifth wedding anniversary, "Who of us then thought five years would see me the Mother of three little ones. Not *I*, of that I am sure and yet so it is and today finds me a loved, loving and happy wife and Mother *even* with *three babies*."[47] Marital happiness had proved much more durable than she expected. Twelve years later she wrote her husband from the Philadelphia Exposition, on what may have been her first major trip east since leaving Michigan on her wedding day: "If only you were here with me I dont know how much I would give for I think of and miss and wish for you almost every moment . . . ."[48] The trip to Philadelphia was apparently planned after she arrived in Michigan for a family visit. Her husband, who was a farmer, approved: "I am verry glad you concluded to go on to the exposition for I know you would not have been satisfied if you had not have went and I should not either. . . . I am begining to count up the days when you will return home again the time is begining to seem long . . . . We are getting allong verry nicely so I hope you will not worry a bout Home affairs . . . . I will close with much love from all to all and a large portion to your self."[49]

Ten years of marriage found Jane Burnett vowing to her husband Wellington that she would never be separated from him again: "you dont' know how I feel when I am away from you, never again will I leave you   I thought when I came up here it would be so very different from S. Cruz. I should feel contented because I would be at home but it is only home where you are—I look forward to meeting you with the most exquisite pleasure—."[50]

Robert Burdette wrote his wife of fifteen years, less than five months before his death, "I dream of you, think of you, long for you, and thank the dear Heavenly for every memory of you, my darling. So dear, so sweet, so helpful you have been and you are to me, my own heart's treasure. The day is tender with its thoughts of you. The waking hours that sometimes come to make the night long, no longer come with dread and with tossings. For my darling comes with them, making them gentle with her caresses; with memories of her tenderness; with whispers of courage and hope, with words of love; with all that she has been and is."[51] Clara Burdette expressed a similarly fervent devotion: "Rob dear—you have been such a sweet, loving husband—I am afraid I do not tell you so often when we are together but my devotion and constant thought of you must testify to what I fail to put into words sometimes."[52]

Lincoln Clark left an unusually steady and continuous record of marital feelings. In 1841, after five years of marriage, he testified: "Most cordially do I reciprocate all the affectionate sentiments of your letter—no business, duty, or exercise of thought and feeling affords me half the delight that it does to love you; to be deprived of that high and heavenly boon, would be to deprive me of all that to me would be worth living for; unless it were for the good of my soul and my little ones . . . ."[53] In 1843, he confessed: "But I can tell you that I love you, and I know that will please you better than philosophy, and is by far the most agreable to me to communicate. *Indeed* and in truth I have thought you much more beautiful of *late* than I ever yet saw you."[54] Lincoln highly valued his wife and commented appreciatively on almost every aspect of her wifely role at one time or another. In 1846, he wrote: "I can say enough I am sure to remind you of my unfailing love, and that is a satisfaction to me at least to record. When those we love are *from* us, nothing is more natural than for us to embody them in their virtues, in other words in their kindness in what they have done for us. Often, often do I think of your pleasant and patient face as you were doing and studying what you should do for me when I was suffering on a bed of sickness, and your constant efforts to strengthen me in every duty, temporal and religious, and if I ever achieve any good fortune or eternity, the credit will be as much yours as mine."[55]

Lincoln, more emotionally expressive than his wife, seemed to accept this difference, albeit with some longing for more words of love on her side. He wrote that his wife Julia *"is my jewel. . . .* they [say] that old man love a lady. Now this is more than you have said or wrote for a long time—well I suppose you have *thought* it, and I must be satisfied with that."[56] In 1848, he closed one letter, "I have stated many facts for the benefit of others since I left, and now one for yours—I *love Julia, Julia Annah; Mine.*"[57] At the end of his twelfth year of marriage, he observed, "I wander over the world and see a thousand strange and interesting sights, but nothing at last is half so interesting as to think of you and write to you except to be *with* you . . . ."[58]

No relationship is without some strain and Lincoln expressed occasional discontent. In 1849, for example, "Murmuring I fear is my sin—I try to see and avoid it—sometimes I feel a little impatient towards you, and the worst of it is that it is without cause; but it is always evanescent. . . ."[59] Nonetheless, in 1852, after almost sixteen years of marriage, Lincoln seemed as extravagantly romantic as ever, "I wish you were here—and sure enough I would praise you and pet you *like* a baby. I know you are sweet, and an accomplished lady—were you here I should be proud of you—I see others—I have seen you—and I believe I have the reputation of being a judge of ladies."[60] He closed this letter, "I think of you every day, and every hour of the day, and almost every minute of the hours." Julia sent him a letter to which he responded in April of the sixteenth year of their marriage: "I always love to read what you have to say, especially that you have not ceased to exercise your youthful feelings in affection and devotion."[61] In 1858, with a stunning consistency and constancy after twenty-two years of marriage, he wrote: "I am always rejoiced to hear from you. I wish I could write a long letter—a large part of it should be taken up in telling you how much I love you, and how desolate I am without you. The image made up of black hai[r], bright black eyes and an intelligent sweet face is almost always before me—but I had rather have the living reality. . . ."[62]

Expressions of marital satisfaction, love, happiness, and longing after ten, twelve, even fifteen years of marriage confound the association of Victorian marriage with stiffness and distance. Certainly some marriages could be characterized as cold, formal, and indifferent, but others appear to be warm, open, and loving. The safest and most compelling generalization concerning the nineteenth-century middle-class marriage is that some were emotionally close and intensely satisfying to both partners, some were distant and less affectionate, while still others were actively troubled.

Obviously, expressions of love were not reserved for courtship letters alone. To the contrary, avowals of both physical and emotional intimacy were not uncommon in nineteenth-century marital exchanges. But these moments of intimacy were often affirmations of uncontested love, based upon taken-for-granted emotional commitment and long-standing resolution of emotional doubt. Hawthorne eloquently described the emotional tone of the "relatively satisfied" marriage: "Oh, what happiness, when we shall be able to look forward to an illimitable time in each other's society—when a day or two of absence will be far more infrequent than the days which we spend together now. Then a quiet will settle down upon us, a passionate quiet, which is the consummation of happiness."[63] There was a quiet—sometimes a passionate quiet—to the relatively contented marital exchange that contrasted sharply to the disquiet, upheavals, emotional crests and valleys of even the most loving courtship.

Courtship letters were full of crisis and the emotional swings from agony to joy which usually accompanied the drama of romantic doubt. Marital letters were more even, less melodramatic, and their emotional range was often narrower. Generally, there was a striking contrast between the content of a couple's dialogue on paper before and after marriage. Courtship letters often dwelt upon the feelings of the sender and receiver. Marital letters varied, but were as a whole much less introspective. Spouses discussed children, friends, relatives, business and finances, domestic tasks, future material goals, and the many responsibilities of everyday life. They spent significantly less time and energy (compared with courtship letters) discussing their feelings about themselves or for each other.

Marriage in the nineteenth century was supposed to be based upon achieving some satisfactory level of emotional and psychic identification before the ceremony. When couples reached this cultural goal, each partner could feel relatively assured of the other's romantic commitment. In fact, marital exchanges about love seem almost to be affirmations of past accomplishment. The ritualistic cycle of doubt, testing, and reassurance which was central in nineteenth-century courtship letters largely disappeared from marital exchanges.

Courtship testing was a definitional ceremony of commitment that the married couple (at least ideally) superseded by the time of the marriage ceremony itself. There was one important exception to this contrast in courtship and marriage letters. In troubled marriages, where either partner had some hope of change, the "premarital" rituals of testing might reappear. As a rule, however, marital letters exhibited strikingly less emotional upheaval.

The nineteenth-century middle-class couple did not conflate romantic love with excitement and drama. They understood and generally accepted that marriage meant an exchange of some of the liveliness and fervor of courtship for a calmer, more predictable, even more boring relationship. This did not mean, however, that marital commitment was expected automatically to dispel romantic love. Nineteenth-century couples who married for love were not easily reconciled to its loss.

🐝 Victorian couples were forced to confront the contradictions that adhered to the cluster of ideas that defined nineteenth-century marriage. The dominant justification for marriage was emotional. Yet, the conception of love as a mysterious agency beyond human control made the recovery of "lost" love exceedingly difficult. Furthermore, divorce in the nineteenth century was not yet conceived of as a wholly legitimate solution to the problem of marital disaffection. Though marriage was based upon individual choice and the ability to develop mutual identification and empathy—in other words, romantic love—divorce was largely sanctioned for failure to perform culturally prescribed duties at socially acceptable levels.[64] Community standards held sway in the performance of the duties of husbandly and wifely roles, at odds with the essentially private, voluntaristic, and personally expressive notions of choosing a marital partner.

While marriage was supposed to be a voluntary act based upon personal choice, the obligations of husband and wife were conceived as prescribed duties, socially mandated, and only involuntarily dissolved. Victorian marriage was thought to be rooted in romantic love, yet by at least 1830, that same love was understood to be unwillable and essentially out of individual control. Furthermore, nineteenth-century men and women approached love as "free and autonomous" individuals (however mythic a conception) and then entered marriage as a relation of dependency. Is it any wonder that marriage was simultaneously desired and yet feared by both unmarried men and women (but especially women)?[65]

The tensions and contradictions within the nineteenth-century concept of marriage can be seen most vividly in the troubled marriage.[66] Not unexpectedly, the troubled marriage was characterized by expressions of insecurity about love and frustrations over dependency. A sense of their inability to recover the emotions that brought them together haunted the exchanges of conflicted couples such as Dorothea and Charles Lummis. Dorothea Lummis, whose marriage to Charles eventually ended in divorce, struggled for many years to save her marriage. Constantly straining to understand the nature of her marital problems and how to cure them, she illuminates many of the cultural contradictions in nineteenth-century marriage.

In the midst of her medical school training in Boston, Dorothea's husband accepted a newspaper job in her hometown of Chillicothe, Ohio. She did not accompany him to Ohio, choosing instead to remain in Boston to finish her degree. This was a painful decision for them both, and Dorothea experienced a range of emotions from sorrow to regret and self-recrimination in the weeks that followed. She confessed, "Ah, sweet boy I miss you so. I miss even your cross looks, and I had almost rather be awfully unhappy with you than have larks here. How did Howells know that 'women loved more, when they had been made to suffer.' I wonder but its true. I guess its sort 'o stupid and dog-like to, but its impossible to be otherwise, but I guess men don't, do they. They just get mad and go off. Which is much more sensible."[67] Whatever truth there was to Dorothea's comparison of men and women, she suffered greatly. Still hurting from past emotional wounds, Dorothea recounted one of the most bitter quarrels of their marriage: "I think that after all, faulty as I plainly am, you hardly did me justice that night alas! for me memorable—when you . . . told me that 'you didn't think women amounted to much except as an appendage—' Things like that have such a wicked life , , , I dont believe you stopped to think what a bitter blow that was to my sweetest hopes and most loyal and tender wishes or you would have spared me, would you dear? for after all wives and husbands have faults and will need forbearance often."[68]

After separating, Dorothea was determined to establish a new equilibrium in her marriage. She vowed to change some of her "horrid ways."[69] In her zeal for reform she even partially recanted her decision to stay in medical school: "Had I known then, what it would cost us, or even that you wanted me with you, I would have given it up . . . ."[70] Separated by almost half a continent, she struggled to patch the relationship together. Wary of criticism at this juncture, she cautioned, "It seems to me that allowing either inward or outward criticism to grow into habit no matter how much cause is given—is fatal to all joy or happiness and makes life a burden and ambition worthless."[71] But Dorothea was not adverse to self-criticism. Desperate to save her marriage, and beset by guilt and self-recrimination, she accused herself of failing to be a "true" wife. "It isn't my faith in you that is gone, but in my own power to be a support and inspirer to you, tho' I would truly live or die to prove capable of it. O, I love you my sweet, I want you, I *do* think you are good and I do want to be your true wife in every way. It is the one thing I care most for and the sorrow of all my days and wakeful nights is, that I can't be good enough to keep your admiration and respect. Let me have a little to keep now, wont you and see if I cant win more and more . . . ."[72]

Within the contradictions and paradoxes of nineteenth-century conceptions of marriage, almost the only area thought to be under individual control was the wifely or husbandly role performance. When the problems in a marital relationship could be located in the performance of a spousal role, people like Dorothea believed that a remedy might be found. Thus, she blamed their marital difficulties on her failure to live up to wifely ideals. Apparently this conception of individual responsibility in marriage was widely held. In a sample of divorces in the last half of the nineteenth century, sex-role obligations provided the most powerful and oft-used measuring stick of public marital failure.[73]

While role performance in a marriage could be defined, measured, mandated, and thus theoretically changed, romantic love was more intractable. The "romantic" definition of love between spouses, unlike earlier views of marital love, was antithetical to any sense of love as an act of will. Love was defined as something beyond individual control. What was theoretically controllable in middle-class conceptions of marital relationships was role behavior. Thus, Victorian ideas of marriage contained a masked but inherent conflict between the spontaneity, irrationality, deeply felt, and ultimately ungovernable conception of romantic love and the prescribed duties of husband and wife which were presumed to be rational goals within the grasp of the average man and woman.[74]

Dorothea grappled with this tension in her own troubled marriage. On the one hand, she vowed to be a better wife. On the other hand, she suggested divorce if Charles no longer loved her, because she could think of no remedy for lovelessness. Dorothea expressed frustration and helplessness in the face of Charles's fading love. Writing of their past romance, she lamented, "Too bad isn't it, that that light that never was on sea or land, must fade into the light of common day. If it was always so with everybody, it would not be so hard, but some people *do* keep it until real old age, and I *hate it* that we cant, if anybody does, for I honestly believe [we] were as happy lovers as ever walked the earth, at least I was."[75] She hoped the love that remained would be their "stay and with its help maybe we can after all work out our own married salvation."[76] But with all her determination and strength of will, she believed that she could not command love where it did not exist. She finally confessed her underlying apprehension: that if her husband's love had disappeared, she would be defenseless: "I can not live in such fond, close and sacred relation, unless I am both loved and respected and if I fail—I fail. If I lost you life would lose its sweetness, but it would not be so ignoble as to cling to you, when your . . . heart and love was not mine . . . ."[77] She clung to the hope that "there is way down under all the work and hurry and worry, still a little bit of romance left . . . ."[78]

Dorothea Lummis yearned to create a stable marriage, but she was willing to dissolve her union because of the very purity of her belief in love. "We married—and rightly enough—because we loved, and I have never regretted it, but I did not expect when I entered for a college course to be separated from you, nor did you. When it seemed best for you to go west, I didn't think of going but I would have done so had I known you wanted me. But we neither of [us] then could foresee the consequences. But were you a bit sour or anything when you wrote me that 'you could get along just as well without me' and that 'it would always irritate you to have me around among your things,' case if it is really so, how can you look forward with any pleasure to spending so much of your life with me as to have a home and children. We can't get back any of our lost happiness unless we both really want to live together can we? You see in some way, though I too have been separated from you, the romance hasnt died out to any extent, and so I am afraid I will be a pretty big bore to you—I guess I could give you up if you wished it, but I am afraid I couldnt go on living with you daily and find that I bored and irritated you."[79]

Dorothea was willing to divorce Charles, not because she no longer loved him, nor even because she had unreasonable expectations of marriage itself. She was one who wrote of marriage as a state of comfort and contentment, not bliss. What she found intolerable was the idea that after his love had died for her, the marriage would continue out of his sense of social duty. In other eras, and other times, and for other individuals in this period, social obligations, economic need, and moral duty would have been a perfectly acceptable, even desirable reason to sustain a marital relationship. Dorothea was repelled because she was committed to the unvarnished logic of romantic love as a completely voluntary relation. She was willing to extend that logic to the obligations of husband and wife within the institution itself, perhaps because her education and temperament gave her a measure of economic independence unusual for most nineteenth-century women.

During the mid-Victorian period, roughly 1830–70, the idea that a voluntary union might be voluntarily undone was inhibited by the still powerful sense of the marital relationship as a set of duties externally imposed upon husband and wife. Later in the century, the idea of marriage as a continuous act of voluntarism became more thinkable. Nonetheless, divorce in late nineteenth-century courts was still argued on the basis of a failure to fulfill sex-role obligations and therefore failure to meet the required level of social performance within marriage. The grounds of voluntarism that Dorothea Lummis expressed, "I dont pine to give you up, but I shall never hang on to you when it gets to be a mere outside duty to you," await the twentieth century to be fully released.[80]

Dorothea's purity of commitment to the voluntaristic and uncontrollable implications of romantic love within marriage was unusual. The logic of the belief in romantic love was readily ameliorated by competing conceptions of spousal obligations. When it came to the problems of marital commitment and dissolution, dominant elements of nineteenth-century culture struggled to disguise and even perhaps repress the essentially voluntary and individualistic elements in the construct of romantic love.[81]

While there is no evidence that divorce was considered in the next marriage, the problems of the voluntaristic conceptions of romantic love operating within the duty-bound context of marriage were again apparent. Charles and Emma Watts's relationship was deeply conflicted. Though assuming the appearance of indifference, anger surfaced on the wife's side to give mute testimony to the emotional trauma of marital failure. The Wattses' marriage ended in his death, but not before he had fled to the Yukon in search of fortune and also perhaps away from his unhappy wife. The strain in this relationship was more apparent by what was not said: Charles constantly complained of his wife's negligence in letter writing, and in fact her silence appeared determinedly hostile.

In the spring of 1898, after managing a makeshift lodge, restaurant, stable, store, and storage space on Bennett Lake in British Columbia, Charles moved to Dawson, the heart of the Yukon Territory, to try to strike it rich quick. The Klondike gold rush was on in earnest and Charles reported that while 10,000 men had already arrived, not 1,000 would earn enough to buy a ticket home. He observed cynically, "It is the greatest over estimated and over advertised country on the face of the earth. The result of the Alaska boom from its inception to the present time will cause more hardships, more heart-aches, more destitution and more financial embarrassment . . . and time will prove it."[82]

In his next extant letter, a homesick and weary Charles vowed to return home, no matter how he fared. "I cannot stand being away so long," he wrote. "This is the only life we have to live, and if we live let it be together."[83] Restless and dissatisfied, he moved to Eldorado City (20 miles from Dawson), where he tried his hand at running a hotel, while speculating on the side in mining claims. He wrote on November 15 that it was 40 below zero, but "At zero we run about out doors, coats unbuttoned and bare-handed."[84] Charles commented frequently on the weather. He once observed that in order to prevent his ink from freezing, he had to keep "the ink bottle sitting on the stove so I can keep writing."[85]

Past transgressions probably haunted this relationship, because Charles insisted, quite unexpectedly, "I have been here 5 *months* and have never been out of the house of an evening after supper but *twice* in that time except to church or Literary Club which meets each Saturday evening. If I can only keep that up after I return home again won't I be a daisy?"[86] He and his wife apparently quarrelled frequently over his evening activities away from home, and he may have been offering her a symbolic olive branch. There is no evidence, however, that his wife responded to his promise of reformation or was even writing him at the time.

In May, Charles communicated a revealing message to his wife through a letter to his daughter. Apparently some hint of scandalous behavior on his part had reached the home front. He protested: "Mr. Rawlings knows nothing about me that he is not welcome to tell the world. My life since coming to Alaska has been an open book. Have eaten *every meal* at a table in a big dining room and *slept every night* in a room where there were *10* beds. Have lived in one house since August lst last and never spent but *2 evenings* in that time away from home. So you see I am willing *all* should learn and know if they think there is a Negro in the fence."[87] Though never plainly stated, his emphasis on the bedroom and his whereabouts at night probably meant that the rumor involved his relationships with other women.

Rumors of extramarital affairs were not the only sources of tension in this marriage. His wife took in laundry to supplement the family income. Charles was skeptical about the economic necessity of his wife's laundry work, but perhaps this was a rationalization of his own failure to provide. "I guess I am mistaken," he told his daughter, "but it does seem to be as though you could get along with what I send you, what the 'Oregonian' sends you [he wrote newspaper stories for them as a freelance correspondent], and milking without Mama having to *wash—wash*—If she had a thousand a year for pin money she would still hunt up some kids' old duds to 'do up.' Some of these days remember she wont *have* to do it unless she insists."[88]

Charles returned home to his wife and daughter in the summer of 1899. His wife shunned him almost completely for he told his daughter disconsolately in early September, as he prepared to leave for the Yukon again: "Life has had many bitter disappointments and as I leave you for so far away, I will try and profit over them. I dont know when I will see you again, and how I would like to kiss you just once more, I feel hurt and can hardly see to write. . . . My visit out during the summer was a cruel disappointment and only that I longed to see you—I would be sorry that I went. I killed the time as best I could—not from choice but just to kill."[89] For the remainder of the year Charles corresponded only with his daughter, although he occasionally expressed tenderness toward his wife, as in his direction to "Kiss Mamma."[90]

In spite of their difficulties, Charles communicated both respect and admiration for his wife through their daughter. Such sentiments probably reflected some degree of sincerity along with a calculated attempt to return to his wife's good graces. In an especially nostalgic and reflective mood, Charles wrote his daughter that while he was perusing his scrapbook, he read among other notices "one of a marriage that was solemized by Rev. S.C. Adams 20 years ago day after tomorrow. Mama has been all these long years a dear, true, noble good wife. Papa perhaps cannot say the same. She never had but one fault—that of a natural scold, but perhaps I was or (we were) in a measure to blame for that. Let us both overlook it anyway and perhaps then she will see life in a broader channel. Then when I come home again for a visit she will ferget to scold, and will try and make my visit a pleasant one. A better woman never lived if she would only wake up some morning and forget that 1 fault."[91] Though accepting some blame for their troubled marriage, Charles suggests that his wife's nagging was the major stumbling block to potential reconciliation. However unlikely a solution, Charles Watts died before his theory could be tested.

The Wattses' marriage was deeply troubled. Charles absented himself from home, apparently as his own makeshift solution to the extreme tensions between them. In response, Emma Watts cloaked herself in a mantle of silence. Her husband dispatched a stream of correspondence with little or no encouragement from her. Returning home in hopes of a summer reconciliation, he encountered a wall of passive resistance as she pointedly ignored him. Charles, at least, had not completely given up on their marriage, but Emma's anger was so intense that after his death, according to family tradition, "grandpa's name was 'verboten' around grandma."[92]

Serious deficiencies in Charles's husbandly role performance, especially as a provider and domestic companion, may account for his wife's anger and hostility. But these problems assume an added dimension because of the difficulty of reviving romantic love. This couple had no idea how to repair the emotional damage in their relationship. Charles could only suggest that his wife stop nagging him—a behavior she might be able to control, but a solution that was hardly responsive to her anger and emotional distancing.

The dilemma of the troubled relationship was the absence of any cultural means to restore or reconstitute splintered love; only role performance—stop nagging or be a better provider—was governable and thus capable of rational direction in the conceptions of nineteenth-century marriage. The strain of this belief system was that though love had become the central cultural rationale for marriage, there was no basis in the conception of love itself for "repairing" or "fixing" it. Romantic love remained the "unmoved mover" in nineteenth-century Americans' lexicon of male-female relationships.

🐾 Elkanah and Mary Walker's marriage was conflicted from the outset. Mary was anxious and uneasy about her husband's affections from the moment she agreed to marry him two days after they had met.[93] On their honeymoon trip across the continent to take up their missionary duties in Oregon, Mary complained to her diary: "Rested well on my ground bedstead I should fll [feel] mh [much] btr [better] if W. wd [would] only treat me with more cordiality. It is so hd [hard] to pls [please] him. I almost despair of ever being able. If I stir it is forwardness if I am still it is inactivity. I keep trying to please but sometimes feel as if it is no use. I am almost certain that more is expected of me than can be had of one woman. I feel that if I have strength to do anything it must come from God."[94] Mary's response to her marriage was uneven and often perplexed. Some moments found her positive and hopeful in that first year, as when she remarked, "Find I am becoming every day more fondly attached to my husband. Indeed he seems every day to become increasingly kind and I am more and more confident of my ability to please him and make him happy . . . ."[95]

More often, however, Mary was discouraged and unhappy. Though she crossed out the following diary entry, the feelings she expressed were not so easily revised. "He reproached again this morning most severely on account of some ingentilities of which I had indeed been guilty. He almost said had he not supposed me more accomplished I had not been his wife. I am almost in despair and without hope of his ever being pleased or satisfied with [me]. I do not know what course to pursue. . . . I never intended to be the wife of a man that did not love and respect me from his heart and not from a stern sense of duty and this I fear yes I have much reason to *fear* is all that secures me that share of kindness that I receive. I know that for the most part I am kindly treated. But I am but too often treated as tho my convenience and wishes were not to be regarded. The thought often occurs I am glad my friends do not witness and can never know."[96] Mary had expectations for marriage that went beyond mere kindness and duty on her husband's part. Furthermore, these expectations must have been the norm in her social circle, for she was relieved that her friends could not witness what she interpreted as her marital failure and humiliation.

Mary held a companionate ideal of marriage, one in which a husband was supposed to regard the wishes of his wife and consider her convenience and well-being. She was deeply disappointed that she received only dutiful kindness from her spouse. "I am tempted to exclaim wo is me that I am a wife," she continued, "Better have lived and died a miserable old maid with no one to share . . . my misfortune. But it is too late. O may he who in his providence has suffered me to become a wife bless me in that relation and enable me to discharge every duty with Christian discretion and propriety."[97] Mary turned to religious solace, but was quickly drawn back to the source of her anguish: "Disappointed not because he is not as good as I anticipated but because I have not gained that place in his heart that I fondly expected and which I think a wife ought to possess."[98] Though the motives for their union were not romantic ones, Mary felt cheated by the absence of romantic attention, tenderness, and emotional empathy. That evening, she shared these feelings with her husband who laughed, reassured her of his love, and apologized for his unintentionally harsh criticism. She was only partially convinced, however, and commented with half-hearted enthusiasm: "I think I will try to feel better."[99]

Soon after, Mary decided to try an experiment to improve her communication with Elkanah. She wrote letters to her husband in her diary which she intended him to read. Hoping to open her heart to him by this device, she pushed to achieve a more understanding relationship. She addressed him in her diary: "I find it in vain to expect my Journal will escape your eye and indeed why should I wish to have it? Certainly my mind knows no seeter solace than the privilege of unbosoming itself to you. It frequently happens that when I think of much I wish to say to you, you are either so much fatigued, so drowsy or so busy that I find no convenient opportunity till what I would have said is forgotten. I have therefore determined to address my journal to you. I shall at all times address you with the unrestrained freedom of a fond and confiding wife. When therefore you have leisure and inclination to know my heart you may here find it ready for converse."[100]

During this experiment, Mary shared her feelings of spiritual unworthiness and sought her husband's help in achieving a more devout frame of mind.[101] In other entries, she expressed anxiety over her husband's health;

she worried over whether she was too indulgent a mother; at several points she described her progress in learning the Indian language. But she confessed toward the end of her experiment that she was not always successful at revealing her feelings to him, even on paper. "I proposed when I commenced this journal to let you know my heart. But I think I do not always do it."[102] By the end of the journal-book that she began "My Dear Husband," she had abandoned the plan of communicating to her husband through her diary. She wrote disconsolately: "Feel lonely tonight. Wish husband would converse about something. Fear we are not that society for one another which men and wife ought to be. Wish I could find what it is prevents our being."[103]

Mary and Elkanah may have eventually reached Mary's goal: "that society for one another which men and wife ought to be." In 1841 Elkanah hinted that he and Mary had finally created some loving rapprochement: "I often wished during the day I had started for home as I want to see my wife and children very much. . . . No man has a wife more deserving of the warmest affection then I have. She is gaining very fast . . . I love her more and more daily and I hope she will continue to improve as much in her manner as she has of late and will continue to make the same exertions to please me. If she does I shall consider myself one of the happiest of men."[104] Whatever new "exertions" Mary was making to please her husband, she too was happy with the results: "The assurance I receive of the love and esteem of my dear husband is to me a source of gratitude and consolation. May we both grow more and more deserving of mutual regard."[105] It may be that Mary and Elkanah developed the romantic love she had missed earlier in their relationship. Mary stopped complaining in her diary of his critical and unresponsive attitude toward her. Six years later she even accused herself of elevating love for her husband above her love for God.[106]

This marriage began without romantic love. But after several years of tension and struggle, the Walkers probably achieved some measure of emotional empathy. Though their engagement was based on very practical grounds, the ideals of nineteenth-century marriage haunted the early years of this union. Mary expected more than a mere dutiful role performance from her husband. She despaired in the first years of their marriage, not because Elkanah neglected his external role obligations, but because the relationship of husband and wife meant something more emotional to her. In nineteenth-century views of marriage, external role duty was supposed to be outweighed by love.

In another troubled marriage, emotional progress took the opposite turn. The marriage of James and Mary Hague developed into a cold formality and contained hostility, after beginning with both warmth and apparent passion. Three years of marriage had not yet dimmed James's enthusiasm: "It does me good to get such warm sweetheart letters from my girl, just as loving and nice as three or four years ago when she didn't know me as well as now and hadn't found out what an irritable cross-grained old fellow I can be sometimes. I am glad you love me as much as you did then in spite of better acquaintance—And, dear girl, I love you as much as ever too; and more if possible; and this morning when I got your letter I just wanted to take you in my arms and put my face against yours and give you a sweet old fashioned hug." [107]

Nonetheless, unresolved tensions in their courtship were an early source of friction in their marriage. Married in his mid-thirties, significantly later than the nineteenth-century male norm, James already had an established identity as a mining engineer. An experienced traveler, he was also adept at living the life of an independent man-about-town. For these or other reasons, he was never as emotionally dependent as Mary desired. After three years of marriage, she did not yet feel emotionally settled in her relationship. Still testing, Mary's wifely expressions of insecurity drew a reassuring response from him: "You think I don't miss you because I am not complaining all the time of your absence. Well, I don't complain—partly because you are having the good time at home that you have been so long looking forward too, and partly because, thanks to my friends and acquantances here, my time passes pleasantly—but if you think for that reason, that I don't prize my wife and baby and home you dont know me much. . . . There's nothing in the world I think of, or love more, than your own dear self and Marian; There is nothing half so essential to my happiness as you and your company and your love; and to have a happy home of our own is my chief desire." [108]

But between 1875 and 1892 a deepening chill settled on their relationship. The correspondence they exchanged in 1892, she in Europe, he in America, was formal, cold, and at times contained a barely concealed rancor. At one point he forbade his family to leave Europe after Mary expressed her wish to return to New York. She commented that she did not cry, "because I have grown tough with years and experience . . . ." [109] She told him dryly, however, that their daughter "went down to the bottom." [110] Eventually moved by his children's pleas, James granted Mary permission to return to America. Perhaps sensing that she could drive a harder bargain, Mary did not immediately grasp the opportunity to return home. She insisted that her return was conditional upon being able to move into a house as well as promise of a more enthusiastic reception: "I hope you have decided in this way with cheerfulness, and that we may feel sure of a welcome when we get back." [111] She remained unconvinced that her husband would be warmly inclined to her when they were re-united. "Of all things I do not wish to feel adrift and uncared for, as I did the first year after I went back from Europe in 1879. I might as well stay here the rest of my life as to go through with such a year again. I generally had a quiet cry after I had got to bed then, when no one would see or hear me—It was a weary forlorn experience."[112] This incident ap-peared to be more typical than exceptional in their later relationship.

Though Mary won James's consent, she correctly perceived that he was not wholeheartedly in favor of her return. He revealed his hesitation when he suggested to his wife that their daughter "would plunge into wild frivolities if she returned to New York."[113] In defending her daugh-ter, Mary indirectly commented on her own life by rejoining that a woman had only two or three years of fun and enjoyment when she was "fresh and young and full of spirits—After that her little day is over, and she must think of other things, unless she is very wealthy—which Marian has not the slightest prospect of being. You judge from a man's standpoint, for his season can go on for years, and a girls is so soon over."[114] Mary detected considerable resistance to her return: "The way in which you tell us we can come home reminds me of the ways of childhood. 'Have your own way and make me hate you'—is between the lines—though I will not make any more comments—but the disappointment which came with reading them was most grievous."[115]

Strain and unconcealed disappointment permeated this relationship. Mary wrote James she had been uncomfortably cold for several weeks. "I don't know if it is because of the chilly tone of your letters that I am so cold—I think it would do my heart good if you could express a little pleasure at our returning home. As it is I am constantly tormented, first with the feeling that you do not want us to come, and then with the distress of the children that all their plans of life must be broken up and changed, if we remain here."[116] She yearned almost wistfully for him to express an "affectionate desire" to live with her.[117] Though Mary Hague eventually moved into her house and received a more enthusiastic en-dorsement of her return to America, she remained dissatisfed with the emotional distance in their marriage.[118]

Returning to America in the fall of 1892, Mary and her children were back in Europe in the fall of 1894. Reading between the lines, it appears as if James was anxious to hurry them off. Mary recognized his ploy: "I was entirely surprised when I received your letter of the twentieth, to find you had not then started for California and had set no time for your departure. From what you said all the time, I supposed of course you would be going within a week, certainly . . . ." She remarked caustically:

"I thought you were in such a hurry to go that you could hardly wait to see us off."[119] These European vignettes vividly portray the Hagues' estrangement.

An episode preceding her death four years later also highlights the difficulties in this relationship. Mary Hague reported that she had suffered "years of misery" and characterized herself as an exceptionally weak woman.[120] In 1898 her doctor, a woman, recommended that she have a hysterectomy. James gave his consent, after earlier refusing permission for a similar operation.[121] Mary was joyous: "If only this could have been done just after Will was born, what years of misery I should have been saved; but I didnt know such things could be done then, or even ten years ago."[122] Buoyant and confident, she predicted in her last letter to him before the operation that she would soon be "light of foot and of heart, without this depressing weakness to weigh like lead on my spirits. I have never been really myself since 1871."[123] This was a surprising calculation, especially so since Mary was engaged in 1871. Speaking twenty-seven years later, she chose her engagement to mark the beginning of her downward spiral of ill-health. In fact, during her courtship Mary was often ill, and used her various ailments (whatever their origin) to dramatize or symbolize her courtship anxiety and premarital doubt over James's love and commitment to her.

Issues of control and autonomy were primary gradients in the Hagues' troubled marriage. Mary's marital frustrations underlined her sense of economic dependence and the narrow range of her personal control. The day she entered the hospital for her hysterectomy she wrote, "I regret the great expense of the performance; if only my own money had not been taken away, I should not trouble you about it."[124] Mary's money—$3,000—was lost in an El Paso bank investment many years earlier.[125] The Hagues had accumulated assets totaling $100,000 at the time she wrote this letter.[126] Nonetheless, losing $3,000 eight years earlier was still psychologically galling to Mary Hague. Though she had a great deal of money at her disposal, this small inheritance loomed as a painful symbol of her lost autonomy.

The issue of female autonomy was highlighted in a troubled nineteenth-century marriage—for female dependency was then a burden, and an irritation to both husband and wife. Husbands were supposed to provide a socially acceptable level of economic support for wives and children. Middle-class women expected (though did not always receive) the economic support which was an undisputed husbandly duty in nineteenth-century American culture. In terms of the sex-role notions of the day, neither spouse expected women to be economically autonomous.[127] Yet in a troubled marriage, women's economic vulnerability could become a central tension.

In a troubled marriage, emotional vulnerability was also at issue. Married couples were usually distressed or dissatisfied by the loss of romantic love. There was, however, little either partner could do to control the other's feelings if they accepted the dominant cultural anatomy of love. The mystique of love as an irrational force of mysterious comings and goings only contributed to the feelings of dependence that may have been especially acute for wives in troubled marriages. Dying soon after her operation, Mary Hague expressed to the end of her life resentment over her financial and emotional vulnerability.

🍒 A troubled marriage could begin with an intense romantic bond. Harriet Russell Strong's separation from her husband a little more than two years after their marriage evoked this impassioned cry: "How can I live three *months*—long *long* months—without you my precious one, when this week has been a year? I dare not look ahead, my heart almost fails me if I do. Your loving words of cheer and comfort 'keep up good heart,' come right home to me, and I pray God for your dear sake to strengthen me. It is as if a part of myself in reality had gone and only a miserable fraction left!" [128] Both husband and wife yearned for the other with passionate intensity. Charles Strong declared: "It seems every moment that I never loved you half so much before. I resolve many times a day never to be again from you for any length of time . . . ." [129]

Their love continued to burn brightly, for on their first anniversary, Harriet exclaimed to her absent husband: "how happy your wife is and after this first year of her married life she loves you so *much* as she did when you first called her your wife, which *love* has been increasing each hour and you continually grow nearer and dearer, God bless you my husband: What more can I say? How fervently I thank Him for giving me so dear and good a husband!" She closed this letter, "may God always keep your heart full of love for your devoted 'year old' wife." [130] Unfortunately, Harriet's prayer was prescient, for sometime in the next nine years the Strongs' relationship soured.

Dating the origin of the Strongs' disaffection is impossible but a significant break occurred in 1874 when Harriet left the house Charles built for his family at the Kernville, California, mine where he was superintendent. [131] From the time she left in 1874 until his death in 1883, the Strongs never lived together on any permanent basis. There is no evidence that divorce was seriously considered by either spouse. Separation appeared to be the Strongs' "solution" to their marital tensions.

Few of Harriet's marital letters to Charles have survived. Charles burned the letters in his possession before he killed himself just days before his twentieth wedding anniversary, and Harriet may have destroyed others herself. Some, however, remain. One of Harriet's most revealing letters may not even have been mailed.[132] Whether or not Charles read this letter, Harriet clearly exposes her version of their marital story. Alongside a narrative of her matrimonial past, Harriet released a torrential flood of feeling, exposing the deepest recesses of her emotional response to Charles: "Every thing I do and say to you or write you is more or less hampered by what I know you think of me—as the flowers open out and give fresh perfume and beauty in the sun—so does the heart warm and expand to love—and nobody's more than mine—if I have any left or ever had   I can almost hear you think—) I have always dreamed and wished for and prayed for, the wealth of human love wh- belongs to a good woman—It was mine for a few months only—why no longer—I shall never know in this world—I have always believed it would be mine again—until last May when you came—I saw and felt—never—never—and my heart died— and afterwards lived a dying three months—wh- more than all else took me nearly to death's door. God spare my worst enemy from a life without hope—God has given me both heart and brain—and all—all—I gave to you—A great pity—marriage without love is better—and let it come afterwards. Through revilings—through cursings by those who wanted you—I have loved on and said not one word to mortal—thank God! I shall never say one word unless my heart gets beyond my will—and breaks—I might then—I ask nothing—you have told me too many times you have no love to give me—never again ask for what should be mine— So live on and try to do for my children—but I cannot do my utmost for am but half a being without the sustaining satisfying love of my husband. . . ."[133]

After thirteen years of marriage this wife was desolated by her husband's coldness and tormented by the emotional emptiness that yawned between them. At the core of Harriet's despair was her complete helplessness to recover Charles's love. To her mind, love came inexplicably and left the same way. She was thus rendered vulnerable by the very passive conception of love as a complete mystery so dominant in Victorian culture. During courtship, basking in the intense glow of their mutual regard, she had celebrated love's mystery. "How glad I am *love* is indescribable, I would dislike to think any thing that *could be fathomed,* could effect *you* and *me* as it does. Completely revolutionizing our whole actions, feelings habits etc. Yes. I am satisfied that we don't know what it *is*. Yet are made so perfectly happy by it."[134] What she could not fathom before marriage was that *not* knowing "what it *is*" would eventually make her perfectly miserable. In the face of her husband's insistence that he had no love to give her, she found to her despair that she must acquiesce.

Her unhappiness was especially galling because love slipped away inexplicably. Underlining her own lack of control was the fact that feelings of love lived on in some marriages—"Every happy couple—is a torture to me—I try to believe no one is loved and there is no such thing as love—but my own heart makes that a falsehood—."[135] She ached for her husband's affection, passionately yearned for love's return, but finally had no way to reclaim it. The very concept of love that she and other Victorians subscribed to precluded rational choice: "All I can wish or ask of you at this late day—is to remember how my heart aches from the past wounds—and I pray and beg you spare me as many heart pangs as is possible; that is all—some day you will know what you have thrown away and discarded and treated as worthless—To think *I* who have been loved always should live an unloved wife!"[136]

It was the will-lessness imputed to the experience of romantic love that made it so troublesome as the basis of marriage. For Victorians, who expected their mates to love them for a lifetime, marriages in which love vanished left them in a peculiar bind. They lived in a culture which was still reluctant to legitimate divorce, yet no longer saw marriage wholly in terms of duty and spousal role obligations. Harriet pined: "Oh for a home— where 'Joy is duty and love is law.' "[137] She longed to be surrounded by love so unquestioned, it might seem like law. Perhaps she coveted the power to demand obedience to a law of love. A law is at least potentially enforceable, so is a duty. They are injunctions to obey which are usually recognized as binding. By contrast romantic love was conceived as lawless by middle-class Victorians. Ironically, love was *not* bound by the act of marriage in the romantic world view. Though she wished to bind his love to her, she recognized her utter powerlessness.

But another conception of love in the American past once required love as a marital duty. In Puritan New England, love was seen as a rational and predictable emotion, not an involuntary feeling.[138] By entering into the marriage contract itself, seventeenth-century New Englanders agreed to exercise the will to love. After marriage, a partner was obligated to love a spouse because love was an intentional emotion that could be directed by choice.

Victorian marital advisers who defended the necessity of indissoluble marriages kept this Puritan tradition alive, conceiving of love as an obligation. Like the Puritan clergy generations earlier, they claimed that marital happiness was generally within the partners' capability to determine.[139]

It is no surprise that these advisers, fearful of marital instability at a time of increasing divorce, emphasized love's reasonableness.[140] For like their Puritan forebears, they recognized that only if love was a reasonable feeling could it also be demanded as an obligation of the marriage contract.

These traditional marital advisers attempted to dissolve one of the deepest tensions in nineteenth-century marriage by eliminating the romantic opposition of love and duty. Subsuming love under the category of duty, they tried to harness love to the power and virtue of self-discipline— but they failed. Though traditional marital advisers might urge the contrary, love was not conceived as a marital duty by the Victorian middle classes.

Charles appears to be a dutiful husband without (from Harriet's perspective) being a loving one. He worried constantly about his ability to fulfill his role as provider in the insecure position of manager of precious metal mines, an often risky venture capital business. Dedicated, as much as one can be at a distance, to fulfilling the role demands of husband and father, he defined "my life's duties—which are to provide for the five blessed Beings who are of right looking to me for food raiment and such comforts as they need and deserve."[141] Moreover, he intently followed the health of his children and wife, providing money for extended vacations in Santa Cruz and other health resorts. He sent his family gifts, spared little expense to satisfy their material desires, and proudly wrote his family regularly.[142] In fact, he sometimes sounded like a model husband. He wrote that he would bear their separation cheerfully "with the hope that it will result in some happiness to us and our blessed little dependent lovable loved and loving souls. . . . When the weather lets up and can get some business started then it will not seem so wretched and forlorn as now. If you can only get along comfortably and none of you get down sick we will patiently wait the developments of another year—and trust for it to bring us something akin to our hearts longings."[143] Harriet's deep disappointment and frustration in her marriage focused on his inability to love her, not on any failure on his part to meet his masculine role obligations of faithful provider, concerned parent, and mindful husband. The problem in marriages such as the Strongs' was precisely the distance that gradually loomed between love and duty.

The line between duty or obligation and love could sometimes be blurred. For example, Harriet complained that Charles forgot both her birthday and his daughter's. While her complaint may be based on more than an expectation of duty, his response was phrased exclusively in those terms. Feeling the sting, he denied that his actions reflected any want of concern and thought for them, but only originated in overwork.[144] The implication was that the lack of leisure in his family's service was the cause of his neglect and indeed proved his constant attention to their needs.

Unhappy marriages often present "two versions of reality rather than two people in conflict," one literary biographer wisely concluded. Her idea is that marriage is a "subjectivist fiction with two points of view often deeply in conflict, sometimes fortuitously congruent."[145] Even though hundreds of Charles's letters have been preserved, his version of their marital reality is more difficult to ascertain than Harriet's and must be read by indirection. Though sincere, charges of romantic rejection obviously reflect Harriet's version of reality. He sometimes seemed more content with a marriage bound by obligations, cheerfully emphasizing his responsibilities as provider and parent. But Charles also manifested a quiet alienation. Damning by faint praise, he described ten years of marriage as a domestic career "Interspersed with some pleasant days and many anxious ones and so I expect will continue to the end. . . ." Unwilling to bow to a single romantic convention in a ten-year anniversary greeting, Charles signed off in the parental role as "Ever Aff Papa."[146]

Ultimately, however, his and her versions of reality coincided in their continued physical separation. Charles's intermittent talk of reuniting his family and complaints about their separate abodes must be contrasted to his actual behavior. He was more intent upon reunion in words than deeds. During June and July of 1877 he suggested that his wife and family join him in Lake Tahoe.[147] After Harriet accepted his invitation, he retreated, claiming that the expense, time limitations, and work responsibility now made it impossible.[148]

He followed a similar pattern earlier in the year, suggesting they might reunite in Galena, if the mine he was superintending proved to be a reliable source of income.[149] Again when Harriet took his proposal seriously, he rebuffed her. "Yours of Sunday has just come—as to your coming up to see me a *day or two* is a little more than I could well stand. All things considered you had best make your arrangements to go to Santa Cruz as soon as it will do—."[150] A bad cold and other long-standing nervous disorders may have played a part in his rather tactless rejection. But, for whatever reason, if even a "day or two" was more of Harriet than Charles could "well stand," her version of a marriage of thwarted love and dashed hopes would seem to be reinforced. Soon after dismissing Harriet's visit, he again openly yearned to have his family with him and twenty days later he proffered another invitation, but his sincerity is at least questionable since at his behest his wife had made other plans.[151] At a safe distance Charles longed for his family but demonstrated conflict about an actual visit.

Charles kept plans alive for a permanent reunion but was no doubt aware that the prospects were dim: "truth is we have no business to live so apart. If it is absolutely necessary that I should live here then here should *all mine* be with me and I very much wish now that I had carried out my plan I made last Spring of building a little comfortable place so that you could have come say about the first of September and stay as I stay and go when we would all go together. But afterthoughts seldom help much, but when another Spring comes all our plans if we are so we can command any plans must be for *union,* and stop this way *of doing—*."[152] The following spring found them no closer to living together, however, and the only barrier would seem to have been of their own making.[153]

After a Christmas visit from Charles in 1877, Harriet commented on the condition of American marriage in an informal essay preserved as her jottings on the back of two of his letters. She bitterly indicted the lack of opportunity and intellectual encouragement of married women. Obviously reflecting upon her own situation, she characterized men and women as separated by a vast chasm. She claimed that women's lives were severely circumscribed by family cares, a wife's only stimulation being occasional social evenings out with her husband.[154] Her thoughts on marriage were uncompromisingly negative: "I do assert that marriage as a rule is the grave of every high thought and noble motive in life—not because the women of Am. have no ambition or mental capacity for progress, but because they do not have the sympathy and encourgagements of their husbands in efforts to improve their minds—."[155] She described a wife who devoted herself to her daughters and after they had matured, the woman, an embarrassment to her family, was left at home while father and daughters took advantage of their social opportunities. Harriet commented that "she is as much shut out from the pleasures . . . of life as if in a Turkish Harem."[156] Though she recognized the "many noble souls that this kind of life cannot keep down and dwarf" yet she also characterized these women as the chosen few and declared dejectedly, "The average Am. women amts to very little."[157] Harriet most likely counted herself among them.

She felt constricted, burdened, neglected, almost vanquished in a marital relationship that left her dissatisfied on practically every level. She blamed a woman's spousal role for her plight, and a canon of domesticity that took all of a woman's mental and physical energy and left her with ashes. Harriet's complete condemnation of woman's role within marriage was extreme. Victorian women expressed role dissatisfactions but usually in less universal terms. In fact, there was a mixed pattern of disquiet on the part of both sexes.

Men might also complain about their marital burdens or discontents.

For example, in the summer of 1878 Charles confronted a financial crisis. He believed the mine that employed him had failed; he was not able to meet his payroll and his men thought that he had betrayed them.[158] Still at his post but facing a bleak prospect of saving the enterprise, unsure of his future, perhaps not even drawing a salary, Charles felt the pinch of Harriet's dependency on him. "When I look back and think of the amt. of gold money I have *worked for* in the last 6 years over $30,000 and where it has gone it takes all the hope out of me about the future. . . . but to work for it by hard days works often under the most trying circumstances and then spend it all and keep in debt besides—Its little good looking back but as it would seem theres nothing to look to for the future we must look somewhere—."[159] In effect Charles blamed Harriet for spending *his* hard-earned money. But this was not a constant theme in their marriage. In flush times he was relatively generous and untroubled by Harriet's lack of penury.

Though economic and emotional dependency was often the crux of tensions in a troubled Victorian marriage, the nature or extent of discontent was rarely constant within the same relationship. It would be unwise to be absolute about nineteenth-century role dissatisfactions. There was usually a mixed and highly contextual pattern of marital disquiet on the part of both sexes.[160] Nonetheless, whatever difficulties beset the troubled Victorian marriage, the absence of love was a dominant motif. American Victorians in this study do not appear reconciled or at peace with a loveless marital relationship. The longing for intimacy and identification, and the hunger for love were neither readily extinguished by time nor easily satisfied by the routines of married life. On the envelope of the last letter she received before his suicide, Harriet scribbled, "My Last letter from him—My All—" and though he may not have loved her as she wished, his final communication ended, "Your aff[ectionate] husband."[161]

Men and women often continued to feel a powerful identification and empathy for each other after marriage. In fact, romantic love was seen— inside and outside marriage—as one of the essential ingredients of marital happiness. No one suggested any circumstance that might mitigate its loss. While nineteenth-century courting couples often cheerfully expected that the drama, excitement, and novelty of courtship would wane in their future marital relations, they feared the absence of love. This fear was not unfounded. Romantic love was conceived as a mysterious force; there was almost no cultural sense of how to regulate its comings and goings. What was thought to be under individual control in nineteenth-century marriage was the role performance of husband and wife. Role duty was conceived as willable, therefore the individual could be held accountable both in his own mind and in a social sense. Because romantic love was believed to be essentially beyond individual control, it was extremely difficult to hold the individual accountable for its loss. Understandably then, romantic love was not conceived of as a duty but was nonetheless joined to a marital institution steeped in dutiful human relations.

Since this cultural mismatch, American marriage has struggled to accommodate this contradiction. The root of the problem lies in how to conceptualize the limits of autonomy of the "free" individual who chooses to enter into a romantically inspired contract. The difficulty of involuntarily maintaining a relationship voluntarily engaged reaches beyond contract law into the realm of the cultural commitment to romantic love as the raison d'être of marital relations. Romantic love was based upon the "fiction" of the independent self, acting as a "free" agent in terms of personal needs. Yet within marriage, the economic dependence of women, the entanglements of family, and the whole web of the social fabric acted to challenge the underlying premise of nineteenth-century American romantic love: atomistic individual freedom. Nineteenth-century culture applauded the application of the ideal of individual freedom to a variety of social situations, but fought against its application within marriage, clinging to older traditions of social responsibility tied to spousal role obligations. It is only in the twentieth century that the romantic "logic" of choosing a mate, already in place by the mid-nineteenth century, was fully extended to the conception of dissolving a marriage. This eased some of the older polarities but left untouched the problem of how to hold the individual accountable for the supposedly unwillable, uncontrollable romantic emotions that formed the basis of the modern companionate marriage.

8

Not for God Only

Patriarchy, Religion, and Romantic Love

H ISTORIANS OF THE American family agree that sometime in the eighteenth century a dramatic transformation began to occur in the relationships of husband and wife, parent and child. Robert Griswold describes this change as a movement from patriarchy to companionship "predicated on the notion of domestic equality between husbands and wives."[1] Another family historian, Daniel Blake Smith, characterizes the shift as

> a movement away from the well-ordered, father-dominated family of the colonial era—with its emphasis on parental control, obedience, and restraint of emotions—toward a strikingly affectionate, self-consciously private family environment in which children became the center of indulgent attention and were expected to marry for reasons of romance and companionship rather than parental design and economic interest.[2]

Carl Degler argues that the "modern" American family emerged in the years between the American Revolution and about 1830. One characteristic of this transformation was the "affection and mutual respect between the partners, both at the time of family formation and in the course of its life. The woman in the marriage enjoyed an increasing degree of influence or autonomy within the family."[3] Mary Beth Norton concludes that the late eighteenth century witnessed "a series of advances for women, in some respects at least."[4] Nancy Cott also suggests that the years from

1780 to 1830 were ones of decisive change within the American family. She too concludes that the increasing success of women divorce petitioners in the eighteenth century "may have signified a retreat from hierarchical models and an advance toward ideals of complementarity in the prevailing conception of the marriage relationship."[5] Jan Lewis also locates significant change in Virginia family values sometime between the end of the American Revolution and 1830. She argues the Virginia gentry turned inward and "exhibited characteristics of a 'modern' personality," seeking warmth, affection, and love in their increasingly idealized family relationships.[6]

In examining the decline of "patriarchy" in the American family, the meaning of the term must be carefully delineated, for its application has sometimes been more impressionistic and rhetorical than enlightening. At times, patriarchy has been employed to express disapproval of any and all aspects of male control or power in social life. Such broad usage has reduced the serviceability of the term. Moreover, indiscriminate overuse has given it a sloganeering quality. Finding no satisfactory substitute for the word itself but recognizing the need for a narrower construction, patriarchy is defined as the husband's culturally given rights of familial privilege and, more specifically, men's household control over women and children.[7]

Though historians are arriving at a consensus that the last half of the eighteenth and early part of the nineteenth century were years in which male-female relationships among the native-born middle-classes became less patriarchal and more companionate, they are uncertain as to the causes of this change. Daniel Blake Smith claims that American "historians have not succeeded in moving beyond the identification of these changes toward a persuasive explanation of causes."[8] Randolph Trumbach also suggests that English historians lack persuasive explanations of this shift in household power among the English aristocracy in the eighteenth century: "The causes of this egalitarian movement are obscure. The rise of a market economy, limited monarchy, Christianity, and kindred structures may all together have produced this great change, but certainly none of them alone was responsible."[9] There were also many factors in America which contributed to the egalitarian movement in domestic relationships. These included the patriotic status of domestic household manufacture (done by women) before and during the Revolutionary War, increasing educational opportunities for women, more positive Protestant ministerial formulations of traditional female images, republican ideology, Enlightenment psychology and pedagogy, challenges to the Calvinistic conception of God and morality, the rising status of mother and child, and various women's reform movements.[10]

Though often described as part of the shift in household power, romantic love is one factor that has been overlooked in explaining the egalitarian movement within the American family.[11] At least a portion of patriarchal authority rested upon customs, traditions, and roles that discouraged intense emotional attachments and suppressed individuality.[12] Relating to each other as romantic selves, men and women foreshortened their emotional distance through empathy and personal identification. Therefore, by bridging the gender gap and encouraging men's empathy in women's lives, romantic love weakened certain aspects of patriarchal family relations.[13] Though economic power remained in the male domain well into the twentieth century, and thus a significant component of the patriarchal household is only now being challenged, non-economic forms of power began to be distributed more evenly in the native-born middle-class household by at least the 1830s, and probably before.[14]

John Marquis expressed a sentiment that was common among nineteenth-century men who were "in love." "[I]f I only succeed in making you happy here then will I be truly happy."[15] It was not that altruism increased mysteriously when couples fell in love, but rather that men identified their self-interest with women and vice-versa. After four years of marriage, Marquis, in the midst of a trip west to search for land to buy and a place to relocate his family, expressed both a physical and spiritual pleasure in his relationship to his wife: "I do not think Neeta that I ever saw the time when I felt such a burning desire to be loved. I feel as if it would make my heart beat very fast just to see your hand writing. . . . I used to feel then that it would be a great pleasure to see you and it was a pleasure to *hope* that I might even someday have you all for my own and thus I think there was a degree of passion which inflamed [at this thought] made me desire you. Now I feel that you are my own wife, that you are a part of myself and seems hard that I cannot fold you in my arms or nestle my head in your bosom and feel your soft arms around me. I do not look upon you as one to minister to my desires but as a very precious gift from my Heavenly Father, given when I was ungrateful and undeserving and which I have never properly appreciated or been sufficiently thankful for. And I think of you as the Mother of our darlings Rossie and Bartie."[16]

His letter reveals several important threads in the unraveling of the patriarchal household in native-born middle-class America. The first was sexual. There was a rich reciprocity in John's physical desire which connected directly to his identification with his wife. The second was service. He declared that he did not consider her as someone whose primary purpose was to minister to his physical needs. Thus he explicitly renounced one of the chief functions of women in a patriarchal household.

But what alternative conceptions of women's role were available to him? Given that women were still excluded from the world of career and commerce in all but the narrowest of ways, husbands could not expect wives to make equivalent economic contributions to the family. It was apparently not enough for nineteenth-century men to think of women as the bearers and rearers of children. John conceptualized his wife's role in more individual terms. He saw Neeta as "a very precious gift from my Heavenly Father." The nature of her gift to him was romantic: she brought him the gift of self.[17]

Before they were married, Neeta worried over her poverty and felt she had little to offer him. John's response was unambiguous: "instead of thinking 'the man who takes Neeta for better or worse will get a dowerless bride' I must think that he will find himself the passion of a 'mine of wealth' just in the possession of her *dear self* . . . ."[18] Violet Blair approached the same issue from the opposite direction. She told her fiancé that she was beautiful, young, graceful, intellectual, refined, spotless in reputation, and from a fine family with considerable wealth. Then she asked quite seriously: "now what can a man give me?" Her answer was simply: "Love—."[19] Albert Janin expressed the same appreciation of Violet Blair. What he most valued was "your giving your precious self to me."[20] When male and female met as romantic selves, to be liked or disliked, loved or hated, on the basis of personality and self-expression, they were moving toward a more egalitarian basis of private family relationships.

Hierarchical gender distinctions nonetheless remained powerful in the nineteenth-century middle-class household. Whatever the rising status of female domesticity, motherhood, and the ideology of separate but equal spheres, family relationships were still heavily laced with the assumption that women should serve the needs of men. John Marquis pondered whether he had been worthy of his wife's love, "and I often feel that I have not been as watchful for your happiness and have not shielded you from as much care, vexation, and disappointment as I could have done, and I am afraid that I have made you think sometimes that married life was a hard path to follow and married happiness a myth. Especially have I blamed myself for the discomfort and care and labor you were subjected to a[t] Mr. T—."[21] John's guilty admission concerning Neeta's care, vexation, and labor indicates that a husband's empathy may not have lightened a wife's household burdens. What it changed was the context of her work.

There was obviously room for considerable slippage between the intentions of courtship and the reality of marriage. The gap between romantic expectations and practice could be wide and was affected by many variables. The important imbalance in economic power between husband

and wife was untouched by romantic ideals. Moreover, romantic love probably did not reduce women's work load in the nineteenth-century middle-class household. But then even so powerful a material force as industrialization shifted the physical burden but did not reduce the time middle-class women spent on household tasks.[22]

Nonetheless, women gained more status, standing, and power through the medium of affection and self-expression in their relationships with men. Though impossible to measure precisely, a husband's care and concern for his wife's happiness appears to be a significant mitigating factor in household affairs.[23] Married less than two years, Lincoln Clark admonished his wife: "Don't work like a slave and break your slender back, and create pains all over you, mind this—I wish I could oversee you all the time."[24] Nine years later, Lincoln was still worried that his wife was working too hard. He asked her to find others to do her work or leave it momentarily "undone": "I have some concern for you—fear it was too cold for you when you came down especially as you had not been out before—I wish you to be more cautious than you have been and not break your back while it is weak—if you cant get others to do the work let it go undone at least for a while—you know my business is to give counsel, and I think it ought to be followed."[25]

James Bell expressed the same concern before marriage: "I never wanted to marry you till I could give you a home where you would not have to work like a slave I have know[n] of many instances, where a couple's happiness were destroyed just by this constant toiling."[26] Wellington Burnett asked his wife to join him in Sacramento "unless you wish to remain. It was not my wish that you should go or remain Darling except for your own good."[27] Such expressions of concern and cooperation were echoed by James Blair, another young husband: "Mary my Pet my heart is almost bursting with love and affection for you and your little babe. . . . I have not a thought that does not aim at some pleasure happiness or comfort for you, indeed my whole happiness now is in contemplating something that will tend to give you pleasure and happiness . . . ."[28] Two years later, James Blair was as solicitous as ever of his wife's well-being: "nothing grieves so much to know that you wanted anything and that I had not supplied it in time for it is my great aim and pleasure to give you every comfort and pleasure that this world can afford."[29] James quite consciously connected a man's love for his wife and his willingness to satisfy *her* needs and desires. The death of a lady on board the steamer he had been sailing to San Francisco prompted this comment on the deceased woman's husband: "I never saw a man more in love and devoted to his wife than he was." Why? "He was all attention to her wants throughout the voyage."[30]

James also harped on a theme in men's solicitude for women that was constant and commonplace. "Mary my pet you must walk out every fair day and gradually increase the exercise until you regain your strength for the baby and my sake." He also urged her to "seek the company of your sister's and all your friends in and out of the house . . . ."[31] Nineteenth-century middle-class men in romantic love were very concerned for the health of their lovers and wives, as a courting man illustrated: "Your 'headaches' I don't like. Can't they be stopped? I cannot bear ought else but perfect health and happiness for you Darling. I think you need a room with a fire-place and then some gentle out of door exercise everyday which you must have. I *must* have you perfectly well."[32]

Wellington Burnett expressed similar anxiety over the health of his wife. "Do you feel any stronger than you did. I do hope so. . . . Let the care of your health be your first and last object in your actions Darling. I do so wish you to be happy."[33] Seven years later, Wellington was still expressing intense concern for his wife's health and happiness. "I hope that you are getting strong again, and well—Do not do anything to fatigue yourself—I hope that the children are not recovering at the expense of your health—you have said nothing about yourself in any of your letters—I wish that you would— I want to hear about *you*—How do you feel? Tell me—If you can come home well, I shall be happy indeed—*You* are all in all to me, in this World. . . ."[34] Though men's concern for women's health was probably tied to practical self-interest in their ability to deliver services in the home, those concerns were often expressed within the empathetic context of romantic love. As Eldred Simkins admonished his fiancée, "Remember you cannot take too much care of yourself."[35]

Masculine attention to women's health, at times almost obsessive, reflected the deeper changes being rung on women's traditional promise to "love, honor, and obey" and men's traditional vow to "love, honor, and cherish." James Blair, perhaps unconsciously, substituted a very different verb in the traditional male ceremonial litany: "My struggle my Mary has been to provide for you and your little cherub and my pleasure and pride will be to see you enjoy what I can afford to give which shall be all I have—I pledged my soul in the presence of God and man to love honor and *serve* you my Mary. I have done so with my whole heart thus far in life . . . ."[36] His pledge to serve her was a quite unpatriarchal notion that had significance even in the conceptualization.

One generation later that love-sick correspondent Albert Janin was still ringing changes on the traditional wedding vows. Albert challenged the female vow to obey: "remember, I do not share the old-fogy notion that a wife must obey and yield to her husband as her lord and master, nor any such nonsense. You and I have got along most delightfully as friends. You like me and I love you ardently and devotedly. What more is needed to ensure our happiness as husband and wife."[37] The ever-irrepressible Violet Blair had a few rules in mind to ensure their conjugal happiness: "don't do anything or say anything that will shock me in any way . . . . Don't mind it if I tell you things that I don't like. . . . Never pretend that ancestors don't matter. . . . Don't say that I am prejudiced because I pride myself on my impartiality. . . . Never try to work on my feelings for I know men too well not to see through that and it provokes me—never treat even the most trivial remark I may make with disrespect . . . never read or ask to read any letter, part of a letter, or note written to me or by me, unless I particularly request you to do so. . . . Now Bertie if you don't like what I have said, let us break off at once—."[38] Such impudence, not to say self-confidence, was unusual, but the point is that nineteenth-century women did not normally cast men in the role of authoritative Victorian patriarch, nor did they define wifely duty primarily in terms of obedience and submission.

In another incident with similar implications, Violet's closest girl-friend, Minna, refused to marry a suitor named Guthrie who believed "The man the head of the woman!" Albert's comment was "What colossal bosh when applied to the 19th century!"[39] Certainly men such as Albert retained considerable economic power in their household relationships. What was changing was their conceptualization of the legitimate uses of that power. While it would be naive to assume that ideas of equality encompassed all there was to power relationships in the nineteenth-century family, it would be equally naive to assume economic control defined the sum total of family power.

Along with many of his male contemporaries, Albert Janin believed in a companionate ideal of male-female relationships, not a patriarchal one.[40] He longed to please the woman he loved and vehemently insisted that he would do all in his power to make her happy.[41] He wanted the whole world to minister to her needs and declared his intentions: "I swear to you that there is not an earthly hope, or scheme of ambition, or habit or feeling of any kind that I would not cheerfully suppress or abandon at your bidding, except my love for you, which is the greatest comfort I have ever known. It controls my every thought and action, it is the arbiter of my destiny, the best part of my nature, and I shall never relinquish it as long as I live—in fact I could not if I would. Now judge, oh ruler of my heart, what power you possess over me, and put me to the test if you doubt me . . . ."[42] When the companionate ideal was emotionally charged with romantic love, it was a powerful counterbalance to male dominance in nineteenth-century male-female relationships.

Rulers of the heart had an uncertain, but nonetheless real tenure during the reign of romantic love, as Albert Janin illustrated in his marriage. Forced to withdraw his bid for a Democratic party nomination to run for Congress, Albert asked his wife to move to New Orleans to establish his residency there for political purposes. Having already leased a house, he was nonetheless willing to give it up if that was her wish.[43] Hearing rumors that her mother-in-law predicted she would return to New Orleans in January, Violet reminded Albert that she "intended to live in Washington every winter. . . . If you choose to go there of course I will not interfere—you are your own master but you are not mine. . . ."[44]

Violet Blair Janin was a beautiful, spoilt southern belle who lived at the height of fashion and in Washington society. A vain and flirtatious woman, she was also extremely assertive in her relationships with men. Violet insisted upon her prerogatives as both a lover and a wife with the reckless abandon of confidence, youth, and a privileged past. In fact, Albert's mother urged him to be more assertive with his new wife. Violet's response was defiant: "Violet Blair Never had a master and never will have one— One single attempt to control me in any thing and we part forever . . . Marriage is an experiment and if it don't work well it is best to smash it, from my point of view— I suppose you will think me very unamiable, but I was provoked that your Mother should say who you should visit— I look upon you as my property and resent any interference on the part of anyone . . . ."[45] In this instance, Violet turned the patriarchal assumption upside down, claiming her husband was her property. She usually approached her relationship with Albert in a more egalitarian spirit, however, insisting on her rights, but also recognizing his.[46]

Throughout her courtship with Albert, Violet expressed a deep concern about maintaining her power over him after marriage. A few months before their marriage she wrote: "I have got an idea that you will be so devoted to business etc. that in five years after we are married you will have ceased to love me, at least to love me as you do now—I fear that when your romance is over, you will forget the little attentions I like to receive from people—Now this is a much stronger reason against marriage than my love of flirtation . . . I know you are a devoted lover but can your love stand the test of five years of my caprices—."[47] Violet understood that romantic love was the source of her power in relationships with men. She was genuinely reluctant to marry for fear of losing the romantic power she possessed over Albert and her myriad suitors: "I know you are a devoted lover . . . yet I have a great many fears about the future, the first evidence of indifference or selfishness on your part will make me hate you as entirely and perfectly as if you were the greatest monster alive— I know a man who was devoted a longer time to a lady than you have been to me, yet now he won't do anything she says and is intensely disagreeable. . . ."[48] Most women did not demonstrate their concern for power in male-female relationships as self-consciously as Violet Blair, but most, in their courtship behavior, repeatedly tested the romantic love of their partner in a ritual which strengthened their emotional hold on men before marriage.

Similarly, courting men in Victorian America most often asserted their hold over women in romantic terms. At least before marriage, men did not normally claim a patriarchal prerogative but rather an emotional one.[49] Nathaniel Hawthorne pondered his "awe" of Sophia Peabody but noted: "And then it is singular, too, that this awe (or whatever it be) does not prevent me from feeling that it is I who have the charge of you, and that my Dove is to follow my guidance and do my bidding. Am I not very bold to say this? And will not you rebel? Oh no; because I possess this power only so far as I love you. My love gives me the right, and your love consents to it."[50] Love was not the same taskmaster in male-female relationships as a patriarchal birthright. Though rulers of the heart could be tyrants, they could also desire "nothing but to be with thee, and to toil for thee, and to make thee a happy wife, wherein would consist his own heavenliest happiness."[51]

Rulers of the heart wished to govern by consent after marriage, as Lincoln Clark explained: "I can not bid you do that in which your *heart* is not, for it is that which I wish to command in all things, not the will."[52] Another husband expressed similar sentiments, though perhaps with less altruistic motives: "I ought to 'mold' you. But you know how I feel about that. I loved you for your very self. I could not bear to lay deliberate hands upon your nature and undertake to change it to my will. Development must be spontaneous to have any charms for me in one I honor and love as my wife."[53] "Molding" could be exercised with great subtlety in male-female relationships, however, and Nathaniel Wheeler most likely pressured Clara, his wife, to resign from a teaching post she held alongside his own because their dual careers threatened his sense of manhood.[54] Nonetheless, Nathaniel scorned a friend whose ideal of the married man was "one who is coarse, sharp, unscrupulous in his business, but keeps a place somewhere he calls 'home,' where he is superior lord, where he takes leave of his other self, and is willing to be amused by the kind offices of wife and children in return giving them what they want, and not swearing at them very often. . . . It's a very low ideal."[55]

The amelioration of patriarchy in the middle-class household was full of contradictions and ambiguities. A man would occasionally claim that love made a woman his property. Eldred Simkins wrote Eliza Trescot in 1863, "Besides, my own, by your own consent you have become my *property*."[56] Yet a few lines later Eldred insisted that he was "governed" by

Eliza and "that to be governed by those we love is the sweetest sacrifice of liberty." Eight days later, he asserted his right to command her after marriage: "If you were only *mine* and I had the right to order! you *should* see me but *alas*! I have not . . . ."[57] Yet, eight months later he promised he would "ever strive to make that present comprise your happiness and contentment."[58]

George Bomford kept vital information hidden from his wife, as he admitted: "and although on many trying occasions—of past years—you may in the retrospect—think that you had not the full share of my confidence—yet believe me—I always made you acquainted where I thought it was just and necessary. . . ." His justification for the exercise of such paternalistic authority, however, was companionate and empathetic: "but I would not add to your load of cares and anxieties—on occasions where you could do no good—, but have indirectly had your opinions—without creating alarm—you therefore if you do not approve my course—on these occasions—cannot but appreciate—the affect[ion] and sincerity of the motives . . . ."[59] Perhaps to compensate, but certainly to please, George gave his wife carte blanche to choose a house in the city: "and whatever may be your wishes and desires—they will be mine—."[60] Though there was no consistency in behavior, middle-class men in romantic love most often expressed a companionate ideal. This meant they frequently placed at least an intellectual priority on sharing over commanding or being served in their romantic male-female relationships. No claim is being made that masculine intentions were faithfully practiced or that the desire to please women extended beyond the bounds of romantic love.[61] But the fact remains that nineteenth-century men emphasized their responsibility for women's happiness, and put at least some of their intentions into practice *while they were in love*.

Wellington Burnett wrote: "I think of you by day and by night, and my greatest and purest wish is that you should be *happy*."[62] Very often nineteenth-century men linked their absorption in women's happiness to characteristics of romantic love. Thus Burnett associated obsessive thoughts of his loved one and concern for her happiness. James Bell also linked several qualities of romantic love with his wish to make his fiancée's life a pleasant and joyous one. "God grant me a safe return," he told her, "that we may share our thoughts together. That our lives may be blended together in peace, and harmony . . . . I would make life pleasant, and joyous for *your* sake, for one that is *dearer* than all others. Yes, *dearer, lovlier,* more *beautiful* and *better* in my eyes, and to my heart than *all* others I've ever met. To day you seem dearer than ever before, I look back over the vista of the past with gladness, in many respects, one in particular, That is, that we have been true to our love, With me absence has strengthened it and made it better, yet I have regretted it *very, very* much."[63] The properties of romantic love that James expressed—the power of absence to strengthen love, abundant praise and positive affirmation of the loved one, and the desire to share completely her thoughts—were all joined to his wish to make her happy. Men's concerns for a woman's happiness sprang from one of the central characteristics of romantic love, the identification of selves across gender boundaries. For, as James Blair revealed, "There is even about me a melancholy pleasure in feeling my heart and mind so entirely usurped by you and yours—I have not a thought that does not aim at some pleasure happiness or comfort for you, indeed my whole happiness now is in contemplating something that will tend to give you pleasure and happiness . . . ."[64]

When men's fixation on women's happiness is connected to romantic love, its role in the decline of patriarchy can be more easily understood. Romantic love did not cast a mysterious spell over men, rather through the practice of self-disclosure and self-discovery they came to blend their inner life with the women they loved. This created a unique association in which men aimed at their own happiness in seeking to give pleasure, comfort, and joy to their lovers and wives. The movement toward more egalitarian household relationships, at least as it was influenced by romantic love, was not some act of collective enlightenment. It was rather the result of the mutual sympathy between men and women created over and over in romantically oriented courtships.

The significance of romantic love is perhaps nowhere more important and yet more difficult to trace than in the realm of religious sensibility and world view. From the beginnings of the more religiously inspired English settlements in the New World, Americans have phrased the language of human relationships in religious terms and their relationship to God in human terms. The poetic New England minister Edward Taylor used the most sensual of metaphors in private meditations on his interaction with God. As one historian notes, Taylor "wished Christ to 'kiss me with the Kisse of his mouth,' for Christ is the 'Fairest among Women,' one whose 'Flame' of 'rich Grace' 'do ravish with delight.' . . . Taylor's Christ was the consoling wife-mother, the virgin drawing her lover to the baptism of purity, both the one seduced and the awe-inspiring, magnificently endowed seducer, as well as the fertilizing sperm."[65] Taylor addressed God in images that vividly illustrate the language of religious emotion in seventeenth-century America: "Oh! that thy love might overflow my Heart!/To fire the same with Love: for Love I would. . . . Lord, blow the Coal: Thy Love Enflame in mee."[66]

Laurel Thatcher Ulrich, in her study of colonial women, also emphasizes the metaphorical use of male-female relationships in early American
religious writing: "Over and over again New Englanders heard the love
of man and wife compared to the bond between Christ and the Church.
Although the analogy obviously ratified the authority of men over women,
ministers seldom explored this implication, preferring to draw upon the
emotional dimension of marriage to personalize the believer's relationship
with Christ."[67] Historian Edmund Morgan, in his classic study of the
Puritan family, also chronicles the extensive use of analogies in New England religious writing which compare the relationship of the believer to
God with that of the wife to husband. As Morgan affirms: "The metaphorical imagery of Puritan theological works, especially those of Thomas
Hooker, displays a singular sensitivity to the warmth of conjugal love.
The relation of husband and wife furnished the usual metaphor by which
the relation of Christ and the believer was designated."[68]

For example, Puritan divine Cotton Mather explained in a sermon
titled *The Mystical Marriage* that the covenant of grace or conversion "appears under the Character of a Marriage, because from this Time, there
is an Union, and not only a Legal Union, but also a Vital Union, between
the Redeemer and the Believer." Mather did not hesitate to compare conjugal coupling to spiritual regeneration: "the Married Couple becomes
One Flesh, so the Redeemer and the Believer become One . . . ."[69] Cler-
gyman John Davenport wrote to a Lady Mary Vere, whose husband had
just died: "The relation which once you had to this earthly husband is
ended, and ceaseth in his death, but the relation you have to our heavenly
husband remayneth inviolable . . . ."[70] Margaret Winthrop told her husband in 1630, "therefore my good Husban chere up thy hart in god and
in the expectation of his favors and blessings in this thy change. . . . as
for me his most unworthy servant I will cleave to my Husban Crist as
neere as I can though my infirmytes be great he is able to heale them
and wil not forsake me in the time of neede. I know I shall have thy
prayers to god for me that what is wanting in thy presence may be sup-
plyed by the comfort of gods spirit."[71] The Reverend Peter Bulkeley preached
"that with the Christian and his God it was 'as it is betweene man and
wife, though shee be foolish, passionate, and wilfull; yet these doe not
breake the Covenant of marriage, so long as shee remaineth faithfull.' "[72]
And the eminent John Cotton of Boston's First Church used even more
sensual allusions in another sermon: "And looke what affection is between
Husband and Wife, hath there been the like affection in your soules towards
the Lord Jesus Christ? Have you a strong and hearty desire to meet him
in the bed of loves, when ever you come to the Congregation, and desire
you to have the seeds of his grace shed abroad in your hearts, and bring

forth the fruits of grace to him, and desire that you may be for him, and for none other? . . . Wil you accept him as your friend and as your Husband? . . ."[73] These examples illustrate that the seventeenth-century language of religious affection was infused with the metaphors of conjugal relationships, and had an affinity with the later language of romantic love.

The canon of religious affections in eighteenth-century America was also an important Christian antecedent to nineteenth-century expressions of romantic love. In the early eighteenth century, Jonathan Edwards, perhaps America's greatest religious thinker, struggled with the question of what constituted true religion at a time of widespread religious revivals and passionate controversy over the nature of the religious experience.[74] Edwards insisted that true religion was only "comprehended in *love,* the chief of the affections, and fountain of all other affections." He wrote that "the essence of all true religion lies in holy love," explaining, "For love is not only one of the affections, but it is the first and chief of the affections, and the fountain of all the affections."[75]

Building upon this basic premise, Edwards identified an anatomy of emotions which resulted from this "fervent love to God." They included: "a dread of God's displeasure, gratitude to God for his goodness, complacence and joy in God when God is graciously and sensibly present, and grief when he is absent, and a joyful hope when a future enjoyment of God is expected, and fervent zeal for the glory of God." Significantly, if the romantic self is substituted for God in Edwards's anatomy of religious emotions, the result is an accurate description of the emotional characteristics of the nineteenth-century American experiences of romantic love. Edwards himself recognized the significant parallel between human love and the love of God when he commented, "And in like manner, from a fervent love to men, will arise all other virtuous affections toward men."[76]

Edwards also held Christ up as an example of true religion because of his "tender and affectionate heart"; he portrayed Christ as someone of extreme emotions, emphasizing his "agonies," "strong crying and tears," "exclamations," "desire," "compassion," "affections," and "ardency."[77] Edwards's emotional Christ was presented as a model of true religion for eighteenth-century men and women. Drawing the parallel between religious and earthly love even closer, Edwards also employed "the holy love and joy of the saints on earth" as a model for the religion of the saints in Heaven. Finally, Edwards virtually shattered the distinction between religious affections in heaven and on earth when he suggested that "The love and joy of the saints on earth, is the beginning and dawning of the light, life, and blessedness of heaven, and is like their love and joy there. . . ."[78] Edwards linked human and divine love by his refusal to make an absolute distinction between religious affections on earth and in heaven.

Though Sydney Ahlstrom claims in his magisterial survey of American religious history that "Edwards expounded a theological tradition that had deep popular roots," Edwards's effects on American cultural life are still a subject of debate.[79] But whatever the effect of the Edwardsean canon, by the nineteenth century native-born Americans within the Protestant tradition had a language of religious emotions readily available to express more earthly love. Small wonder, therefore, that nineteenth-century romantic love was described by lovers in religious terms and concepts: sacred, holy, hallowed, worship, salvation, sacraments, altar, and so on.[80] The metaphorical comparisons of human and godly love were long-standing. The cultural significance of nineteenth-century romantic expressions was not that the descriptions of earthly love were phrased in religious vocabulary, but rather that the metaphorical distance between God and man—the bedrock of an earlier world view—was collapsing.

Native-born men and women in romantic love were becoming more God-like to each other. Not all men and women, certainly. Even within a middle-class framework it is impossible to assess the quantitative dimension of such profound change because nothing so central to people's world view affected everyone at the same rate or with equal intensity.[81] The complexities must be respected. Some people continued to believe in the traditional Judeo-Christian God even as they made the selfhood of the loved one their central symbol of ultimate significance. God was not abandoned by these nineteenth-century couples, He was just moved to a less central location in the cosmos.

The danger of human affection, especially in male-female relationships, was feared by America's more religious settlers from the outset. John Cotton warned, "let such as have wives look at them not for their own ends, but to bee better fitted for Gods service, and bring them nearer to God, and then wee so have wives, as if wee had them not . . . ."[82] Puritan ministers repeatedly warned against loving a spouse inordinately, urging moderation in all human love. As Edmund Morgan explains: "The highest love of all Christians was reserved for God himself; and since human beings, husbands and wives, were only the creatures of God, they could not take his place. . . . To prize them too highly was to upset the order to creation and descend to idolatry."[83] Even the letters of John and Margaret Winthrop, often cited as early examples of the most sentimental conjugal love, indicate a firm respect for the principle of limited human affection as well as emotional commitment to God as first and primary to all others.[84]

This tradition or world view stretched well into the nineteenth century. Thus in the 1820s, Anna Harrison's suitor could write with some assurance, secure in the time-honored boundary between secular and religious affection: "Does not all Heaven approve of our happiness, for it has been given us by our Heavenly Father and all he does is approved there. There was joy in Heaven when he called us by his Holy Spirit from darkness to his glorious light. Will it not add to the brilliancy of the crown of your rejoicing that since I have known you I have gone on more vigorously in the Christian course? . . . I do not believe my feelings for you are too earthly, though I know from the word of truth that the heart is deceitful above all things and desperately wicked and am conscious of it from the teachings of the Holy Spirit. I believe our earthly happiness cannot be too great when our enjoyment of it leads to higher feelings of gratitude to him who has given us these good gifts to provoke each other unto good works—when it leads us to greater love for our Savior and more obedience to the Holy Spirit. May God grant these things may be so, yea that we may do all things for the Glory of God."[85] He placed secular happiness in romantic love a distant second to the spiritual fulfillment of his Christian relationship to God. And he argued that those who were disappointed in marriage expected too much of their happiness to originate in human affections: "Those who have been disappointed in not finding the married state one of unmingled, uninterrupted enjoyment did not estimate it by that standard which we do—the Holy Scriptures, nor did they seek their happiness from Him who is the only source of it."[86]

Ordinary Americans in humble circumstances were apparently even more tenacious than their more affluent countrymen in observing the principles of "limited" human love. One cultural historian concluded in his study of the popular mind: "Common people of the mid-nineteenth century portrayed God as jealous and as a deity not to be mocked by man's idolatrous love of his fellows. The popular didactic poet Martin Tupper, who was read widely both in America and in England, could warn in his poem 'Of Marriage,' 'Take heed lest she love thee before God; that she be not an idolater.' "[87] A man of modest means, writing in his diary in 1850, epitomized the time-honored order of primary emotional commitment to God: "O 'tis delightful to dwell upon this heaven born principle—love.—Love to God first. Love to my chosen companion. Love to my children."[88] By contrast, John Marquis asked his wife to "believe that my heart beats only for you[,] our babies[,] and our god."[89] God came last in Marquis's revealing sequence of emotional loyalties. While God was not dead or even dying in middle-class sensibilities, it was a momentous shift from "Love to God first" to an order of emotional priorities in which god (uncapitalized) receded to third place.

What happened was the realization of the worst fears of early American religious leaders: human affections were beginning to be placed above religious ones as nineteenth-century men and women felt a more powerful pull from their relationships on earth than their relationships to Heaven. Of course, this was a matter of degree. Religious sentiment abounded in nineteenth-century middle-class private expression, but the personhood of the loved one, by the 1830s, had become a powerful rival to God as the individual's central symbol of ultimate significance. More and more middle-class individuals—propelled by romantic love—were worshipping in the new temple of individual selfhood.

Such a momentous shift may be captured in before-and-after snapshots of historical experience. However, one would expect the process of transition to be particularly elusive. In fact, many couples either easily rearranged their emotional priorities or inherited the new order and found nothing remarkable in the romantic "theology of Selfhood."[90] Others, however, not so sanguine, struggled to keep God first in their lives. In both cases, a portion of this process of change is revealed and the deeper connections between secularization and romantic love can be examined.

Samuel Francis Smith was a theologically conservative Baptist minister. Mary Smith was a pious Massachusetts Congregationalist who eventually became his wife. During their courtship they both struggled against the secular influence of romantic love. At the beginning of their correspondence in January 1834, they agreed to "breath the spirit of heaven" in all their subjects "whatever they shall be."[91] True to their commitment, Samuel wrote in late January: "Your letters my dear Mary be assured do give me unmingled gratification. I trust they are not merely causes of pure delight to my own heart, which is comparatively of little importance, but what is more to be desired, occasions of gratitude and prayer to the Father of Spirits . . . . Surely our correspondence then will not have been in vain, if it leads us to send up but one more burst of praise—to feel but one more thrill of humble yet adoring gratitude in our daily devotions."[92]

While Samuel was keeping his eyes on heaven, Mary responded to the allure of more earthly feelings. To Samuel's breath of heaven, she answered: "It is to me a rich consolation to know that I have your prayers. I need them—*indeed*. I *do need* them. I desire to be more holy—more heavenly: more weaned from earth. I am in *continual* danger of doing wrong—My affections too strongly fix upon things [that] 'must perish with their using—.' "[93] Samuel, understanding clearly that her affections fixed upon him, offered her a paragraph from his Sunday sermon as an antidote.[94]

Perhaps inspired by Samuel's "cure," Mary renewed her effort to fix her gaze on heaven. She asked Samuel on the upcoming occasion of her twenty-first birthday to "unite your cries with mine, that I may be enabled to lay aside 'every weight' every sin and press boldly onward. That I may the rest of my days, whether few or many—serve God—and 'be clothed in his likeness' imitating him who ever 'went about doing good.' " [95] But Mary's thoughts soon turned in a more earthly direction for she wrote on February 10, "If I am not with you my friend, in 'presence' I am often (alas!) I fear too often with you in 'heart.' Oh pray for me that my affections may not dwell *too* much upon earthly *treasures;* but may be strongly, most strongly placed upon Him 'who is worthy of them all'—." [96] Earthly treasures were on both their minds, for ten days later Samuel proposed that they set aside April 1 "to ask for the direction of the Holy Spirit" on the question of marriage. [97]

In his next letter, Samuel expressed a typical desire to be with her, except that the scene he pictured (in this post-Freudian age) is difficult to accept at face value: "And now, my beloved, if I had you with me at the moment, we would stop just here in the letter and fall down before God, and praise and adore him for all his goodness—." Though it is tempting to the twentieth-century mind to read something more than religious worship into Samuel's urge to fall down with Mary, he most likely intended nothing more. Obviously struggling himself, he again included a traditional warning against putting too much importance on human affection: "I beg you, my dearest Mary, see that you do not worship any image of clay—and pray that I may be kept from similar idolatry." [98]

When Mary asked Samuel for reassurance that she was the "girl of your *undivided* affection," Samuel's response was passionately affirmative. [99] Something important was happening between them. In a letter written the next day, March 14, and out of his usual time sequence, Samuel again urged that their affections be set upon God, but added quietly, "I hope I shall by and by learn to love you more rationally and less passionately." [100]

Mary responded in a highly charged letter that was full of self-recriminations. After receiving two reassuring letters from Samuel, she was ready to escalate the emotional stakes and put Samuel's emotional commitment to an even severer test. "There dwells not one on this earth who possesses so entirely, so unreservedly—my whole affections as yourself,—My hand too should that *cure* be wanted I would give you as freely and as happily as I have given my heart. You *may never ask it*—it may be a boon not worth your possessing. I know not. . . ." [101]

Mary appeared unwilling to leave her future to the device of mutual prayers and thus pressured Samuel with a far more human expedient. Self-criticism such as "it [marriage] may be a boon not worth your possessing" almost forced Samuel into a more affirmative rejoinder. And expressions such as "Cast me not from your bosom, I pray. I beseech— love me trust me still" cannot be easily ignored. In revealing so much of herself and expressing such intense emotional commitment to Samuel, Mary was issuing an imperious summons to him to express his own emotional stance toward their relationship. Her expressions of affection were urgent and highly charged: "Tis not the cold passion which the passing crowd call friendship—it is something more—yea *nothing less* than a deep settled permanent and yet lively ardor—would lead me to offer every thing, all I have, at its shrine. As *well* as I love my parents—as well as I love my connections to friends—Yet all *all* could I resign most willingly—most happily for *your* sake. My affection for them dwindles into comparative insignificance when I think of what I bear towards you. Cast me not from your bosom, I pray, I beseech—love me trust me still and believe me when I say all I have is *your own*. To use a borrowed expression which speaks my heart 'had I as much proof that I loved Jesus Christ as I have of my love to you I should prize it more than rubies?' "[102]

Mary inched close to the edge of blasphemy when she admitted she was more certain that she loved Samuel than Christ. Samuel's response was positive, bordering on elation, as he rejoiced in the certainty that "I may address you as my own, almost as securely and as sweetly as if the *final negotiations* had been completed."[103] But later he warned her not to set him above Christ for, he reminded her, they would both eventually sit at Christ's feet.[104]

Samuel formally proposed on April 3, including a quite unromantic but characteristically Puritanical injunction that "The union herein proposed, remember, my dear, *may* be the most disastrous act in our lives—that shall set upon all that remains to us a seal of wo—I would have this not forgotten, but I cannot bear to dwell upon it."[105] With a renewed commitment to the division between earthly and heavenly love, Mary accepted his proposal of marriage in her April 7 letter: "*It is* my '*sweet*' *love in the honest sincerity of my whole soul, that I solemnly pledge to you my heart and hand.* And *most* happily, yea, triumphantly *do I accept yours as the richest earthly boon.*"[106] Though Samuel was deeply gratified by her response, he nonetheless suggested that she should instruct her parents to stop them, if they did not believe the marriage was God's will. His reasons were revealing: "my attachment, love, you know is such that I probably should go forward at any rate—blind, almost to the voice of God . . . ."[107]

The secular precipice of romantic love was looming for them both and Mary hinted that she was near the edge in a letter written one month after their engagement was sealed: "Why, *why is it* my dear Francis that I love you as I do? Is it to be a snare to me—a source of unhappiness through my short pilgrimage? *I pray not.* But is there no danger. I am sure there is—for our heavenly Father will surely chasten us if we make an idol of any one of his merciful gifts."[108] In the intervening weeks Samuel traveled from Maine to Massachusetts to see Mary, and in the first letter written after their separation, Mary was clinging to her traditional religious priorities by a slender thread. " 'I would fain love earthly beings less'—or *one* at least. I would not make you an idol—indeed I *would* not. But oh—if I ever had one you *are that one most certainly.*"[109]

The secular shift in Mary's emotional hierarchy, despite all her best efforts, is revealed by the clever trap she laid for Samuel in what might have been an unconscious attempt on her part to bring his feelings in line with her own. "Oh I do feel it such a luxury to hide myself a little time in my own room and think of God and heaven and your precious self! But what would you think dearest of my piety—if I should tell you that you engage the chief of my thoughts and attentions even then? But it is thus And what *shall* I *do?*"[110] Mary put Samuel in the uncomfortable position of either rebuking her for thinking of him when she was alone; chastising her for being engaged by him rather than God; or accepting his own priority over God. Samuel managed to escape her trap, responding that he was the worst person to ask because he "covets" all her thoughts. Furthermore, he declared that the thoughts she had of him alone with her God for five minutes were even more precious to him than "12 times 5 minutes at any other time." He told her he also often had thoughts of her in religious moments, but then slipped in the warning, "While therefore we love, and must still love each other with a pure heart *fervently,* May we have grace to love Christ more."[111] Samuel escaped the unhappy prospect of alienating her or his God.

Then the "death letter" arrived. Mary's letter to Samuel (in the customary time sequence of their exchange) was one day late, and arrived sealed in black wax. Samuel leapt to the extraordinary conclusion that Mary was dead. He "reasoned" that this was her last letter to him before she died, forwarded by a kind friend who had included a postscript to inform him of her demise. "[B]ut my trembling heart at once suggested," he wrote, "—'Well this is the end of my sweet intercourse—my dear Mary is no more—' You will not wonder, my blessed, still spared love, that I now feel you to be dearer and lovlier than ever."[112] But then Samuel added, "Well, my dear, with all my deficiencies, 'such as I' am 'give I' myself 'unto thee'—and to God and the World."[113] He gave himself in a revealing order: Mary first, then God, followed by the world. Samuel did not openly recognize his emotional shift. But in a July 3 letter, he hinted at this reordering: "It would greatly cheer me, beloved, if you could but be here today, tomorrow, *every* day—For every day, I see some new reason to long for your presence. If I could at all times so strongly anticipate communion with God, and if the anticipation were so sweet of being with him as is the anticipation of being with yourself, I should charge my heart to leap and sing for joy."[114]

A seesaw struggle waged in Samuel's heart between God and Mary.

One moment he lamented the fact that he loved her twice as much as he loved Christ, the next he was encouraged by the "warfare between the spirit and the flesh" because it showed that "we have some knowledge of our own hearts, and some desire to bring them into conformity to God."[115] Striving to bring his heart into conformity with his intellectual commitment, Samuel philosophized: "I cannot but believe, my dear Mary, that the enjoyment of God has in it so much to fill the hungerings and thirstings of the soul, that we should both be willing to forgo the fulfillment of our loveliest anticipations, for the sake of experiencing it."[116]

As their wedding day approached, this couple was full of the expectation of bliss and anxious over the possibilities of pain and disappointment, but neither appeared willing to forego "the fulfillment of our loveliest anticipation." The warfare of flesh and spirit still waged in Samuel's heart: "indeed I am beginning to be almost seriously afraid I shall put you in the place of God as the guide and the object of all I do . . . . We shall do no wrong, dear one, by loving each other with our *whole hearts,* if we only love each other 'in Christ.' "[117] This was a clever compromise, to be sure, but if it was no longer wrong for man and wife to love each other with their *"whole hearts,"* the older universe of religious meaning with God at the absolute center was in jeopardy. In fact, Samuel was afraid he had already put Mary "in the place of God."

🐦 This new theology of the self was illustrated and confirmed by many different correspondents. Nathaniel Hawthorne referred to Sophia's love letters as his "spiritual food" and remarked, "I keep them to be the treasure of my still and secret hours, such hours as pious people spend in prayer; and the communion which my spirit then holds with yours has something of religion in it."[118] The "something of religion" was romantic love. Hawthorne marveled at his feelings of love, "But I leave the mystery here. Sometime or other, it may be made plainer to me. But methinks it converts my love into religion."[119] Hawthorne underscored the seriousness of his claim two months later: "Oh, my Dove, I have really thought sometimes, that God gave you to me to be the salvation of my soul."[120] In almost any variety of Judeo-Christian theology, only God had the power to save a mortal soul. To call Sophia his salvation was to attribute special powers to romantic love formerly reserved in religious circles for the Almighty alone. Without any fanfare, Hawthorne substituted human love for God's love as the source of his salvation, and attributed the substitution itself to God. Though this was no doubt heretical within a nineteenth-century Christian world view, yet the romantic self had already become such a fundamental spiritual category to some people that

Hawthorne, unlike Samuel and Mary, no longer even perceived a conflict between an all-consuming earthly love and religious values.

Hawthorne's expressions of love to Sophia are a vivid example of the way earlier American religious analogies between the relationship of husband to wife and Christ to the believer were completely reversed in middle-class expressions of romantic love. Puritan divines used the relationship of husband and wife as a metaphor to illuminate and explain the pivotal relationship of God and believer. Nineteenth-century native-born men and women used the relationship of God and believer as a metaphor to illuminate and explain the pivotal relationship of men and women in romantic love. Thus, when Hawthorne remarked: "Oh dearest, blessedest Dove, I never felt sure of going to Heaven, till I knew that you loved me," he was reversing a long-standing Christian metaphorical order by making the believer's assurance of a heavenly afterlife based on a relationship with Christ a secondary concern, illuminating his primary "spiritual" relationship with Sophia. Romantic love had become a form of salvation that was at the very least a rival to and for some perhaps a substitute for Christian salvation.

Hawthorne's sense of spiritual rebirth in romantic love was expressed in the basic structure and language of Christian salvation. "Indeed, we are but shadows—we are not endowed with real life, and all that seems most real about us is but the thinnest substance of a dream—till the heart is touched. That touch creates us—then we begin to be—thereby we are beings of reality, and inheritors of eternity."[121] What is most striking about this passage is that identical words could have been used in a Christian pulpit to refer to religious salvation; instead Hawthorne was speaking of romantic love and a human lover in a nineteenth-century love letter.

A new husband such as Wellington Burnett described his wife's letters as "holy," and continued, "You are all purity and sacred in my eyes darling."[122] "I love you," he told her, "as ardently and with as holy a love as my nature is capable of."[123] Another courting male, Nathaniel Wheeler, described the progress of his romantic relationship as "too sacred" ever to recapitulate fully.[124] But four years later, he reminisced about their Sabbaths together, calling their romantic communion a "sacrament": "Today I'm sighing for the old days, when long hours were ours, when both silence and speech were golden, and the long bright Sabbath afternoon drew gently to its close while we read, or talked, or dreamed,—together. Then the solemn twilight hour, like a sacrament binding us together, and leaving its blessing for the morrow."[125]

Dorothea Lummis also used the sacramental metaphor to describe her love: "Ah Carl, I thought once in that miserable time, of Love's sweet sacraments and wondered if they were all over purely for us forever. But it cannot be. O! if you were here, or I with you, we might put our dear tired selves at rest, and then perhaps love would bless us again."[126] Dorothea again called theirs a "sacred relation" and in a moment of hopeful anticipation, exclaimed: "The love that yet remains strong in our hearts must be our stay and with its help maybe we can after all work out our own married salvation."[127] Robert Burdette claimed: "You love me. Darling as I have already written you, that is spiritualizing my life from day to day."[128] Continuing the substitution of the lover in romantic love for the divine in traditional religious worship, Burdette acknowledged: "When I pray, I whisper your dear name as the Romanist dwells on Mary's. You are all of my thought and oh, so much of my religion."[129]

There are many examples of the Christian metaphors that both informed and structured the meaning of romantic love. Eldred Simkins exclaimed, "For oh! tis a thing most exquisite and holy to be truly loved. . . ."[130] His fiancée described her vision of their future together: "Life to us will not be stale, flat and unprofitable. No Eldred, we will have higher than earthly aims, I hope of hearing these blessed words, 'Servant, well done!' "[131] This prompted Eldred to remark that guarding and cherishing her "would not be an earthly aim."[132] Later Eldred compared himself to Christ's "disciples listening to the promises of the future Eliseum. . . ." He followed this with an extended analogy between himself and Christian in the allegorical *Pilgrim's Progress:* "I have a far more difficult task than he—Christian didn't love! his love was Heaven itself and he could afford to wait and toil for that—I will too one of these days—and yet Christian was all impatience and joy when he beheld the promised land and pressed forward on his way with great eagerness. So you see you strengthen my side of the question and be not surprised if I like Christian, sigh for that sweet rest and happiness that once was strong enough to tempt even Angels down from Heaven itself."[133] Eldred's goal was to rest in Eliza's arms, a far more earthly aim than that of Bunyan's allegorical sojourner.

One woman wrote to her husband in a private poem on married love, "Thou art my church and thou my book of psalms!" She also addressed him: "Toi! qu'en mon coeur seul je nomme, /je te vois partout et, en toi, je vois Dieu." ["You! who alone has my heart, /I see you everywhere, and in you, I see God."][134] Clara Wheeler Baker described her lover's attention: "Your love is worship—that an angel in Heaven might envy me."[135] She characterized her romantic feelings in terms of the sacred: " 'This hour' is church time but I am not attending divine service preferring to pay my devotions to my Lover."[136] Clara also paid devotions to her lover in church: "With a letter in my pocket to Robert what else could I do through the service but think of him. And do you know, dear, except for the twilight hour—you always seem nearer to me in church than any where else." [137]

One significant aspect of the religious investiture of romantic love by nineteenth-century middle-class men and women was the sacralization of sexuality. Kisses were "holy," marriage beds were "sacred," embraces were "devoted." [138] Sex in nineteenth-century America was more than merely acceptable: for the native-born middle classes, sex was the most important "sacrament" of the newly ascendant God of Self. Robert Burdette, that most ardent of Victorian lovers, eloquently expressed the religious dimension of nineteenth-century sexuality: "And my soul thrills with tenderness when it prays for you, My Violet. And so all my Love is tender and reverent; it mingles with my religion so that my prayers and my kisses meet." [139] Prayers and kisses met in many nineteenth-century men's and women's view of passion as a form of worship in romantic relationships. Nineteenth-century middle-class sexuality was embedded in a complex transmutation of Christian imagery, structure, and belief into the symbols of romantic love. Within the new religion of romantic love, sex was construed as a sacred act. [140]

The emphasis on romantic love as sacramental was not merely metaphorical convenience. The process of mutual identification encouraged by nineteenth-century courtship nurtured self-expression, self-knowledge and thus self-development. Consequently, the use of the metaphor of salvation or rebirth had an experiential dimension in the deepening and expanding awareness of selfhood that was created through romantic love. Charles Strong described his feelings in the first bloom of romance: "Was thro' breakfast just in time for church. A large congregation, good sermon and good music, all of which sounded new to me, because I was a new being just made." [141] Charles might have used similar language to describe Christian rebirth, but he was actually giving an account of his romantic feelings.

Charles also referred to his fiancée as "The one Being of all others" and employed the relatively common phrase, "the Idol of my heart." [142] In the former case, he portrayed Harriet in language that might have denoted the Supreme Being. In the latter instance, he described a feeling that blatantly violated a basic Judeo-Christian religious principle. [143] Yet Charles appeared quite guilt-free and unselfconscious of any incongruity between sitting in church and feeling reborn in romantic love or idolizing his eventual wife. And he could still thank the "Bountiful Giver" "for blessing me with such a love on earth." God had not left Charles Strong's universe of meaning, but for many like him, human love had become centrally sustaining to individual identity. Romantic love was a sacred experience to lovers such as Charles, who repeatedly understood and gave this experience meaningful shape with the metaphors of Christian conversion and rebirth. Charles recalled: "You remember the little brick office where we *first met* and those 'earnest eyes' were upon me. That office is sacred to us now—McConnell has it. I went in there yesterday 'where we were born.' "[144]

🐾 The shift from a God-centered to a person-centered universe of meaning occurred in fits and starts. Though some nineteenth-century men and women were already worshipping at the altar of human selfhood by the 1830s, others were expressing a restraint of earthly love that would have met with approval two hundred years earlier. For some it was a struggle, for others it was a *fait accompli*. And still others reflected the transition by their strain. They acknowledged the older standards, sometimes fought to maintain them, but reluctantly admitted the shift in their own personal emotional priorities.[145] Few abandoned God or religion, but they began to find ultimate fulfillment and the central symbol of their life's meaning in the new theology of the romantic self.

The correspondence of Lincoln Clark illustrates the shift from the older religious *mentalité* to a matrix of meaning in which his wife took the place of God as his central symbol of ultimate significance. This happened not because of some abstract disintegration of religious institutions or values but because romantic love created an "other" in human form that answered deep needs of personal identity and rivaled God as a center of personal if not always cosmic gravity. God remained a viable symbol in individual lives, predominantly in terms of the world "out there"; on the personal level the center of gravity was shifting to human personality in nineteenth-century America.

Thus, very early in their marriage, Lincoln wrote that he hoped his last letter had reached Julia so that "it may serve to assure the beloved wife of my bosom that I am incapable of even the colour of neglect to one who has given so many evidences of deep and devoted affection which will ever be regarded as a blessing inferior only to that which laid down its life for the redemption of the object of its regard."[146] At this point Lincoln's emotional hierarchy was still traditional and God-centered.

Lincoln, however, was involved in a process that was definitely secularizing his emotional priorities. After several years, he hinted that his relationship with his wife was growing in equivalency to his religious sympathies. Julia did not approve of writing letters on the Sabbath, but he risked her disapproval more than once. In 1841 he defended himself with the idea that writing to her on the Sabbath was his sacred duty: "Perhaps you will be so much surprised that I should devote a *portion* even of this sacred day in this way, that my letter will be less acceptable than it otherwise would be; but I do not see how I can defer this necessary, pleasant, and I think I may say sacred duty."[147] Two years later he referred to his family as his "little domestic church," and he continued, "I remember you all, all!: and as when a boy I followed the reaper and gathered up the hand-fuls, and bound them together for the granary, so tho' going from you, I follow you, gather you all in the band of love, baptise you with its warm pledges . . . ." Yet he clung to a traditional religious order, however narrowly, praying to the "Father above" for his family's spiritual welfare.[148]

After ten years of marriage, Lincoln confessed to Julia: "and be assured that I love you as I fear with all my heart which you know is against one of the commandments."[149] Writing on the Sabbath a year later, he indicated that the emotional shift to a new center of gravity was complete: "I believe that I have often told you that I love the coming of this day—and more perhaps now that I am *alone* than when you were with me, for it gives me uninterrupted time to hold communion with one I know I love and the *Great One* who *I hope I love.*"[150] Romantic love had the power to invest the spouse, fiancée, or friend with such emotional significance that religious relationships assumed a more peripheral role in the personal psyche. Romantic relationships had an immediacy and a physical presence which created intense competition with the love of a more abstract religious being. Thus, Lincoln felt his love for Julia, whom he "knew" he loved, more strongly than he was able to feel his love for God, whom he only "hoped" he loved.

The emotional distance between Lincoln's Sabbath confession and the next furtive note is unmistakable. In 1823, Mary Garland worried to her husband: "I hope you will during my absence fill my seat at Church, and in your devotion remember one—who ever remembers you at a throne of Grace, as it is the sabbath I cannot conscientiously devote more time to you so I bid you my Dear Husband for the present Adieu."[151] Mary obviously remained entrenched in a system of meaning in which her relationship with God (as conceptualized by a strict Sabbatarian rule) was primary. Thus she was chastened in writing a letter to her husband on Sunday.

Examples of very pious men and women who succumbed emotionally to the new religion of personal love while intellectually recognizing the priority of the older theological framework of religious meaning are not hard to find. The pious missionary, Mary Walker, wrote in her journal in 1847, after a long struggle to achieve a romantically satisfying marriage: "But I find myself inclined to depend on something else rather than on God. My heart is full of idols. In trouble I am prone to cast myself into the arms of my husband rather than my Saviour."[152] Emily Lovell also saw her husband as an idol and her love as worship. In doing so, however, she expressed self-conscious pangs of guilt, "but if I am wicked enough to worship you dear, as I do, I feel that I shall be made to suffer."[153]

Ironically, in the courtship of James Bell and Augusta Hallock, romantic love contributed to Bell's Christian "conversion." Supernatural priorities, however, were only temporarily paramount in James's universe of meaning. Deepening romantic love later challenged and compromised the very religious order that it had originally inspired. On May 27, 1858, James invited his girlfriend Gusta to write him about his piety. He was responding to her anxiety over his lack of religious conviction. Though James was unchurched and unrepentant, the religious language and religiously inspired metaphors for romantic feelings were so pervasive that he pledged his love for Gusta through Christian imagery: "I love you as truly and devotedly as man can love woman. Wherever I have been you have been the *star* that I have turned to with love and delight. My absence from you has only srenghtend that love. It has woven itsselfe into my very being and 'Dearest' believe me when I say that I could no more forget the being who has the keeping of my heart than a true Christian can forget his God."[154]

In late 1858 James's thoughts turned to marriage and he asked Gusta what she thought about the subject.[155] She responded warily in her next letter, telling him she was in no hurry but when he began to think about marriage, he should let her know "so I can *too* . . . ."[156] Subsequently James observed that though they loved each other enough, he was not yet ready to tie the knot.[157] But Gusta's thoughts remained focused on her potential marriage partner, and she worried more earnestly than ever about his lack of faith. She expressed serious reservations about marrying him in his current unconverted state. "Now I'll tell you a candid fact, and its this—while you are a joy, a happiness, to me—you are the greatest trouble I'v got in the world. I have passed some of the soberest, saddest hours of my life, thinking of you, and myself,—of the Future.—Its your want of religion—I not only tremble for you, but *myself* . . . ."[158] She drove home her point forcefully, characterizing her own value system and emotional needs as most basically at odds with his. It appeared she was reevaluating her commitment to their future together. She wrote: "*You* to whom I *should* look for such advise could not give it. You to whom my every sorrowing thought should be confided and to whom I should look, to point and lead the way, *could not feel for me*. In every thing else you could; but that which lies nearest my heart would find no unison in yours. May the time never come when I shall look to you for religious sympathy and *look in vain*[.] I speak strong, but its as I feel."[159]

Troubled, anxious, and perplexed, James remarked that he could not find the motivation to pursue religious experience. Fearing that ending his relationship with Gusta would be better for her, he still fiercely hoped to continue it. "I think that the great love I have for you; will always prompt me to do that which will promote your happiness. I shall have to agree with you when you say that there can be no sympathy between us in such matters. I believe that I have done wrong in trying to win you for a companion through this life. I am afraid that I shall fail on my part . . . . Some times I am tempted to say that we had better stop where we are. But I cant bear the thought of *that:* I have loved you so long and so much that I can't think of such a thing."[160]

Under the emotional pressure of losing her, James experienced a religious awakening. By the time her next letter reached him, James found God. He commented on his feelings toward her after his religious conversion: "I love you with a better, purer, a holier, and less earthly *love* than I did before."[161] Gusta was ecstatic and admitted she had been determined to end their relationship before word of his new-found faith arrived. She claimed she would have given him up because she was afraid of losing her own soul.[162] At this point Gusta's relationship with God (in her terms) was firmly anchored at the center of her system of emotional priorities.

James's religious life continued to develop and in early 1861 he decided to be baptized and join the church.[163] Later in the same year, he entered the momentous arena of war, volunteering to fight for the northern cause in the Civil War. It was not until early 1862 that he began to hint at a possible conflict between his earthly and heavenly love: "I have your picture with me, and when I feal sad, or lonly, I take it; and my Bible to a quiet place. Is that wrong?"[164]

James was with the Army of the Potomac, the main fighting force in the eastern theater of the Civil War. By the end of 1862 he was aware that he and Gusta might never meet again, but he was comforted by the belief that "we *will* meet on those beautiful shores where *so many* are waiting for loved ones. Yes Gusta, if I go *first* I will *wait* for you there on the other side of the dark waters and the chain that may be snapped here, will be united in more beautiful, and brighter links. There. is a thrill of joy from my soul, that lingers for hours, when I think of that better world, where we can live free from sin, and wickedness. I dont write this to complain of this world, or my condition, no, no, for I love the world, tis beautiful to me to much so. perhaps for my own good."[165] James hinted that though Heaven was a comforting abstraction, his heart was riveted on her and their earthly future together.

His feelings of romantic love intensified at the close of 1862. After surviving the bloody battle of Fredericksburg, he was even more romantically inclined. "Gusta I *do* love you *so* much. My heart is *running over* with sweet loving thoughts of *you*. Confused, and tumultious are the fealings that sweep oer me. Yet amid all the varied emotions of my soul, I feal *very happy*. Yes, more than happy, (if that is possible) I feal so because you love me, and the thought comes to me, am I not loveing you *to much*."[166]

While James expressed flickerings of conflict between his secular and religious love, Gusta remained relatively sanguine about her emotional priorities. She appeared to feel little threat from romantic love in her religious life. But in April and May of 1863, she suddenly confessed, "James I have been thinking of you a great deal today, perhaps not any more tho' than I do *other* days. And I have many times been afraid that I thought *too* much of you. Such things sometimes happen you know. I must guard against letting this great love that I have for you,—interfering with the love that I owe to God."[167] This very religious young woman was experiencing the challenge of romantic love to her formerly secure God-centered universe of meaning. At the end of April, while expressing the magnitude of her love for James, she barely managed an aside regarding her love for God: "I was thinking today how much I *did* love you—A few years ago I would not have thought it possible for me to love anyone as I do now. *Then* I loved you the best of any one, and I do *now*, but *that* love has enlarged, untill its magnitude *astonishes* me. And James I am glad that it is so. I would not have it less. Aside from the love I owe to God;—the best, the dearest thoughts and emotions of my heart I give to *you* . . . ."[168]

The shift to the personhood of the beloved as the object of ultimate significance occurred for James and Gusta in the middle of May 1863. James described the power of romantic love in elegant simplicity: "Sometimes I think that it is best, that we cant tell the great love of our hearts, that a portion is left to be told by the glance of the eye, and that quiet happiness that comes to us in the presance of those we love. Gusta, I do wrong in loving you as I do, I love you best of all, and every thing. Perhaps this may pain you, but I find it impossible to resist the tide that sweeps over me."[169] Gusta responded by a similar confession and then an attack upon her own feelings. "As I read it over and over I asked myself is it possible that you and I think and feel so much alike? How your words come home to me when you say—'Gusta I do wrong in loving you as I do. I love you best of all and every thing.' James I am doing the *same* thing—I *know* I am, and it has troubled me a *great deal*. We must *not* do this. It is wrong and God will not look with favor upon us. *He*, claims, deserves, and should have our best highest and purest affections,—Our great love for each other must not interfere in the least with that which we owe to our Savior—He who died for Sinners—for *us*. Our God, and our Country *first* What I have just said will apply to myself more than to you. I guess."[170]

Gusta's intellectual commitment to a God-centered universe of meaning remained firm and clear, but her emotional priorities drifted in a secular direction nonetheless. In early August 1863 she attempted to pressure her emotions back in line with her intellectual priorities: "Tonight I would return from all my wanderings. I would live near to my Savior. . . . Life soon goes out. It dissolves as a vapor, and we are *gone* but if God, be our portion, how blissful our condition—how *glorious* the exchange How kind is Death to open the gates to that beautiful city."[171] Unfortunately, Gusta was soon to test her own rhetoric about the kindness of death.

James was wounded in late September and wrote Gusta from Stanton Hospital in Washington, D.C., where she managed to reach him before he died. Some twist of fate brought Gusta to his side—after years of separation—and thus enabled her to witness his dying words. What follows is probably her record, written on a little scrap of paper in blue ink. She was at the edge of James's hospital bed on Monday night, October 5, 1863, when he spoke to her:

> Oh I am *so* glad you've come!
> I *knew* you'd come
> I had rather see *you* than any
> body else
> Tell Eddie and Annie to *live*
> for Heaven. Tell all my Ill. [Illinois] friends
> to *live* for Heaven
> We'll meet in Heaven. I'll
> wait for you there. You know
> I *told* you that I'd wait for
> you there.
> Tell Mate that I wanted to
> see her oh—oh how much
> Tell Asher that I love him
> even in death—and pray for him
> It looks light. O Lord take my spirit
> Yes I know you, You
> are Gusta. *my* Gusta.
> Yes take—take my bible. Keep
> it. Tues. morn 6 AM
> Gusta kiss me,—kiss me
> closer.—You will love

me *always* wont you Gusta?
X X X O I'm *so so* thankful
If there should be any mis
take—good bye again

On the back side of this little scrap of paper, her record continues:

Gusta do you want to
live? If you do—you can
by going round this room
Gusta forgive all my
sins. I left it all to your
judgment
Whats the matter Dave? [172]

James Bell's dying words spoke plainly to his faith in a religious afterlife. His conception of the ultimate end of existence was Christian, but
mingled within and weaving in and out of his expression in the last moments of mortal life was also the credo of romantic love. Heaven seemed
worthwhile because he would meet Gusta there, but in a reunion that was
so modeled on his image of their earthly meeting as to barely qualify as
spiritual. Still, on Monday night James retained a grip on his Christian
frame of reference. He commended his spirit to the Lord, and gave Gusta
his Bible.

Death worked on its own timetable. James regained consciousness again
at dawn and as life was ebbing away, he asked Gusta to forgive all his
sins. In this conscious or perhaps semiconscious deathbed request, Gusta
had taken the place of God. She claimed his best and highest affection—
so much so that he asked her to do something quintessentially God-like.
James gave the power of forgiveness and perhaps even final judgment to
Gusta: "Gusta forgive all my sins. I left it all to your judgment . . . ."
The symbolic substitution of the romantic self for the Judeo-Christian
God could hardly be clearer.

But if for some this was an easy transition, for others it was as profound as it was painful. Thus, Gusta Hallock was still struggling two
years after James Bell's death to understand the meaning of the powerful
love she experienced for him. After his death, she wrote sporadically to a
soldier she met in the hospital where James died. In a rough draft of one
of her letters to this soldier, she crossed out certain thoughts which (after
she penned them) were presumably either so personal or painful that she
did not want to share them with her correspondent. Her deletions (which
are in italics in what follows) are poignant testimony to the emotional

clash within Gusta of two world views and the cost to one woman who
was caught in the transition from a traditional sacred cosmos to a newer
religion of personal love:

> 14 years ago in the heart of a school girl of 15, there commenced an
> affection which lived through the changes of 12 long years and then went
> out beneath—the dark shadows of the Death Angel. *That love seemed a*
> *part of my very existance—it was an idolatry—God saw it. 'His ways are not*
> *our ways.' How infinite is he in wisdom and mercy. I have bowed in submis-*
> *sion to the dreadful blow.*[173]

🎜 Narratives of romantic love present dramatic evidence of the de-
sacralization of consciousness in American life. However, this evidence
may also be interpreted, not as a decrease, but as a shift in religious meaning. Though Americans may have invested less emotion in an earlier Judeo-Christian world view, this does not necessarily mean that they became
less religious as a consequence.[174] Rather the evidence suggests that many
reinvested more religious significance in human love. Propelled by that
love, Gusta became a symbol of ultimate meaning in James's living and
dying. However heretical from an earlier Christian perspective, this religious investment in human love actually relied heavily upon Christian
structures and symbols, often substituting functions—for example, the relationship of man and woman for that of God and believer—but changing
little else. Historian Ralph Henry Gabriel's observation is richly suggestive
in this light: "Though such influences cannot be measured it is a reasonable guess that Christianity had as much to do with giving romanticism
its dominant position in the climate of opinion as did literary and artistic
importations from across the Atlantic."[175]

Romantic love made inroads in nineteenth century American cultural
life not because it challenged Christianity but because it absorbed its basic
functions, used its basic language, and retained its basic structure and
world view. The "Romantic Self" became a most powerful God and romantic love became the new salvation alongside Judeo-Christian concepts
of God and theology which retained a prominent place in the shifting
cosmology. Heaven and hell remained the dominant conceptions of an
afterlife; and nineteenth-century men and women still went to church and
worshipped in traditional ways. But the universe of ultimate meaning had
begun to shift to include the "Self" as a symbol of ultimate significance,
and some of the burdens of human existence that were formerly sustained
by religious institutions were now also beginning to be laid upon the
agency of human love.

John Milton, in the seventeenth century, penned a revealing and justly famous line in Book Four of *Paradise Lost:* "He for God only, she for God in him." Though "she for God in him" was an idea that was not altogether acceptable to more religiously minded seventeenth-century American Protestants, it still expressed a prevailing sense of patriarchy and hierarchy in male-female relationships that is useful to contrast to the nineteenth-century view.[176] Within the context of the nineteenth-century religion of romantic love, Milton's lines must be changed to reflect a different emotional logic: "he for God in her, she for God in him." This more symmetrical formulation neatly summarizes the role of romantic love in ameliorating patriarchy and in creating a new vision of the individual's place in the cosmos. Nineteenth-century middle-class men no longer saw themselves in Milton's phrase, "for God only." Under the aegis of romantic love (as well as a host of other influences) men and women were discovering God in each other.

🐝 Still captives of the Victorian dichotomy between public and private matters, historians have been slow to recognize that the transformation of male-female relationships is more than a topic of private consequence. Romantic love shaped the contours of American history as surely as technology and finance. Not only gender and sexuality but basic conceptions of what it meant to be a person grew from and were nurtured by the romantic ethos. Romantic love changed individual feelings, perceptions, and behavior. It is at this level that large social trends actually transform society and at this level that all "isms" become viable social forces. Romantic love has touched many dimensions of American society—dividing loyalties, demanding alterations of expression, channeling personal energies to and from public life, encouraging intense communion and collective distance. Unstable, dynamic, but tightly harnessed to the demands of a new social order, nineteenth-century romantic love was an essential component of the American commitment to self-expression with profound consequences for both public and private life.

NOTES

Note on Use of Sources and Citation Form

The use of *[sic]* in intimate letters is distracting for the reader. The love letters in this study contain myriad grammatical or spelling errors, but these mistakes need not be highlighted through such an intrusive device. Therefore, *[sic]* will not be used to mark errors in quotations from letters.

No silent corrections have been made. Correspondents sometimes wrote in extreme haste and also exhibited various levels of expressive skills. The letters take their flavor and tone from correspondents' elegant turns of phrase as well as their most awkward grammar and spelling. To edit these letters would be an unnecessary and artificial intervention between the reader and the letter writer. An edited letter, however convenient for contemporary publication, is a subtle distortion of the experience of letter writing and reading. Two exceptions, however, have been made. First, a word is deleted if it was inadvertently repeated in a text. Second, ampersand signs varied so much across correspondence that "and" has been substituted for the polyglot forms of abbreviation. In addition, unconscious "corrections" undoubtedly have crept into the text. Psychological studies in perception have repeatedly demonstrated the human tendency to fill in and even reformulate information when presented with incomplete or incorrect visual or aural patterns.

Furthermore, as anyone who has read a stack of manuscript letters can attest, deciphering the difference between a flyspeck and a period, a capital and a lower-case letter, an ink stain and a deliberate punctuation mark is at times like a Rorschach test. What one reader sees as a period, another reads as a comma; and another, a paper mark. The subjective element involved in deciphering hand-written punctuation is present in these quotations. Accurate word transcription also has a subjective element as cursive writing and alphabet formation vary, even within the practice of a single correspondent. Other readers may differ in their interpretation of some of the punctuation and spelling in these manuscripts. Thus, quotations are avoided in which the meaning may be changed by doubtful handwriting.

Occasionally, a correspondent's nickname is employed because the context warrants a more informal designation. Augusta Hallock, for example, is most often referred to as Gusta, because this is usually more appropriate than Augusta for discussion of intimate relationships. However, the initials of formal first names are always used in the specific letter citation.

All emphasis and italics in correspondence follow the original, and only added emphasis is noted in a citation.

Short titles are used in endnotes, but in the Advice-Book Bibliography, complete titles are given as they appear in the *National Union Catalogue.* M.D. or other titles appear in endnotes when they appear in the *NUC.* M.D. is inserted in the Bibliography when the author has been identified as a physician.

Introduction

1. Simone de Beauvoir, *The Second Sex,* trans. H. M. Parshley (New York, 1953; rpt., 1974), 712–43.

2. D. Lummis to C. Lummis, [Feb. 28–March 2, 1884], Dorothea Rhodes Lummis Moore Collection, Henry E. Huntington Library, hereafter cited as Moore Collection.

Chapter 1. The "Pen Is the Tongue of the Absent"

1. Theocritus, Jr. (pseud.), *Dictionary of Love* (New York, 1858), 146.

2. E. Simkins to E. Trescot, Jan. 1, Feb. 25, 1864; E. Trescot to E. Simkins, Dec. 22, 1863, Eldred Simkins Collection, Henry E. Huntington Library, hereafter cited as Simkins Collection.

3. A. Hallock to J. Bell, Oct. 14, 1862, James Alvin Bell Collection, Henry E. Huntington Library, hereafter cited as Bell Collection.

4. Ibid., Oct. 2, 1862.

5. L. Clark to J. Clark, June 1, 1843, Lincoln Clark Collection, Henry E. Huntington Library, hereafter cited as Clark Collection.

6. W. Burnett to J. Burnett, Nov. 28, 1859, Wellington Cleveland Burnett Collection, Henry E. Huntington Library, hereafter cited as Burnett Collection.

7. See Marilyn Ferris Motz, *True Sisterhood: Michigan Women and Their Kin, 1820–1920* (Albany, 1983), 53, 64; Alice P. Kenney, " 'Evidences of Regard': Three Generations of American Love Letters," *Bulletin of the New York Public Library* 76(1972), 94; Ellen K. Rothman, *Hands and Hearts: A History of Courtship in America* (New York, 1984), 10; Daniel J. Boorstin, *The Americans: The Democratic Experience* (New York, 1973), 130–36; Wayne E. Fuller, *The American Mail: Enlarger of the Common Life* (Chicago, 1972); Wesley Everett Rich, *The History of the United States Post Office to the Year 1829* (Cambridge, Mass., 1924); Peter T. Rohrbach and Lowell S. Newman, *American Issue: The U.S. Postage Stamp, 1842–1869* (Washington, D.C., 1984).

8. Anonymous, *The Art of Good Behavior; and Letter Writer on Love, Courtship, and Marriage* (New York, 1846), 78.

9. The high cultural value placed on love letters, coupled with the common practice of saving them, suggests that this category of letter is widely available for historical study, Victorian censorship notwithstanding.

10. General conduct manuals did not always include a discussion of letter writing. On letter-writers and letter writing, see the Introduction by Brian W. Downs to Samuel Richardson, *Familiar Letters on Important Occasions* (London, 1741; rpt., 1928), ix–xxvi; Jean Robertson, *The Art of Letter Writing* (Liverpool, 1942); Alvin F. Harlow, *Old Post Bags: The Story of the Sending of a Letter in Ancient and Modern Times* (New York, 1928); Ruth Perry, *Women, Letters, and the Novel* (New York, 1980), 66–67, 88–91; Norman Fruman, "Some Principles of Epistolary Interpretation," *Centrum New Series* 1(Fall 1981), 93–106.

11. Robertson, *Art of Letter Writing*, 66.

12. Anonymous, *A New Letter-Writer, For the Use of Ladies* (Philadelphia, 186–), iii. This was bound with *A New Letter-Writer, For the Use of Gentlemen.* In this edition the preface to the ladies' letter-writer seems addressed to both sexes.

13. See, in order of letter heading, Emily Thornwell, *The Lady's Guide to Perfect Gentility* (New York, 1856), 167; Anonymous, *A New Letter-Writer, For the Use of Ladies,* 46; ibid., 47; An American Gentleman, *Good Behavior for Young Gentlemen: Founded on Principles of Common Sense, and the Usages of Good American Society* (Rochester, N.Y., 1848), 36; Anonymous, *The Ladies' and Gentleman's Model Letter-Writer: A Complete Guide to Correspondence on All Subjects with Household and Commercial Forms* (New York, 1871), 36.

14. Few, if any, love letters in this sample were etiquette book copy. The one kind of letter that approached the etiquette book models in length and style was the first letter of a correspondence. When extant it was often a formal request to begin a correspondence and might have been copied, though no evidence exists to prove that any in this sample were not original. It is possible that other samplings may turn up caches of etiquette-book model love letters, but they are unlikely to be found in large numbers written by middle-class correspondents.

15. Karen Halttunen, *Confidence Men and Painted Women: A Study of Middle-Class Culture in America, 1830–1870* (New Haven, 1982), 129.

16. Anonymous, *A New Letter-Writer, For the Use of Ladies,* vi.

17. Thornwell, *The Lady's Guide,* 161.

18. Anonymous, *The Art of Good Behavior,* 35.

19. Anonymous, *A New Letter-Writer, For the Use of Ladies,* xi–xii.

20. Ibid., vi–vii. The preface was written for both ladies and gentlemen. See fn. 12.

21. Ibid., 78, 90.

22. Anonymous, *The Ladies' and Gentleman's Model Letter-Writer,* 38–39, 100.

23. Anonymous, *The Art of Good Behavior,* title page.

24. [Mrs. Eliza Farrar] By a Lady, *The Young Lady's Friend* (Boston, 1837), 290–91.

25. Though many commentators have noted that literacy was widespread in America by the early 19th century, not enough attention has been paid to how

the powers of articulation functioned as a strategy to distinguish class, status, and social worth.

26. John Marquis to Neeta Haile, Nov. 11, 1865. All quotations in this paragraph and the next are from the Nov. 11th letter. Neeta Marquis Collection, Henry E. Huntington Library, hereafter cited as Marquis Collection.

27. Ibid.; Anonymous, *The Art of Good Behavior,* was aimed at "those who have not enjoyed the advantages of fashionable life . . . ." The editor included five pages of poetry in this style, intentionally offering readers an alternative to "Roses are red . . ." (118–25).

28. An American Gentleman, *Good Behavior,* 40.

29. Peter Gay, *Education of the Senses,* vol. 1, *The Bourgeois Experience: Victoria to Freud* (New York, 1984), 404–5, 418–22, 445, 452–60; Daniel Scott Smith, "Family Limitation, Sexual Control, and Domestic Feminism in Victorian America," in *A Heritage of Her Own,* Nancy F. Cott and Elizabeth H. Pleck, eds. (New York, 1979), 222–45. Smith remarks in a 1979 postscript to his essay, "The distinction between public and private and the corresponding allocation of men to the former and women to the latter is a central (if not *the* central) theme in the history of gender" (239–40).

30. [Samuel Roberts Wells], *How to Behave; or A Pocket Manual of Republican Etiquette* (New York, 1858), 113.

31. Anonymous, *The Art of Good Behavior,* 35.

32. Burton J. Bledstein titles one of his chapters "Space and Words," in *The Culture of Professionalism: The Middle Class and the Development of Higher Education in America* (New York, 1976), 46–79. He insightfully brings together these two central aspects of middle-class experience, emphasizing that words were the social currency of mid-Victorians, and also underlining the importance of enclosing and defining private space in Victorian culture. See esp. 61–65, 70–75.

33. The separation of public and private life had class implications for nineteenth-century Americans. Paul E. Johnson suggests that "Perhaps more than any other act, the removal of workmen from the homes of employers created an autonomous working class." See *A Shopkeeper's Millennium: Society and Revivals in Rochester, New York, 1815–1837* (New York, 1978), 106. As the door to more prosperous homes was shut tight against the world, classes were created by a literal separation of owner and worker. The differentiation of space set up barriers to communication which contributed to the social and cultural gap between different kinds of workers. Home was now separated from business and in this act of separation it became possible to build barriers based upon individual control of self and environment, reflected in one's involvement in different social settings. Middle-class formation fed off this growing social and cultural segmentation.

Mary P. Ryan documents the emergence of a middle class by mid-19th century in *Cradle of the Middle Class: The Family in Oneida County, New York, 1790–1865* (New York, 1981). But as she notes, class formation in American life is still a "dimly understood" phenomenon (xiii and 13–15). See Ryan for a succinct summary of the strategies by which native-born families reproduced their middle-class standing in the subsequent generation (184–85). See also Stuart M. Blumin's insightful and suggestive essay, "The Hypothesis of Middle-Class Formation in Nineteenth-Century America: A Critique and Some Proposals," *American Historical Review* 90 (April 1985), 305, 299–338; Anthony F. C. Wallace, *Rockdale: The Growth of an American Village in the Early Industrial Revolution* (New York, 1972), 44–65; Michael B. Katz, by contrast, argues for a two-class division in this period in "Social Class in North American Urban History," *Journal of Interdisciplinary History* 11 (Spring 1981), 579–605.

34. Evidence for these generalizations on romantic love can be found in Chapters 2, 3, and the last half of both Chapters 4 and 8.

35. Halttunen, *Confidence Men and Painted Women,* 55, 104, 107–10. The phrase "social geography" is Halttunen's. See esp. 104, 107, 110.

36. François Peyre-Ferry, *The Art of Epistolary Composition* (Middletown, Conn., 1826), 30.

37. L. Clark to J. Clark, May 13, 1847, Clark Collection. There are few adequate histories of letter-writing to draw upon. See George A. Saintsbury, *A Letter Book: Selected with an Introduction on the History and Art of Letter-Writing* (London, 1922); Everett Emerson, ed., *Letters from New England: The Massachusetts Bay Colony, 1629–1638* (Amherst, Mass., 1976).

38. J. Marquis to N. Marquis, April 4, March 29, March 6, Feb. 21, Feb. 17, 1870; Oct. 2, 1866, Marquis Collection.

39. For example, ibid., April 5, 1863.

40. N. Wheeler to C. Bradley, Sept. 13, Oct. 25, Nov. 1, 1874, Nov. 22, 1876, Jan. 2, Feb. 2, 1877, Feb. 3, 1878, Clara Bradley Burdette Collection, Henry E. Huntington Library, hereafter cited as Burdette Collection.

41. J. Hague to M. W. Foote, Nov. 20, Dec. 13, Dec. 31, 1870, James Duncan Hague Collection, Henry E. Huntington Library, hereafter cited as Hague Collection.

42. Ibid., July 4, Aug. 11, 1871.

43. H. Russell to C. Strong, Jan. 14, 1862, [Dec. 1862], Jan. 6, 1863, Harriet Williams Russell Strong Collection, Henry E. Huntington Library, hereafter cited as Strong Collection.

44. C. Strong to H. Russell, Dec. 17, 1862, Jan. 8, Jan. 11, 1863.

45. See Joe L. Dubbert, *A Man's Place: Masculinity in Transition* (Englewood Cliffs, N.J., 1979), 80–121, esp. 97, 115; and Elizabeth H. Pleck and Joseph H. Pleck, eds., *The American Man* (Englewood Cliffs, N.J., 1980), 1–49, esp. 29. Both these early synthetic men's histories were bold, intelligent, and pioneering efforts, but constrained by the limits of published primary sources and the dearth of secondary sources on private male-female relationships. There are also problems of evidence and interpretation on the private side of the male role in Peter Gabriel Filene, *Him/Her/Self: Sex Roles in Modern America* (New York, 1975), 68–71; and David G. Pugh, *Sons of Liberty: The Masculine Mind in Nineteenth-Century America* (Westport, Conn., 1983), 54, 60–61, 79–80.

46. Striking gender differences are claimed for overland trail diaries by both John Mack Faragher, *Women and Men on the Overland Trail* (New Haven, 1979),

see esp. 128–33; and Lillian Schlissel, *Women's Diaries of the Westward Journey* (New York, 1982), 14–15. Though Faragher introduces his book by insisting that men and women "were part of a common culture, that they were, indeed, more alike than different," he actually focuses upon and emphasizes gender contrast. Among his observations are these: women's writing was personalized, men's depersonalized; men wrote tersely and with taciturnity, while women enriched and elaborated their descriptions; men wrote about objects and things, while women wrote about people; men's persona was rigid while women's was more emotional and empathetic (132–33). He observes, "Men hid their feelings from themselves, understood themselves in an incomplete way, but found comfort in the company of their similarly repressed brothers" (133).

If, however, the diaries of the overland trail were often intended to be public documents, as some commentators have suggested, then they may illustrate not so much the paucity of men's emotional life as the ideology of masculine separate spheres in which men were culturally encouraged to hide themselves emotionally in the public sphere. As Schlissel notes in *Women's Diaries,* "Overland diaries were a special kind of diary, often meant to be published in county newspapers or sent to relatives intending to make the same journey the following season" (11). See also Andrew J. Rotter, " 'Matilda for Gods Sake Write': Women and Families on the Argonaut Mind," *California History* 58 (Summer 1979), 130. This may also help to explain the absence of contrast in male and female rural diaries examined by Marilyn Ferris Motz. She found in her rural Michigan sample that both men *and* women wrote "terse accounts of weather and daily activities, with little commentary on emotions or reactions" (Motz, *True Sisterhood,* 77). It may be that male and female language differences receded in genuinely private forms of expression. See Chapter 5 for a fuller treatment of 19th-century gender roles and the ideals of masculinity.

Class variables might offer another plausible explanation of the different patterns of masculine response found in samples of 19th-century private expression. It is possible, though it remains to be proved, that expression weighted on the lower-middle-class side contains more dramatic gender differences than middle-to upper-middle-class evidence. It is also possible that only in activities relating to romantic love were any class of men allowed the full range of emotional expression in 19th-century American culture. Motz takes the latter position in *True Sisterhood,* 80.

47. T[imothy] S[hay] Arthur, *Advice to Young Men on Their Duties and Conduct in Life* (Philadelphia, [1847]), 109–10. See John F. Kasson, "Civility and Rudeness: Urban Etiquette and the Bourgeois Social Order in Nineteenth-Century America," in *Prospects,* Jack Salzman, ed., vol. 9 (New York, 1984), 143–67.

48. This phrase was taken from E. Simkins to E. Trescot, Oct. 26, 1862, Simkins Collection.

49. Ibid., Jan. 8, 1864; see also Aug. 14, 1864.

50. N. Wheeler to C. Bradley, Sept. 13, 1874, Burdette Collection.

51. Ibid., Oct. 25, 1874.

52. E. Lovell to M. Lovell, Nov. 15, 1862, Mansfield Lovell Collection, Henry E. Huntington Library, hereafter cited as Lovell Collection.

53. E. Trescot to E. Simkins, Aug. 13, 1864, Simkins Collection.

54. N. Wheeler to C. Bradley, Jan. 7, 1878, Burdette Collection.

55. J. Bolton to A. Harrison, July 16, 1838, Harrison Family Collection, Robert Alonzo Brock Collection, Henry E. Huntington Library, hereafter cited as Harrison Collection.

56. A. Hallock to J. Bell, June 29, 1862, Bell Collection.

57. Ibid., July 8, 1863.

58. E. Simkins to E. Trescot, Feb. 25, 1864, Simkins Collection.

59. Ibid., Oct. 18, 1864.

60. C. Baker to R. Burdette, Aug. 12, 1898, Burdette Collection.

61. Ibid.

62. S. F. Smith to M. Smith, May 6, 1834, Samuel Francis Smith Collection, Henry E. Huntington Library, hereafter cited as Smith Collection.

63. R. Burdette to C. Baker, Nov. 14, 1898, Burdette Collection.

64. A. Janin to V. Blair, Dec. 18, 1871, Janin Family Collection, Henry E. Huntington Library, hereafter cited as Janin Collection.

65. Ibid., Dec. 9, 1872.

66. N. Hawthorne to S. Peabody, April 30, 1839, Nathaniel Hawthorne Collection, Henry E. Huntington Library, hereafter cited as Hawthorne Collection.

67. Ibid., July 15, 1839.

68. Ibid., April 30, 1839.

69. Ibid., April 21, 1840.

70. Ibid.

71. E. Simkins to E. Trescot, Nov. 2, 1863, Simkins Collection.

72. L. Clark to J. Clark, April 29, 1852, Clark Collection.

73. J. Bell to A. Hallock, [late Sept. 1862], Bell Collection.

74. J. Hague to M. Hague, May 12, 1875, Hague Collection.

75. E. Trescot to E. Simkins, Aug. 23, 1864, Simkins Collection.

76. J. Bolton to A. Harrison, July 16, 1838, Harrison Collection.

77. J. W. North to A. Loomis, July 8, 1848, John Wesley North Collection, Henry E. Huntington Library, hereafter cited as North Collection.

78. Clara Baker noted the length of Robert Burdette's letter in her letter to him of Feb. 23, 1899, Burdette Collection. Since she censored his correspondence, missing pages at the end of letters make it impossible to calculate the length of any censored originals.

79. C. Strong to H. Russell, Dec. 14, 1862, Strong Collection.

80. E. Simkins to E. Trescot, Sept. 28, 1864, Simkins Collection.

81. Ibid.

82. E. Trescot to E. Simkins, Jan. 14, 1864.

83. Ibid., July 12, 1864.

84. Ibid., July 14, 1864.

85. A. Hallock to J. Bell, Jan. 14, 1861, Bell Collection. This dimension of the relationship of Bell and Hallock is explored more fully in Chapter 5.

86. S. F. Smith to M. Smith, August. 10, 1834, Smith Collection.

87. J. Marquis to N. Marquis, Feb. 7, 1870, Marquis Collection.

88. Chapter 2 documents these contentions and analyzes the process more fully.

89. Anonymous, "Female Letter-Writers," *Taits' Edinburgh Magazine* 1 (May 1832), 197–98.

Chapter 2. Falling in Love

1. Peter N. Stearns and Carol Z. Stearns, "Emotionology: Clarifying the History of Emotions and Emotional Standards," *American Historical Review* 90 (Oct. 1985), 813–36.

2. Though romance, romantic love, romanticism, and the romantic movement are often used and related terms, they are not interchangeable. Applied to fiction, romance refers to a particular type of novel, with both high culture and popular culture forms. High culture romance emphasizes allegorical characters and symbolic landscapes more than naturalistic or realistic re-creations of character and culture, as, for example, in *The Scarlet Letter*. See Michael Davitt Bell, *The Development of American Romance: The Sacrifice of Relation* (Chicago, 1980); Richard Chase, *The American Novel and Its Tradition* (Garden City, N.Y., 1957); Perry Miller, "The Romance and the Novel," in *Nature's Nation* (Cambridge, Mass.: 1967), 241–78; Joel Porte, *The Romance in America: Studies in Cooper, Poe, Hawthorne, Melville, and James* (Middletown, Conn., 1969).

Popular culture romance is a formulaic type of fiction, with heroines and heroes who fall in love but must overcome many obstacles finally to consummate their special attraction. See Kay Mussell, *Women's Gothic and Romantic Fiction: A Reference Guide* (Westport, Conn., 1981); John G. Cawelti, *Adventure, Mystery, and Romance: Formula Stories as Art and Popular Culture* (Chicago, 1976); Janice A. Radway, "The Utopian Impulse in Popular Literature: Gothic Romances and 'Feminist' Protest," *American Quarterly* 33 (Summer 1981), 140–62; Leslie Fiedler uses both types of romantic fiction in *Love and Death In the American Novel* (New York, 1960; rev. ed., 1969). See also Janice A. Radway, *Reading the Romance: Women, Patriarchy, and Popular Literature* (Chapel Hill, 1984); Tania Modleski, *Loving with a Vengeance: Mass-Produced Fantasies for Women* (Hamden, Conn., 1982).

The romantic movement describes a particular period of cultural history—essentially the late 18th and early 19th century. Romanticism is the concomitant world view. Definitions of romanticism have been debated fiercely, but it is primarily a term used to designate a new constellation of values celebrating organicism, creative imagination, and the constructive powers of the mind, sense experience, personal growth, and individual diversity—embodied in literary and philosophical works of the 19th century. See Morse Peckham, *The Triumph of Romanticism: Collected Essays* (Columbia, S.C., 1970); Mario Praz, *The Romantic*

Agony, trans. Angus Davidson (London, 1933); Northrop Frye, ed., *Romanticism Reconsidered: Selected Papers from the English Institute* (New York, 1963).

Non-fictional romantic love, while linked to fictional romance, the romantic movement, and romanticism, is a separate phenomenon. Though literature was undoubtedly a key force in its content and transmission, romantic love is embodied in the social world of real as well as fictional life.

3. See, for example, Herman R. Lantz, "Romantic Love in the Pre-Modern Period: A Sociological Commentary," *Journal of Social History* 15 (Spring 1982), 349–70. There is disagreement about the nature of courtly love. See F. X. Newman, ed., *The Meaning of Courtly Love* (Albany, N.Y., 1968); Irving Singer, *The Nature of Love,* vol. 2, *Courtly and Romantic* (Chicago, 1984), 19–36. Two historians who have created controversy around their claims for the presence or absence of romantic love are Lawrence Stone, *The Family, Sex and Marriage in England, 1500–1800* (New York, 1977), and Edward Shorter, *The Making of the Modern Family* (New York, 1975). Two classic literary studies on love are C. S. Lewis, *The Allegory of Love: A Study in Medieval Tradition* (Oxford, Eng., 1936), esp. 1–43, and Denis De Rougemont, *Love in the Western World,* trans. Montgomery Belgion (New York, 1940; rev. ed., 1956), 108–22. While these latter two authors agree on the origins of romantic love, their interpretations of its significance are almost diametrically opposed. Lewis sees romantic love as the source of the empathetic happiness and intimacy in modern marriage; De Rougemont believes romantic love is the ultimate barrier to a realistic commitment to another human being and thus at the heart of what destroys the happiness of the married couple. De Rougemont opposes the Christian to the romantic ideal. Lewis sees romantic love itself as the source of the Christian marriage ideal; thus he discovers no fatal opposition. See also Singer, *Courtly and Romantic,* 283–302.

4. Daniel Scott Smith and Michael S. Hindus, "Premarital Pregnancy in America, 1640–1971: An Overview and Interpretation," *Journal of Interdisciplinary History* 5 (Spring 1975), 537–70; Daniel Scott Smith, "Parental Power and Marriage Patterns: An Analysis of Historical Trends in Hingham, Massachusetts," *Journal of Marriage and the Family* 35 (Aug. 1973), 419–28; Ellen K. Rothman, *Hands and Hearts: A History of Courtship in America* (New York, 1984), 31 35, 103–7.

5. All the definitions and formulations of romantic love in this chapter are based on the experience of 19th-century lovers. I avoided imposing, as much as possible, a priori philosophical and psychological constructs of love drawn from other historical eras. I use the term "identification" not in any technical psychological sense, but rather to describe the "emotional merging" of individual-to-individual or individual-to-group which Philip Gleason has dubbed "the vernacular meaning of the word." See his "Identifying Identity: A Semantic History," *Journal of American History* 69 (March 1983), 910–31, esp. 915–17. I do not take the unity of the self, implicit in such terms as personality and identity, as a given but as an historical construct, central to understanding Victorian romantic love and the profound effects it has had on American social and cultural life.

6. Perry Miller, *The New England Mind: The Seventeenth Century* (New York,

1939), 53–63; Stone, *The Family, Sex, and Marriage,* 224–27, 262–64; Edmund Leites, "The Duty to Desire: Love, Friendship, and Sexuality in Some Puritan Theories of Marriage," *Journal of Social History* 15 (Spring 1982), 383–408; Robert Middlekauff, *The Mathers: Three Generations of Puritan Intellectuals, 1596–1728* (New York, 1971), 3–8.

7. William Haller and Malleville Haller, "The Puritan Art of Love," *Huntington Library Quarterly* 5 (Jan. 1942), 235–72; William Haller, "'Hail, Wedded Love,'" *English Literary History* 13 (June 1946), 79–97; Edmund S. Morgan, *The Puritan Family: Religion and Domestic Relations in Seventeenth-Century New England* (Boston, 1944; rev. ed., 1966), esp. 29–64. Morgan's Puritans diverge from Haller's in terms of the emphasis on controlling human affection.

8. See David E. Stannard, *The Puritan Way of Death: A Study in Religion, Culture, and Social Change* (New York, 1977), 72–91.

9. Kenneth Lockridge, *A New England Town: The First Hundred Years, Dedham, Massachusetts, 1636–1736* (New York, 1970); Philip J. Greven, Jr., *Four Generations: Population, Land, and Family in Colonial Andover, Massachusetts* (Ithaca, 1970); Michael Zuckerman, *Peaceable Kingdoms: New England Towns in the Eighteenth Century* (New York, 1970).

10. Jan Lewis, "Domestic Tranquillity and the Management of Emotion Among the Gentry of Pre-Revolutionary Virginia," *William and Mary Quarterly* 39 (Jan. 1982), 135–49; Michael Zuckerman, "William Byrd's Family," *Perspectives in American History* 12 (1979), 255–311; Rhys Isaac, *The Transformation of Virginia 1740–1790* (Chapel Hill, 1982), 58–138; Jan Lewis, *The Pursuit of Happiness: Family and Values in Jefferson's Virginia* (New York, 1983), 36–39, 44–46.

11. Isaac, *The Transformation of Virginia,* 161–72; Lewis, *The Pursuit of Happiness,* 51; William G. McLoughlin, *Isaac Backus and the American Pietistic Tradition* (Boston, 1967); Edwin S. Gaustad, *The Great Awakening in New England* (New York, 1957); Richard L. Bushman, *From Puritan to Yankee: Character and the Social Order in Connecticut, 1690–1765* (Cambridge, Mass., 1967; rpt., 1970), 196–220.

12. Herman R. Lantz et al., "Pre-Industrial Patterns in the Colonial Family in America: A Content Analysis of Colonial Magazines," *American Sociological Review* 33 (June 1968), 413–26; Herman R. Lantz et al., "The Changing American Family from the Preindustrial to the Industrial Period: A Final Report," *American Sociological Review* 42 (June 1977), 406–21.

13. Robert A. Gross, *The Minutemen and Their World* (New York, 1976), 75–108; Lockridge, *A New England Town,* 167–80; Smith, "Parental Power and Marriage Patterns," 419–28.

14. Peckham, *The Triumph of Romanticism,* 38, my emphasis. I found chapter 3, "The Dilemma of a Century: Four Stages of Romanticism" (36–57), particularly helpful. The romantic movement characterized certain external roles as inherently self-expressive and therefore often viewed them as "natural" rather than as social constructs. The artist was one such "anti-role" in the romantic tradition.

15. Perhaps the most significant conduits of that romantic sense of self or socialization of romantic behavior were literature and music. For the literary

manifestations, see Fiedler, *Love and Death in the American Novel,* 13, 18, 47, 49, 54–55, 58, 100–102; R. W. B. Lewis, *The American Adam: Innocence, Tragedy and Tradition in the Nineteenth Century* (Chicago, 1955); Herbert Ross Brown, *The Sentimental Novel in America, 1789–1860* (Durham, N.C., 1940); Lantz et al., "Preindustrial Patterns," 413–26, and "Changing American Family," 414–15; James D. Hart, *The Popular Book: A History of America's Literary Taste* (New York, 1950; rpt., 1963), 51–84, 180–200. Joyce W. Warren, *The American Narcissus: Individualism and Women in Nineteenth-Century American Fiction* (New Brunswick, N.J., 1984), argues that only men were granted full individuality in 19th-century American fiction (1–19); Mary Kelley strenuously disagrees in *Private Woman, Public Stage: Literary Domesticity in Nineteenth-Century America* (New York, 1984).

The power of romantic love is affirmed and some kind of romantic sensibility and behavior is displayed in popular 19th-century songs. See Nicholas E. Tawa, *A Music for the Millions: Antebellum Democratic Attitudes and the Birth of American Popular Music* (New York, 1984), 70–83, 110–18; Sigmund Spaeth, *Read 'Em and Weep: The Songs You Forgot To Remember* (Garden City, N.Y., 1927).

16. See Chapter 8.

17. The historical study of childrearing has been a major exception because it is a subject area that must concern itself with changing patterns of socialization. See Carl N. Degler, *At Odds: Women and the Family in America from the Revolution to the Present* (New York, 1980), 66–110; Glenn Davis, *Childhood and History in America* (New York, 1976); Jacqueline S. Reinier, "Rearing the Republican Child: Attitudes and Practices in Post-Revolutionary Philadelphia," *William and Mary Quarterly* 3rd ser. 39 (Jan. 1982), 150–51. Robert L. Griswold succinctly and incisively summarizes the current scholarship on the history of American childrearing in *Family and Divorce in California, 1850–1890: Victorian Illusions and Everyday Realities* (Albany, 1982), 141–50; also see Chapter 8, fn. 18 for other relevant citations.

18. A. Janin to V. Blair, Nov. 6, 1871, Janin Collection.

19. Ibid., Dec. 14, 1871.

20. Ibid., Nov. 30, 1871.

21. Ibid., Jan. 4, 1872.

22. A sister to A. Hallock, Feb. 14, 1866, Bell Collection.

23. J. Bell to A. Hallock, March 25, 1860.

24. D. Lummis to C. Lummis, [April 10, 1884], Moore Collection. She quotes him in her letter.

25. N. Hawthorne to S. Peabody, Aug. 26, 1839, Hawthorne Collection.

26. M. Smith to S. F. Smith, Jan. 27, 1834, Smith Collection.

27. S. F. Smith to M. Smith, March 13, March 14, 1834.

28. Ibid., July 3, 1834.

29. M. Smith to S. F. Smith, July 7, 1834.

30. N. Wheeler to C. Bradley, Feb. 13, 1875, Burdette Collection.

31. Ibid., March 12, 1876.

32. Ibid.

33. E. Simkins to E. Trescot, Oct. 26, 1862, Simkins Collection.

34. Ibid., April 15, 1863.

35. Ibid., July 19, 1863.

36. Ibid., Sept. 4, 1863.

37. Ibid.

38. Ibid., Sept. 11, 1863.

39. E. Trescot to E. Simkins, July 14, 1864.

40. Ibid., Sept. 8, 1864.

41. These factors included evangelical Protestantism, education, the communications revolution, increasing mobility, the changing birth-rate and child-oriented family patterns, and industrialization.

42. J. Bell to A. Hallock, Oct. 18, 1858, Bell Collection.

43. Ibid., Sept. 19, 1863.

44. Ibid., May 16, 1859.

45. A. Hallock to J. Bell, Aug. 31, 1862.

46. J. Bell to A. Hallock, Jan. 27, 1862.

47. A. Hallock to J. Bell, Dec. 1, 1861.

48. J. Bell to A. Hallock, Dec. 26, 1862.

49. Ibid., Sept. 19, 1863.

50. Ibid.

51. Anita Clair Fellman and Michael Fellman, *Making Sense of Self: Medical Advice Literature in Late Nineteenth-Century America* (Philadelphia, 1981), 3–21; Daniel Walker Howe, "Victorian Culture in America," in *Victorian America,* D. W. Howe, ed. (Philadelphia, 1976), 3–28, esp. 17–25; John F. Kasson, "Civility and Rudeness: Urban Etiquette and the Bourgeois Social Order in Nineteenth-Century America," *Prospects,* Jack Salzman, ed., vol. 9 (New York, 1984), 143–67; Burton J. Bledstein, *The Culture of Professionalism: The Middle Class and the Development of Higher Education in America* (New York, 1976), 146–47.

52. Warren I. Susman, " 'Personality' and the Making of Twentieth-Century Culture," in *New Directions in American Intellectual History,* John Higham and Paul K. Conkin, eds. (Baltimore, 1979), 212–26. The *Oxford English Dictionary* indicates that in its early usages, personality referred to the contrast between a person and a thing, or to the trinity of the Godhead, or to the actual existence of a person. The modern meaning of personality as a "quality or assemblage of qualities which makes a person what he is as distinct from other people" dates, according to the *O.E.D.,* from the late 18th century.

53. See fn. 82 in Chapter 4; and Karen Halttunen, *Confidence Men and Painted Women: A Study of Middle-Class Culture in America, 1830–1870* (New Haven, 1982), 109–10.

54. Obviously reaching this goal was mitigated by the power of role socialization and ultimately constrained by the paradoxical fact that the romantic self was its own kind of role.

55. Examples of the specific dialogues of courtship can be found in Chapters 5 and 6.

56. The dominant cultural patterns of courtship, including the multiple dimensions of ritual courtship testing, are discussed at length in Chapter 6.

57. S. F. Smith to M. Smith, April 25, 1834; M. Smith to S. F. Smith, April 24[–28], 1834, Smith Collection. Mary's response was written April 28, soon after she received Samuel's letter of the 25th.

58. S. F. Smith to M. Smith, Aug. 1, 1834; M. Smith to S. F. Smith, Aug. 4, 1834.

59. J. Bell to A. Hallock, Feb. 4, 1862, Bell Collection.

60. J. Hague to M. W. Foote, Jan. 19, 1872, Hague Collection; "grouty" was a slang term for peevish or sulky. See Eric Partridge, *A Dictionary of Slang and Unconventional English,* Paul Beale, ed., 8th ed. (London, 1984).

61. J. Marquis to N. Haile, March 10, 1865, Marquis Collection.

62. J. Hague to M. W. Foote, Jan. 19, 1872, Hague Collection.

63. N. Wheeler to C. Bradley, Nov. 16, 1876, Burdette Collection.

64. A. Hallock to J. Bell, Dec. 14, 1861, Bell Collection. Fanny Fern published what today might be called "lifestyle" columns in the *New York Ledger,* a popular monthly magazine, and in weeklies such as the *Saturday Evening Post;* she also wrote novels, essays, and sketches; her most famous collection of sketches was titled *Fern Leaves from Fanny's Port-Folio* (1853). See Frank Luther Mott, *A History of American Magazines, 1850–1865,* vol. 2 (Cambridge, Mass., 1938), 23, 357.

65. J. Marquis to N. Haile, March 10, 1865, Marquis Collection.

66. A. Janin to V. Blair, Nov. 27, 1871, Janin Collection.

67. Ibid., Nov. 30, 1871.

68. Ibid. Quoted in his letter to her Jan. 2, 1872. Albert Janin also criticized her through sarcasm in a March 4, 1862, letter.

69. Ibid., Jan. 15, Jan. 25, Feb. 1, 1872.

70. Ibid., March 7, 1872.

71. N. Hawthorne to S. Peabody, Jan. 13, 1841, Hawthorne Collection.

72. A. Hallock to J. Bell, July 21, 1863, Bell Collection.

73. See Chapter 5 for a fuller discussion of 19th-century sex-role ideals and their application.

74. Quentin Anderson observed that Hawthorne's fiction is "shrewd . . . about the quality of contemporary relationships," in *The Imperial Self: An Essay in American Literary and Cultural History* (New York, 1971), 60. See also Leland S. Person, Jr., "Hawthorne's Love Letters: Writing and Relationship," *American Literature* 59 (May 1987), 211–27.

75. N. Hawthorne to S. Peabody, May 26, 1839, Hawthorne Collection.

76. Ibid., April 6, 1842.

77. Ibid., July 15, 1839.

78. Empathy is a word that was not invented until the early 20th century. See *A New English Dictionary of Historical Principles: Introduction, Supplement, and Bibliography* (Oxford, Eng., 1933), 329. I have used "empathy" sparingly because it does not fully capture the experience that is being described in these 19th-century sources.

79. N. Hawthorne to S. Peabody, April 6, 1840, Hawthorne Collection.

80. Ibid., March 30, April 6, 1840.

81. Ibid., Oct. 10, 1839.

82. Ibid., Nov. 27, 1840.

83. C. Baker to R. Burdette, Aug. 7, 1898, Burdette Collection.

84. Ibid.

85. E. Boynton to W. S. Harbert. Undated letter fragment written sometime prior to Nov. 18, 1870, Elizabeth Morrisson Boynton Harbert Collection, Henry E. Huntington Library, hereafter cited as Harbert Collection.

86. J. Burnett to W. Burnett, Sept. 30, 1856, Burnett Collection.

87. M. Smith to S. F. Smith, March 3, 1834, Smith Collection.

88. Ibid., June 9, 1834.

89. Ibid., July 7, 1834.

90. S. F. Smith to M. Smith, July 10, 1834.

91. A. Hallock to J. Bell, April 25, 1863, Bell Collection.

92. J. Bell to A. Hallock, Aug. 22, 1863.

93. E. Simkins to E. Trescot, Jan. 1, 1864, Simkins Collection.

94. N. Wheeler to C. Bradley, Nov. 16, 1876, Burdette Collection.

95. Ibid., Aug. 2, 1877.

96. A. Janin to V. Blair, Nov. 6, 1871, Janin Collection.

97. Ibid., Nov. 7, 1871.

98. Ibid., Nov. 17, 1871.

99. Ibid., Jan. 4, 1872.

100. Ibid., Dec. 18, 1871.

101. Edward Shorter sees the essence of romantic love as spontaneity and empathy, in *The Making of the Modern Family*. While not incorrect, these characteristics do not fully encompass the particular experience of romantic love in the 19th-century American middle class. Shorter, however, has understood one of the central effects of this romantic experience: "One consequence of such intense emotional exchange has been the dismantling of strict sex-role divisions" (15–16).

102. Chapter 8 presents evidence that gender boundaries—crossed at least temporarily in romantic love—affected private male-female power relationships in the 19th-century household.

103. I found very few *emotional* differences *in private* between men and women when both were immersed in romantic love. The disparity that did exist in 19th-century middle-class male versus female behavior and feelings within the circumference of romantic love stemmed from women's economic dependence and men's relative economic independence. Both sexes were wary, for different reasons, of the economic consequences of emotional involvement, and this proved to be one of the major barriers to 19th-century male-female emotional intimacy. I am aware of Simone de Beauvoir's argument that European and American women experienced different emotions in romantic love than men (Simone de Beauvoir, *The Second Sex*, trans. H. M. Parshley [New York, 1953; rpt., 1974], 712–43). The evidence of this study, however, challenges the idea that the emotions of romantic love in Victorian relationships were rigidly sex-typed. Of course there is no doubt that 19th-century women's economic dependence created differences in their response to courtship and marriage as well as affected women's feelings of emo-

tional vulnerability. Furthermore, de Beauvoir's analysis may be quite correct for post-World War II men and women. Romantic love is not an historical monolith and must be treated with more chronological precision. See Francesca M. Cancian, "The Feminization of Love," *Signs: Journal of Women in Culture and Society* 11 (Summer 1986), 692–709, for an argument that the definition of love in contemporary American culture, emphasizing the verbalization of feelings, has excluded more non-verbal "masculine" styles of expressing love.

104. Part of the process of creating romantic love in the 19th century involved dramatic courtship testing rituals which accounted for some of the emotional fluctuation, anxiety, and intensity of feeling found before marriage. See Chapter 6. Nineteenth-century courtship testing played a major role in the early stages of romantic excitement and fluctuation of feelings. After marriage, as long as romantic love survived, the potential for such courtship drama existed, but in the "relatively untroubled" marriage, the emotional pitch was lower and the emotional fluctuations appeared less intense. See Chapter 7. This did *not* mean that romantic love had necessarily departed from a marriage, but rather that emotional testing had abated. I would argue that if romantic love was not moribund, the emotional response pattern—jealousy, obsessive thoughts of the other, emotional highs and lows, pain of absence, etc.—could be activated by various situational crises in marriage. The workings of romantic love in marriage will necessitate more documentation and consideration in this light.

105. L. Clark to J. Clark, March 28, 1847, Clark Collection.

106. A. Janin to V. Blair, Nov. 20, 1871, Janin Collection.

107. Ibid., Dec. 7, 1871.

108. Ibid. See also Dec. 14, 1871.

109. Ibid., Dec. 14, 1871.

110. Ibid., Jan. 11, 1872.

111. Ibid.

112. Ibid., Jan 25, 1872.

113. Ibid., Feb. 22, 1872. Her response is contained in his letter.

114. Ibid.

115. Ibid.

116. Ibid., Feb. 23, 1872.

117. Ibid., March 2, 1872. He reports her response in his letter.

118. Ibid.

119. Jealousy seems to be the product of the intense identification of lovers which created a heightened awareness of dependence upon another.

120. J. Marquis to N. Haile, Sept. 17, 1865, Marquis Collection.

121. Ibid.

122. See Chapter 6 for detailed examples of courtship anxiety and emotional distress.

123. A. Janin to V. Blair, Nov. 9, 1871.

124. M. Granger to L. Hodge, March 25, 1866, Benjamin Hodge Collection, Henry E. Huntington Library, hereafter cited as Hodge Collection.

125. Lovers who were separated by enough distance to necessitate a protracted

correspondence may have undergone a heightened version of anxiety, disappointment, and need. Nonetheless, romantic love created exceptional vulnerability to another human being, whatever their geographical distance.

126. N. Hawthorne to S. Peabody, Feb. 27, 1842, Hawthorne Collection.

127. N. Wheeler to C. Bradley, Aug. 2, 1877, Burdette Collection.

128. E. Trescot to E. Simkins, Oct. 4, 1864, Simkins Collection.

129. H. Russell to C. Strong, Dec. 20, 1862, Strong Collection.

130. M. Smith to S. F. Smith, July 15, 1834, Smith Collection.

131. N. Wheeler to C. Bradley, Jan. 11, 1877, Burdette Collection.

132. N. Hawthorne to S. Peabody, March 15, 1840, Hawthorne Collection.

133. E. Simkins to E. Trescot, Aug. 14, 1864, Simkins Collection.

134. A. Janin to V. Blair, Nov. 14, 1871, Janin Collection.

135. D. Lummis to C. Lummis [April 16, 1884], Moore Collection.

136. A. Janin to V. Blair, Feb. 1, 1872, Janin Collection.

137. N. Hawthorne to S. Peabody, July 30, 1839, Hawthorne Collection.

138. Ibid., Dec. 18, Dec. 24, 1839.

139. Suzanne Lebsock, *The Free Women of Petersburg: Status and Culture in a Southern Town, 1784–1860* (New York, 1984), 53.

140. A. Hallock to J. Bell, July 21, 1863, Bell Collection.

141. J. Bell to A. Hallock, July 11, 1863.

142. Ibid., March 6, 1863.

143. A. Janin to V. Blair, July 3, 1871, Janin Collection.

144. Ibid., March 24, 1874.

145. M. Smith to S. F. Smith, June 9, 1834, Smith Collection.

146. S. F. Smith to M. Smith, Aug. 15, 1834.

147. N. Hawthorne to S. Peabody, Jan. 3, 1840, Hawthorne Collection.

148. C. Strong to H. Russell, Jan. 1, 1863, Strong Collection.

149. H. Russell to C. Strong, Jan. 7, 1863.

150. L. Hodge to M. Granger, Jan. 9, 1866, Hodge Collection.

151. M. Granger to L. Hodge, March 15, 1866.

152. E. Simkins to E. Trescot, Sullivan's Island, June 24, 1864, Simkins Collection.

153. Ibid., July 10, 1864.

154. N. Hawthorne to S. Peabody, Jan. 1, 1840, Hawthorne Collection.

155. C. Strong to H. Russell, Dec. 14, 1862, Strong Collection.

156. H. Russell to C. Strong, Dec. 14, 1862.

157. J. Bell to A. Hallock, Sept. 19, 1863, Bell Collection.

158. E. Simkins to E. Trescot, Jan. 8, 1864, Simkins Collection.

159. A. Janin to V. Blair, March 24, 1874, Janin Collection.

160. N. Wheeler to C. Bradley, Jan. 10, 1875, Burdette Collection.

161. Alexis de Tocqueville, *Democracy in America,* trans. Henry Reeve; Francis Bowen, ed. (New York, 1838; rev. ed., 2 vols., 1945), vol. 2, 106.

162. Suzanne Lebsock perceptively analyzes the reasons to be skeptical about the beneficial influence of love and the companionate marriage in women's lives in *The Free Women of Petersburg,* 15–53, esp. 32–35, 48–53.

Chapter 3. "Lie Still and Think of the Empire"

1. Captain Frederick Marryat, *A Diary in America, with remarks on its institutions*, vol. 2 (London, 1839), 246–47.

2. Marryat's story is taken seriously by Peter T. Cominos, "Innocent Femina Sensualis in Unconscious Conflict," in *Suffer and Be Still: Women in the Victorian Age*, Martha Vicinus, ed. (Bloomington, 1972), 155–72, esp. 157; and Duncan Crow, *The Victorian Woman* (New York, 1972), 28–29. Other instances of a serious use of this incident are cited in Carl N. Degler, "What Ought To Be and What Was: Women's Sexuality in the Nineteenth Century," *American Historical Review* 79 (Dec. 1974), 1467, fn. 1. It is possible, however, that the young ladies were playing a little joke on Captain Marryat. This is how the piano leg covering is interpreted by Ronald G. Walters in *Primers for Prudery: Sexual Advice to Victorian America* (Englewood Cliffs, N.J., 1974), 2. Marryat was the object of much criticism in the American press as his comments offended his hosts on more than one occasion. On the basis of the hostility he created while in America, and his tactless and blundering style, it is not illogical to suggest that the piano legs were sheathed as a practical joke on an obnoxious English traveler. It is also possible that Marryat, who had a sense of humor, fabricated the incident. The diary has been called the most anti-American of the two hundred or more books written by Englishmen on America from 1836 to 1860. See Jules Zanger, Foreword to Captain Frederick Marryat, *Diary in America* (Bloomington, 1960), 9–32, esp. 9, 26–28, 32. This is an excellent modern edition but includes only the diary itself and not Marryat's remarks and analysis which accompanied most 19th-century editions.

3. Zanger, *Diary in America*, 9–26.

4. Ibid., 23–24.

5. "Repressive hypothesis" is a term coined by Michel Foucault to refer to the belief that sex was hidden, silenced, and condemned in the historical past. The agent of this repression was supposedly the Victorian bourgeoisie, and the modern myth is that contemporary sexual freedom has been wrested from the grip of this repressive heritage. See *The History of Sexuality*, trans. Robert Hurley (New York, 1978; rpt., 1980), 3–34. Historians who have been influenced by the repressive hypothesis are legion but particularly influential or highly visible studies include John S. Haller Jr. and Robin M. Haller, *The Physician and Sexuality in Victorian America* (Urbana, 1974); Barbara Welter, "The Cult of True Womanhood: 1820–1860" *American Quarterly* 18 (Summer 1966), 151–74; G. J. Barker-Benfield, *The Horrors of the Half-Known Life: Male Attitudes Toward Women and Sexuality in Nineteenth-Century America* (New York, 1976); Walters, *Primers for Prudery*, 1–22; Howard Gadlin, "Private Lives and Public Order: A Critical View of the History of Intimate Relations in the U.S.," *Massachusetts Review* 17 (Summer 1976), 304–30, esp. 316; Cominos, "Innocent Femina Sensualis in Unconscious Conflict," 155–72; Crow, *The Victorian Woman*, 19–31. Most of these historians were sensitive to matters of interpretation and methodology but were working within what was an unquestioned paradigm of Victorian culture.

6. Charles E. Rosenberg, "Sexuality, Class and Role in 19th-Century America," *American Quarterly* 25 (May 1973), 131–53, esp. 149–50; also see *No Other Gods: On Science and American Social Thought* (Baltimore, 1976) for a collection of Rosenberg's articles. Nancy F. Cott notes the theoretical distinction between the sexual ideology of female passionlessness and behavior in "Passionlessness: An Interpretation of Victorian Sexual Ideology, 1790–1850," in *A Heritage of Her Own*, Nancy F. Cott and Elizabeth H. Pleck, eds. (New York, 1979), 162–81, esp. 162–63. Carroll Smith-Rosenberg characterizes Victorian America as a society of infinite sexual complexity in "Sex as Symbol in Victorian Purity: An Ethnohistorical Analysis of Jacksonian America," in *Turning Points: Historical and Sociological Essays on the Family, American Journal of Sociology,* 84, Supplement, John Demos and Sarane Spence Boocock, eds. (Chicago, 1978), S212–24; also see *Disorderly Conduct: Visions of Gender in Victorian America* (New York, 1985) for a collection of Carroll Smith-Rosenberg's articles. Daniel Scott Smith observes that historical variation in sexual ideology is undoubtedly greater than changes in sexual behavior in "Family Limitation, Sexual Control and Domestic Feminism in Victorian America," in Cott and Pleck, eds., *A Heritage of Her Own,* 222–45, esp. 234. Michael Gordon concludes that women were not presented as typically asexual in respectable Victorian marriage manuals. He also argues that in these manuals "nonprocreative sex is never *fully* accepted throughout most of the nineteenth century." See Michael Gordon, "From an Unfortunate Necessity to a Cult of Mutual Orgasm: Sex in American Marital Education Literature, 1830–1940," in *Studies in the Sociology of Sex,* James Henslin, ed. (New York, 1971), 53–77 and fn. 63 in my Chapter 4.

7. Since Edmund S. Morgan wrote his classic essay, "The Puritans and Sex," *New England Quarterly* 15 (Dec. 1942), 591–607, historians have been increasingly concerned with the sexual dimensions of historical experience. As both social and cultural historians have placed more emphasis upon the private side of American history, the subject of sexuality has grown in scholarly significance. Indeed, a Dec. 1982 theme issue of *Reviews in American History* titled "The Promise of American History: Progress and Prospects" was dedicated to summarizing "the astonishing quantity of new information and reinterpretation of American history" produced from 1972 to 1982; one of the twenty essays in this issue was devoted exclusively to a discussion of sexuality. See Estelle B. Freedman, "Sexuality in Nineteenth-Century America: Behavior, Ideology, and Politics," *Reviews in American History* 10 (Dec. 1982), 196–215. As the quantity of scholarship on various aspects of 19th-century sexuality is very large, a brief summary follows of the six major concerns of scholarly writing in this area.

The first directs attention to one of the most significant demographic shifts in all of American history—namely, at the beginning of the 19th century the average number of children in a white native-born American family was 7.04 (almost the same as it had been for two centuries of American history) and 100 years later, at the end of the 19th century and long before the supposedly revolutionary birth control pill, the average white native-born American family had only 3.56 children. This dramatic decline in the birth rate has resulted in a large body of scholarship on 19th-century birth control, abortion, and the motivations as well as the interpersonal dynamics of family size reduction. Among the most significant works in this regard are James C. Mohr, *Abortion in America* (New York, 1978); Linda Gordon, *Woman's Body, Woman's Right: A Social History of Birth Control in America* (New York, 1976); James Reed, *From Private Vice to Public Virtue: The Birth Control Movement and American Society Since 1830* (New York, 1978); and Smith, "Family Limitation, Sexual Control, and Domestic Feminism."

The second major area of scholarship on 19th-century sexuality involves quantitative studies of premarital pregnancy rates, marital fertility rates, and divorce. See Daniel Scott Smith and Michael S. Hindus, "Premarital Pregnancy in America, 1640–1971: An Overview and Interpretation," *Journal of Interdisciplinary History* 5 (Spring 1975), 537–70; also Robert V. Wells, "Demographic Change and the Life Cycle of American Families," *Journal of Interdisciplinary History* 2 (Autumn 1971), 273–83. Attempts to move beyond 19th-century ideology to behavior have been undertaken in quantitative studies using family reconstitution or aggregate analysis. See Daniel Scott Smith, "Parental Power and Marriage Patterns: An Analysis of Historical Trends in Hingham, Massachusetts," *Journal of Marriage and the Family* 35 (Aug. 1973), 419–39; Daniel Scott Smith, "The Dating of the American Sexual Revolution: Evidence and Interpretation," in *The American Family in Social-Historical Perspective*, Michael Gordon, ed., 2nd ed. (New York, 1978), 426–38.

The third area of scholarship converges around sexual politics—the social and political organization of sexuality in 19th-century America including utopian experiments, moral reform movements, purity crusades, and other efforts at state control of sexuality. Examples of scholarship along these lines include William Leach, *True Love and Perfect Union: The Feminist Reform of Sex and Society* (New York, 1980); Lawrence Foster, *Religion and Sexuality: Three American Communal Experiments of the Nineteenth Century* (New York, 1981); Hal D. Sears, *The Sex Radicals: Free Love in High Victorian America* (Lawrence, Kan., 1977); Barbara Leslie Epstein, *The Politics of Domesticity: Women, Evangelism, and Temperance in Nineteenth-Century America* (Middletown, Conn., 1981); David J. Pivar, *Purity Crusade: Sexual Morality and Social Control, 1868–1900* (Westport, Conn., 1973).

The fourth category reflects American cultural pluralism and includes the study of sexuality in minority populations, as well as the examination of same-sex love in 19th-century America. The last begins with Carroll Smith-Rosenberg's enormously influential article, "The Female World of Love and Ritual: Relations Between Women in Nineteenth-Century America," which first appeared in *Signs: Journal of Women in Culture and Society* 1 (Autumn 1975), 1–29. For an overview of the history of homosexuality, see Jonathan Katz, ed., *Gay American History: Lesbians and Gay Men in the U.S.A.* (New York, 1976). Historians' interest in sexuality is not confined to white native-born couples as illustrated by works on slave sexuality and family norms such as Herbert G. Gutman, *The Black Family in Slavery and Freedom, 1750–1925* (New York, 1976); Eugene Genovese, *Roll, Jordan, Roll: The World the Slaves Made* (New York, 1972), 458–75. This category also includes works on prostitution, including Anne M. Butler, *Daughters of Joy, Sisters of Misery: Prostitutes in the American West, 1865–1890* (Urbana, 1985); Marion Goldman, *Gold Diggers and Silver Miners: Prostitution and Social Life on the Comstock Lode* (Ann Arbor, 1981); Joel Best, "Careers in Brothel Prostitution: St. Paul, 1865–1883," *Journal of Interdisciplinary History* 12 (Spring 1982), 597–619.

The fifth category of research, and by far the most voluminous, deals with sexual ideology, sexual advice, and didactic non-fiction in 19th-century America, from the pens of doctors, clergymen, and moral reformers. By ideology, I mean authoritative concepts and images used to justify and defend certain definitions of social situations as well as motivate action. Historians have been intrigued with many different dimensions of 19th-century sexual ideology from domesticity to womanhood to the morals of young men. Works in this area include Stephen Nissenbaum, *Sex, Diet, and Debility in Jacksonian America: Sylvester Graham and Health Reform* (Westport, Conn., 1980); Haller and Haller, *The Physician and Sexuality in Victorian America;* Welter, "The Cult of True Womanhood, 1820–1860"; Walters, *Primers for Prudery;* Cott, "Passionlessness: An Interpretation of Victorian Sexual Ideology"; Smith-Rosenberg, "Sex as Symbol in Victorian Purity"; and many others. Evidence in ideological studies supports the repressive hypothesis more strongly than any other scholarship. This data must be confronted alongside evidence of erotic attitudes and behavior in private 19th-century male-female relationships.

The sixth category is the examination of qualitative sources for insight into both private beliefs and behavior. Some scholars working in this area have launched an aggressive and far-reaching attack upon the repressive hypothesis. Several notable efforts to bring qualitative materials to bear upon questions of the behavior and attitudes of 19th-century men and women as they actually lived their lives are the groundbreaking article by Carl N. Degler, "What Ought To Be and What Was," 1479–90; Carl N. Degler, *At Odds: Women and the Family in America from the Revolution to the Present* (New York, 1980); a seminal book by Nancy F. Cott, *The Bonds of Womanhood: "Woman's Sphere" in New England, 1780–1835* (New Haven, 1977); another work with far-reaching impact on the repressive hypothesis, Peter Gay, *Education of the Senses,* vol. 1, *The Bourgeois Experience: Victoria to Freud* (New York, 1984); and one of the first modern studies of American courtship, Ellen K. Rothman, *Hands and Hearts: A History of Courtship in America* (New York, 1984). Rothman rejects the repressive hypothesis in part, particularly in the first half of her book (17–176). Her work covers a broad historical expanse from 1770 to 1920, and as she moves closer to the end of the 19th century, she reimposes conventional wisdom's repressive framework, "liberating" her couples again in the 1920s. This may reflect an effort to reconcile her findings with the standard interpretation of American sexual history which almost always characterizes the twenties as the era of American sexual liberation.

John D'Emilio and Estelle B. Freedman, in *Intimate Matters: A History of Sexuality in America* (New York, 1988), xiv, have, in their words, "attempted to translate this new body of scholarly work," namely, all the areas represented above, "into a synthetic, interpretive narrative." This is the first comprehensive history of sexuality in America.

8. Elizabeth Hampsten, *Read This Only to Yourself: The Private Writings of Midwestern Women, 1880–1910* (Bloomington, 1982), 9; Julie Roy Jeffrey, *Frontier Women: The Trans-Mississippi West, 1840–1880* (New York, 1979), 66–68; Rothman, *Hands and Hearts,* 51–54, 122–24; Degler, "What Ought To Be and What Was," 1479–90, and *At Odds,* 26–51, 249–78; Gay, *Education of the Senses,* 71–144.

9. In attacking the stereotype of Victorian prudery a scholar may be in the position of announcing what everyone claims to have known all along, and yet few completely believe. See, for example, the attitude expressed by Neil McKendrick, "Sex and Married Victorians," review of *Education of the Senses,* vol. 1, *The Bourgeois Experience: Victoria to Freud,* by Peter Gay, in *New York Times Book Review* (Jan. 8, 1984), 1, 35.

10. Carol Zisowitz Stearns and Peter N. Stearns, "Victorian Sexuality: Can Historians Do It Better?" *Journal of Social History* 18 (Summer 1985), 625–34. This lively essay seems to come close, at several points, to embracing a late 20th-century conceit, heaping such significance on the orgasm that all erotic experience is measured by its presence or absence. See 629–32.

11. Estelle B. Freedman and Erna Olafson Hellerstein, "Introduction to Part II," in *Victorian Women: A Documentary Account of Women's Lives in Nineteenth-Century England, France, and the United States,* Erna Olafson Hellerstein, Leslie Parker Hume, and Karen M. Offen, eds. (Stanford, 1981), 118–33, esp. 124–25.

12. Cott, "Passionlessness," 162–81; Rothman, *Hands and Hearts,* 50–51, 234.

13. This tension was not always apparent nor necessarily present; it surfaced in relationships where the taboo against intercourse was violated or where there was an emotional and/or physical struggle over the intercourse taboo. Such courtship tension in the sexual arena was best supported not by my own data sample but by the later 19th-century correspondence quoted in Rothman's *Hands and Hearts,* esp. 126–39 and 232–40. Rothman's interpretation of the sexual dimension of 19th century courtship sometimes coincided, but also diverged from mine. While she emphasized "self-control" in the later Victorian period, I was struck more by expressions of passion and sexual interest. This is not to say that there was no dimension of self-control in my data. However, evidence of restraint and inhibition seems to me to be tied to the violation of the intercourse taboo which, when that boundary was crossed, might be interpreted as a sign of the failure of restraint as much as evidence of its positive operation. Certainly crossing that tabooed boundary before marriage could create tensions and self-doubts in 19th-century correspondents.

14. See Chapter 8, notes 138–44.

15. I believe this is the root of Victorian condemnations of sexual licentiousness and their horror at sexual seduction.

16. Cott, "Passionlessness," 173. Cott also recognized that a bridge of romantic ideology might join the opposition between women's sexual being and her spiritual character.

17. Mrs. E. B. Duffey, *What Women Should Know* (Philadelphia, 1873), 64.

18. J. Marquis to N. Marquis, March 29, 1870, Marquis Collection.

19. B. Meyers to J. Tinsley, Jan. 30, 1844, Tinsley Family Papers, Robert Alonzo Brock Collection.

20. L. Clark to J. Clark, March 17, 1847, Clark Collection.

21. This couple was married Jan. 10, 1867. The quotation is from a letter from A. Baldwin to F. Baldwin, Sept. 5, 1869, Frank Dwight Baldwin Collection, Henry E. Huntington Library, hereafter cited as Baldwin Collection.

22. Ibid., Oct. 1, 1870. "Conflumux" appears to be Alice's private euphemism for male genitalia.

23. Ibid., Dec. 18, 1870. Two slang expressions occur in this passage. Eric Partridge, in *A Dictionary of Slang and Unconventional English,* Paul Beale, ed., 8th ed. (London, 1984), defines "giving someone the mittens" as jilting or dismissing them. According to Partridge, "Long Tom" was established slang for a penis.

24. Her frustration with other aspects of her sex role is discussed in Chapter 5.

25. A. Baldwin to F. Baldwin, June 22, 1873, Baldwin Collection. See Partridge's definition of "bubbies" in *Dictionary of Slang.*

26. Jane Burnett later became an influential clubwoman in San Francisco and was known for her organizational skills. She organized the Young Woman's Christian Association in that city. See Bailey Millard, *History of the San Francisco Bay Region,* vol. 3 (San Francisco, 1924), 31–32. See also Jane Cleveland Burnett, *My Memories of Early California Days,* March 21, 1917, typescript in the "Memories Files," Burnett Collection. There are several extant versions of Jane Burnett's recollections, all dated differently.

27. J. Burnett to W. Burnett, Jan. 9, 1857.

28. Ibid., Jan. 10, 1857.

29. Ibid., Jan. 18, 1857.

30. Ibid., June 2, 1857.

31. Ibid., Dec. 2, 1859.

32. For example, E. Lovell to M. Lovell, Nov. 6, 1862, Lovell Collection.

33. Ibid., Nov. 4, 1862.

34. Ibid., Nov. 6, 1862. Most of the last six lines are in the margins of the letter.

35. Ibid., Nov. 7, 1862. Harold Wentworth and Stuart Berg Flexner, in *Dictionary of American Slang,* Second Supplemented Ed. (New York, 1975), define the tabooed usage of "eat" as "To perform cunnilingus or fellatio on a person." They note that it is usually used as a euphemism, since "it can be considered as having the first meaning" which is simply to regard someone as exceptionally sweet or adorable (170).

36. M. Lovell to E. Lovell, Dec. 12, 1862, Lovell Collection.

37. Mansfield and Emily were married in 1849. See Jon L. Wakelyn, *Biographical Dictionary of the Confederacy* (Westport, Conn., 1977).

38. J. Bolton to A. Harrison, July 16, 1838, Harrison Collection.

39. A. Janin to V. Blair, March 4, 1872, Janin Collection.

40. Ibid., April 26, 1874.

41. J. Marquis to N. Marquis, Oct. 2, 1866, Marquis Collection.

42. J. Hague to M. W. Foote, February 4, 1872, Hague Collection.

43. A. Baldwin to F. Baldwin, June 22, 1873, Baldwin Collection. One reader interpreted this statement as completely ironic, suggesting that Alice was not issuing a teasing sexual invitation, but was actually portraying herself, at that moment, as physically undesirable. While this interpretation is possible, I read the comment differently.

44. See fn. 13 in this chapter.

45. A. Hallock to J. Bell, March 1, 1857, Bell Collection.

46. Ibid., Aug. 8, 1863.

47. J. Bell to A. Hallock, Oct. 2, 1857.

48. A. Hallock to J. Bell, April 3, 1856; misdated, actually 1857.

49. J. Bell to A. Hallock, Nov. 23, 1861.

50. A. Hallock to J. Bell, Dec. 1, 1861.

51. J. Bell to A. Hallock, Dec. 7, 1861.

52. Ibid., Camp California, [undated but probably written around Dec. 17, 1861].

53. A. Hallock to J. Bell, Dec. 31, 1861.

54. J. Bell to A. Hallock, Feb. 4, 1862 (actually Feb. 3; James noted the error in his letter).

55. A. Hallock to J. Bell, March 8, 1861 (actually 1862). This letter is certainly misdated. Internal evidence proves it must have been written after Sept. 1861, when James was mustered into the Union Army.

56. E. Elliott to G. Hallock, [undated but probably written before 1859].

57. L. Hodge to M. Granger, May 20, 1866, Hodge Collection.

58. M. Granger to L. Hodge, Jan. 28, 1866.

59. L. Hodge to M. Granger, Feb. 10, 1867.

60. Ibid.

61. E. Simkins to E. Trescot, Aug. 14, 1864, Simkins Collection. I read this collection soon after its arrival at the Huntington Library. Before the originals were formally catalogued, typescripts of the letters, transcribed by a relative, were made available to me by Mrs. Virginia Rust. My notes were made from the typescript copies.

62. E. Trescot to E. Simkins, July 14, 1864.

63. E. Simkins to E. Trescot, Aug. 14, 1864.

64. Ibid., [undated but written sometime around Sept. 10–12, 1864].

65. Ibid., Oct. 21, 1864.

66. E. Trescot to E. Simkins, Aug. 23, 1864.

67. Ibid., July 14, 1864.

68. Ibid., Sept. 14, 1864.

69. M. Smith to S. F. Smith, July 15, 1834, Smith Collection. Mary's maiden

name was also Smith. All correspondence exchanged before Sept. 16, 1834, were premarital courtship letters. Her marriage on that date did not change her surname.

70. Ibid., July 15, 1834.

71. Ibid.

72. S. F. Smith to M. Smith, July 25, 1834.

73. Ibid., Aug. 10, 1834.

74. Ibid., Aug. 15, 1834.

75. Ibid., Sept. 1, 1834.

76. M. Smith to S. F. Smith, May 8, 1836.

77. Ibid., May 21, 1836.

78. *The National Cyclopaedia of American Biography,* vol. 42 (New York, 1958), 578–79; Turbese Lummis Fiske and Keith Lummis, *Charles F. Lummis: The Man and His West* (Norman, Okla., 1975); Dudley Gordon, *Charles F. Lummis: Crusader in Corduroy* (Los Angeles, 1972); Edwin R. Bingham, *Charles F. Lummis: Editor of the Southwest* (San Marino, 1955).

79. The biographies of Charles Lummis give uneven treatment to his first wife, Dorothea. The work by Fiske and Lummis incorporates many of Dorothea's letters still in the possession of Charles's son, Keith Lummis, and is the fullest and fairest treatment of Dorothea and Charles's relationship. See Fiske and Lummis, *The Man and His West,* 13–15, 41–42, 52, 65–67, 69; also see Bingham, *Charles F. Lummis,* 6–12. Dudley Gordon blames Dorothea's "masculine" character for the break-up of her relationship with Charles, basing his estimation of her personality upon handwriting analysis. See Gordon, *Crusader in Corduroy,* 42–45, 73.

Dorothea's medical career was in an unorthodox branch of medicine, homeopathy, which was more open to women in the 19th century. See Martin Kaufman, *Homeopathy in America: The Rise and Fall of a Medical Heresy* (Baltimore, 1971); Richard Harrison Shryock, "Women in American Medicine," in *Medicine in America: Historical Essays* (Baltimore, 1966), 177–99; Mary Roth Walsh, *"Doctors Wanted: No Women Need Apply": Sexual Barriers in the Medical Profession, 1835–1975* (New Haven, 1977); Regina Morantz-Sanchez, *Sympathy and Science: Women Physicians in American Medicine* (New York, 1985).

80. William L. O'Neill, "Divorce in the Progressive Era," *American Quarterly* 17 (Summer 1965), 203–17. Between 1880 and 1900 the divorce rate in America virtually doubled: in 1880, one divorce for every 21 marriages to one divorce for every 12 marriages in 1900.

81. D. Lummis to C. Lummis, [Sept. 21–23, 1883], Moore Collection.

82. Ibid., [Sept. 24–26, 1883].

83. Ibid., Oct. 8 [actually Oct. 7], 1883.

84. Ibid., [Oct. 11–14, 1883]. I disagree with the dating of this letter. It is catalogued as [Oct. 11 and 13, 1883]. Internal evidence indicates otherwise.

85. Ibid., [Oct. 12, Oct. 15–17, Oct. 18–21, Oct. 23–24, 1883].

86. Ibid., [Oct. 23–24, 1883].

87. Ibid., [Nov. 8–9, 1883].

88. Ibid., [Nov. 1–4, 1883].

89. Ibid., [Nov. 5–7, 1883].

90. Ibid., [Nov. 12–13, 1883].

91. Ibid., [Nov. 8–9, 1883].

92. Ibid., [Dec. 10–11, 1883].

93. Ibid., [Jan. 10–13, 1884]. In a Jan. 18–20, 1884, letter, Dorothea referred to Charles's response to her inquiry: "So you were happy as ever!" She then reaffirmed her admonition to take time for "all love's blessings." She also referred in detail to their last lovemaking in a letter begun Feb. 21, 1884.

94. Ibid., [Feb. 6, 1884].

95. Ibid., [Feb. 22, 1884].

96. Ibid., [April 16, 1884].

97. See Walters, *Primers for Prudery,* 65–78; Welter, "Cult of True Womanhood," 154–58.

98. Cott, "Passionlessness," 163–64, 173–75.

99. R. Burdette to C. Baker, Saturday night, Aug. 20, 1898, Burdette Collection. See Chapter 4 for a detailed treatment of the erotic dimension of the Burdettes' courtship.

100. Ibid., Monday evening, Oct. 24, 1898.

101. Ibid., Monday night, Three Rivers [November 7, 1898].

102. C. Baker to R. Burdette, Oct. 26, 1898.

103. Ibid., Jan. 22, 1899.

104. J. W. North to A. Loomis, June 30, 1848, North Collection. This letter was censored with scissors, and not much of it remains. See Chapter 4 for a discussion of the relation between Victorian sexuality and censorship.

105. Rosenberg, "Sexuality, Class, and Role," 139–40; Gordon, *Woman's Body, Women's Right,* 96, 102, and 106; Sears, *The Sex Radicals,* 22.

106. See my discussion of sexual advice books in Chapter 4.

107. This conclusion is based upon both the pattern of private attitudes found in correspondence between men and women and my reading of public sexual advice.

108. Smith and Hindus, "Premarital Pregnancy in America," 537–70. While both helpful and highly suggestive, one must be cautious in using the premarital *pregnancy* rate to determine rates of premarital *intercourse*. Coition which did not result in conception either for natural or artificial reasons as well as aborted pregnancies are not included in premarital pregnancy counts by definition.

The jump in premarital pregnancy, sometimes as high as 30 percent of all first births in sample 18th-century New England towns, can be explained not only by the disruption of the Revolutionary War, but also by the fact that courtship was changing in the last half of the 18th century from a parental-run system, albeit with strong veto power in the child's hands, to a participant-run system based on romantic love with only minimal veto power in parental hands.

Henry Reed Stiles first propounded a "disequilibrium model" of late 18th-century courtship behavior. He argued that the French and Indian Wars as well as the Revolution itself may have introduced a social instability that "flooded the land with immorality and infidelity," and destroyed the chasteness of bundling. See his *Bundling: Its Origin, Progress and Decline in America* (New York, 1934), 76–77.

A shift probably occurred in late 18th-century courtship from dominant pre-modern external forms of social control based upon community and parental surveillance to reliance on internalized psychological limits based upon romantic love. Perhaps the location of sexual limits also shifted from external to internal control mechanisms with a period of relative confusion and sexual readjustment in the late 18th century. For a general discussion of the pre-industrial to industrial shift in the location of social control see Michael Zuckerman, "The Fabrication of Identity in Early America," *William and Mary Quarterly* 3rd ser. 34 (April 1977), 183–214. See also Nancy F. Cott, "Eighteenth-Century Family and Social Life Revealed in Massachusetts Divorce Records," *Journal of Social History* 10 (Fall 1976), 20–43.

109. N. Hawthorne to S. Hawthorne, April 7, 1856, Hawthorne Collection.

110. D. Lummis to C. Lummis, [Nov. 8–9, 1883], Moore Collection.

111. M. Smith to S. F. Smith, May 15, 1836, Smith Collection.

112. J. Burnett to W. Burnett, Sept. 30, 1856, Burnett Collection.

113. W. Burnett to J. Burnett, June 9, 1857.

114. A. Baldwin to F. Baldwin, Oct. 29, 1870, Baldwin Collection.

115. Ibid., Oct. 28, 1869.

116. Degler, *At Odds,* 178–209.

117. N. Wheeler to C. Wheeler, Aug. 25, 1880, Burdette Collection. Nathaniel's comments in the next three paragraphs are taken from this letter.

118. See Chapter 4 for a fuller treatment of the issue.

119. N. Hawthorne to S. Hawthorne, April 16, 1849, Hawthorne Collection.

120. M. Smith to S. F. Smith, May 15, 1836, Smith Collection. Though "intercourse" as sexual connection was in use by the early 19th century, its meaning in this passage is not exclusively sexual. Mary may be referring simply to intimate communication in general. Ambiguous, but probably not primarily sexual, references to "intercourse" are also found in Chapters 6 and 8.

121. M. Walker, Diary, July 26, 1839, Elkanah Bartlett Walker Collection, Henry E. Huntington Library, hereafter cited as Walker Collection.

122. M. Smith to S. F. Smith, May 15, 1836, Smith Collection.

123. Keith Thomas, "The Double Standard," *Journal of the History of Ideas* 20 (April 1959), 195–216. See esp. 210–13.

124. Gordon, *Woman's Body, Woman's Right,* 100–101, 103–6; Cott, "Passionlessness," 173; Smith, "Family Limitation, Sexual Control, and Domestic Feminism," 235.

125. See fn. 124 above, and also Chapter 4.

126. Degler, *At Odds,* 209–26. See also Reed, *From Private Vice to Public Virtue,* 3–18.

127. L. Clark to J. Clark, Dec. 19, 1851, Clark Collection. The Clarks had six children; their last two died in early childhood.

128. Ibid., Dec. 21, 1851.

129. Ibid., July 29, 1852.

130. Ibid., Feb. 2, 1853. Their mutual apprehension may have stemmed from the fact that her last pregnancy was an accident. This might explain her unhappiness at the time: "I am truly sorry that you suffer so much with sick, sick—" Lincoln commented. "I hope this will be your last trial of the sort, altho you may be doubtful—you must try and have patience this time, long as it may seem. You said you could not get much love into your letter—I do not blame you for that—I can easily understand that sore sickness keeps love in surveilance. But I am not sick, and can truly say that I love you and should like to have you with me all the time." L. Clark to J. Clark, Sept. 23, 1850. This child was born in March 1851 and died in July 1852. See L. Clark to his daughter Catherine Clark, March 21, 1852 and L. Clark to J. Clark, July 27, 1852.

In the aftermath of what may have been an unplanned pregnancy, they both seem wary. Reading between the lines, however, Lincoln appears less apprehensive than Julia. Though he inquired nervously about her "lady conditions" after they slept together, his sexual enthusiasm was undimmed. He wrote on Dec. 29, 1852: "Hope your health will be good—you can be nearly as quiet here as you choose. I have a large pleasant room—and a *bed* of the same kind, bless you." Her interest in sharing a bed appears much weaker than his. See L. Clark to J. Clark, Jan. 9, 1853. After stalling and arguing against making the trip, she traveled to Washington to be with him; Lincoln's uncharacteristically tepid response, close to damning her by faint praise, suggests that the recent and possibly unintended pregnancy affected their sexual relationship. See L. Clark to J. Clark, April 2, 1853.

The letters indicate no other sexual problems either before or after the three-year period beginning from evidence of the unwanted pregnancy in September 1850 and extending through 1853. Lincoln referred neither before 1850 nor after 1853 to Julia's "lady conditions" and Julia was not pregnant again as far as the extant evidence revealed. Though her methods of birth control are vague, clearly the Clarks practiced some form of cooperative family planning, even if not always with complete success.

131. V. Janin to A. Janin, July 2, 1874, Janin Collection.

132. Ibid., Aug. 19, 1874.

133. Ibid., Aug. 22, 1874.

134. Ibid., Oct. 20, 1874.

135. Ibid., Oct. 26, 1874.

136. Ibid., Nov. 10, 1874.

137. Ibid., Nov. 21, 1874.

138. M. Walker, Diary, Dec. 3, 1842, Walker Collection.

139. E. Elliot to A. Hallock, Dec. 30, 1860, Bell Collection.

140. L. Hodge to M. Granger, July 29, 1867, Hodge Collection.

Chapter 4. Secrecy, Sin, and Sexual Enticement

1. Michel Foucault, *The History of Sexuality,* trans. Robert Hurley (New York, 1978; rpt., 1980), 45.

2. Randall Stewart, "Letters to Sophia," *Huntington Library Quarterly* 7 (Aug. 1944), 387–95. Nathaniel Hawthorne, *The Letters, 1813–1843, The Centenary Edition of the Works of Nathaniel Hawthorne,* Thomas Woodson, L. Neal Smith, Norman Holmes Pearson, eds., vol. 15 (Columbus, 1984), 7–9, 9–10, and *The Letters, 1843–1853, The Centenary Edition of the Works of Nathaniel Hawthorne,* vol. 16 (Columbus, 1985). The Centenary editors of Hawthorne's works attempted to restore all passages overmarked by Sophia. Unfortunately, they do not indicate those censored passages that have been successfully recovered, believing that recovery "has removed the necessity of noting them in the text."

3. N. Hawthorne to S. Peabody, July 24, Aug. 26, 1839, quoted in Stewart, "Letters to Sophia," 391–92. These quotations are taken from letters written before they were married.

4. N. Hawthorne to S. Peabody, Dec. 2, 1844; April 16, 1849; Feb. 7, 1856. All are quoted in Stewart, "Letters to Sophia," 394. This group of quotations is taken from letters written after their marriage.

5. N. Hawthorne to S. Peabody, April 21, 1840. Hawthorne Collection. See also N. Hawthorne to S. Peabody, Dec. 18, Aug. 21, May 26, 1839.

6. Ibid., Nov. 20, 1839.

7. Hawthorne burned most of Sophia's love letters, thus actually outdoing his wife in the censorship department. Hawthorne, *The Letters, 1813–1843,* 6; Lou Ann Gaeddert, *A New England Love Story: Nathaniel Hawthorne and Sophia Peabody* (New York, 1980), 82. Only a few of Sophia's courtship letters to Hawthorne are extant.

8. E. Trescot to E. Simkins, Dec. 1, 186[3], Simkins Collection. Though Eldred was less concerned about secrecy, he wrote one half of his Dec. 4, 1863, letter for public perusal, and the other half for her eyes only.

9. A. Baldwin to F. Baldwin, Sept. 15, 1874, Baldwin Collection.

10. L. Clark to J. Clark, Dec. 20, 1846, Clark Collection.

11. A. Hallock to J. Bell, June 2, 1860, Bell Collection.

12. C. Baker to R. Burdette, Jan. 22, 1899, Burdette Collection.

13. R. Burdette to C. Baker, Feb. 7, 1898.

14. Ibid., Oct. 28, 1898.

15. Clara B. Burdette, *Robert J. Burdette: His Message* (Pasadena, 1922), 11. He was born July 20, 1844. Though she is listed on the title page as the editor, this is more a biography than an edited collection of her husband's work. See also Clara Burdette, *The Answer to "Clara, what are you going to do with your life?"* (Pasadena, 1951), 3. She was born July 22, 1855. For more biographical information about her life, see Dorothy Grace Miller, "Within the Bounds of Propriety: Clara Burdette and the Women's Movement" (Ph.d. diss., University of California, Riverside, 1984).

16. Burdette, *Robert J. Burdette: His Message,* 121–62, 241–94.

17. Burdette, *The Answer,* 158–59, 234–41. Excerpts from her speeches reveal a very conventional approach to the culture of her times.

18. R. Burdette to C. Baker, April 27, 1897, Burdette Collection. In a Dec. 28, 1897, letter, Robert confessed that during their last face-to-face encounter, in May 1897, he had yearned to hold her and kiss her lips.

19. Ibid., Dec. 28, 1897. Burdette, *The Answer,* 172.

20. R. Burdette to C. Baker, Jan. 6, 1898, Burdette Collection.

21. Ibid., Jan. 9, 1898.

22. Ibid., Jan. 21, 1898.

23. Ibid., March 15, 1898.

24. Ibid., "The Nunnery," Saturday night, March 19, 1898. Robert repeated this theme many times; for example, Sunday night, March 20, 1898; March 22, 1898 (in all 3 letters written that day).

25. Ibid., Wednesday morning, March 23, 1898.

26. Ibid., Saturday night, March 27, 1898.

27. Ibid., Saturday night, [April 2, 1898].

28. Ibid., April 6, 1898.

29. Ibid., March 29, 1898. I could only find an occasional extant letter from Clara scattered in various places in this collection before Aug. 3, 1898. After that date, Clara's letters were abundant, albeit thoroughly censored, most likely by Clara herself.

30. Ibid., April 9, 1898.

31. Ibid., April 24, 1898 (20 pages), April 13, 1898 (12 pages), April 14, 1898 (12 pages), April 21, 1898 (16 pages + a 4-page appendix).

32. Ibid., April 21, 1898. See also April 18, 1898, beginning "My own White Violet."

33. Ibid., April 25, 1898.

34. Ibid., April 28, 1898, Thursday, 5 p.m.

35. Ibid., April 29, 1898.

36. Ibid., May 1, 1898. In this letter he reported her contention that their "love-making" applied only to the "make-believe" world of letters.

37. Ibid., May 22, 1898.

38. Ibid.

39. Ibid., May 29, 1898.

40. Ibid., June 8, 1898.

41. Ibid., June 13, June 16, July 27, 1898.

42. C. Baker to R. Burdette, Aug. 12, 1898.

43. Ibid., Aug. 16, 1898 (I refer to the earlier of two letters she wrote this day); R. Burdette to C. Baker, Aug. 12, 1898.

44. His proposal came during the period June 21–July 27, 1898, which marked their first opportunity to spend time with each other since the onset of serious courting. R. Burdette to C. Baker, Aug. 13, 1898. A Sunday morning [Aug. 14, 1898] letter described the entire proposal scene in detail. Perhaps their previous

marriages explain their willingness to violate the intercourse taboo before marriage. However, I do not think the Burdettes are an unrepresentative couple so much as an extreme manifestation of quite common patterns.

45. Ibid., Monday night [Aug. 15, 1898].

46. Ibid.

47. Ibid., Tuesday twilight, Aug. 16, 1898.

48. Ibid., Dec. 22, 1898.

49. C. Baker to R. Burdette, Aug. 24, 1898. Though Clara censored Robert's letters, she was not completely successful. He was supposed to pen the erotic passages on separate sheets that could be removed without throwing away the entire letter. Thus in his correspondence, whole pages are missing, particularly at the end of letters. Clara appears to have relied upon the separate-pages strategy, and to have given less rigorous attention to the line-by-line reading of his voluminous correspondence. Interestingly, she was more arduous in censoring her own letters.

50. The first four pages of this letter are missing, presumably censored, but the letter is most certainly written in early Sept. 1898, before she departed for a cross-country trip by train to see Robert again.

51. R. Burdette to C. Baker, Sept. 27, 1898, Broad Street Station; C. Baker to R. Burdette, Oct. 17, 1897 (this letter, dated incorrectly, was certainly written in 1898); C. Baker to R. Burdette, Sunday afternoon, Oct. 23, 1898.

52. R. Burdette to C. Baker, Feb. 18, 1899, Saturday night. This is a 27-page letter written at several sittings; in it, he expressed his moralistic attitudes toward cards. Burdette is described in an undated clipping from a Charleston newspaper quoting the *Genessee Republican*. This notice appeared around Jan. 11, 1899, the date he lectured in Charleston, South Carolina.

53. Ibid., Jan. 19, 1899.

54. Ibid., Feb. 27, 1899; and Burdette, *The Answer*, 183.

55. See Peter Gay, *Education of the Senses*, vol. 1, *The Bourgeois Experience: Victoria to Freud* (New York, 1984), 71–108; Polly Longsworth, ed., *Austin and Mabel: The Amherst Affair and Love Letters of Austin Dickinson and Mabel Loomis Todd* (New York, 1984).

56. For example, the Social Purity movement in the last quarter of the 19th century exercised a power that questioned, monitored, searched out, and attempted to control sexuality. The purity crusaders led a successful campaign against licensed prostitution and expressed in various other efforts at social reform the belief that sexual expression needed to be limited. While it had its latitudinarian elements, censorship was an integral part of this reform movement. See David J. Pivar, *Purity Crusade: Sexual Morality and Social Control, 1868–1900* (Westport, Conn., 1973), 180–85, 232–38, 262. Carl Degler noted that in spite of some of the movement's more liberal features, Anthony Comstock was always welcome among the Social Purists. Degler also pointed out that "one consequence of the Social Purity movement no one had anticipated and one that some may well have regretted" was that it opened up the public discussion of sexuality because "one

could not attack the evils of brothels and prostitution and vice in general without mentioning names, institutions, and practices." See Carl N. Degler, *At Odds: Women and the Family in America from the Revolution to the Present* (New York, 1980), 287, 279–97.

57. The Rev. John Bayley, *Marriage As It Is and As It Should Be* (New York, 1857), 145–46.

58. See, for example, T[imothy] S[hay] Arthur, *Advice to Young Men on Their Duties and Conduct in Life* (Philadelphia, [1847?]), 109–10.

59. Carl N. Degler, "What Ought To Be and What Was: Women's Sexuality in the Nineteenth Century," *American Historical Review* 79 (December 1974), 1467–90; Anita Clair Fellman and Michael Fellman, *Making Sense of Self: Medical Advice Literature in Late Nineteenth-Century America* (Philadelphia, 1981), esp. 95; Charles Rosenberg, "Sexuality, Class and Role in 19th-Century America," *American Quarterly* 25 (May 1973), esp. 133–34, 138–39, 151–53; Carroll Smith-Rosenberg, "Sex as Symbol in Victorian Purity: An Ethnohistorical Analysis of Jacksonian America," in *Turning Points: Historical and Sociological Essays on the Family, American Journal of Sociology,* 84, Supplement, John Demos and Sarane Spence Boocock, eds. (Chicago, 1978), S212.

60. Fellman and Fellman, *Making Sense of Self,* 12–16, 91–92, 135–41; Rosenberg, "Sexuality, Class and Role," 137, 150-53; Smith-Rosenberg, "Sex as Symbol in Victorian Purity," 213, 217–21, 227–34.

61. See the Victorian Advice-Book Bibliography, below, for a listing of the medical and moral advice books that form the basis for these generalizations. This sample is deliberately limited to books written for a lay audience. I did not include articles written for a medical audience and published in professional medical journals. I consider the latter to be a different genre. At the very least the public's exposure to medical journals was limited.

62. The Fellmans use the term "moderate" to delineate a middle-range group of Victorian advisers who advocated moderation in sexual activity. But their criteria included advisers who, in my scheme, would fall into the restrictionist camp. See Fellman and Fellman, *Making Sense of Self,* 95–99.

63. Michael Gordon illustrates the need to categorize and divide 19th-century sexual advisers. His conclusion that "nonprocreative sex was never *fully* accepted throughout most of the nineteenth century" applies to restrictionists but not to moderates and enthusiasts. See his "From an Unfortunate Necessity to a Cult of Mutual Orgasm: Sex in American Marital Education Literature, 1830–1940," in *Studies in the Sociology of Sex,* James M. Henslin, ed. (New York, 1971), 55.

64. Ironically, William Acton, whose famous quote on the absence of female passion is almost obligatory in discussing 19th-century sexuality, is unrepresentative of the position taken by this sample of American medical and moral guide books. See fn. 70 in this chapter. F. Barry Smith also questions Acton's usefulness as a representative adviser, in "Sexuality in Britain, 1800–1900: Some Suggested Revisions," *A Widening Sphere: Changing Roles of Victorian Women,* Martha Vicinus, ed. (Bloomington, 1977), 182–98.

65. Edward B. Foote, M.D., *Plain Home Talk About the Human System—the Habits of Men and Women . . . Embracing Medical Common Sense* (New York, 1870; rev. ed., Chicago, 1896), 621, 618.

66. Dr. H[oratio] R[obinson] Storer, *Is It I? A Book for Every Man* (Boston, 1868), 53, 129.

67. George H. Napheys, *The Physical Life of Woman: Advice to the Maiden, Wife and Mother* (Philadelphia, 1884), 90–92, 96–98.

68. Mrs. E[liza] B[isbee] Duffey, *What Women Should Know* (Philadelphia, 1873), 97, 99.

69. Anonymous, *Satan in Society; By a Physician* (Cincinnati, 1882), 145, 147–49.

70. William Acton, *The Functions and Disorders of the Reproductive Organs* (Philadelphia, 1867), 144. His discussion of "false impotence" is found on 137–48. Acton contrasted the absence of desire—defined as "false"—to "true" impotence, which he considered a physical inability to consummate a sexual relationship.

71. Ibid., 145. All subsequent quotes in the paragraph are also found on 145.

72. J[ohn] H[arvey] Kellogg, *Plain Facts for Old and Young* (Burlington, Iowa, 1882), 265–66; John Cowan, *The Science of a New Life* (New York, 1873), 112, 115. William Leach believes that John Cowan's work is the "best single source for insight into the sexual beliefs of most feminists of the time." See *True Love and Perfect Union: The Feminist Reform of Sex and Society* (New York, 1980), 32. I am highly skeptical of this contention as Cowan, a restrictionist, advocated sex once every three years after a successful procreative act.

73. J[ohn] H[arvey] Kellogg, *Ladies' Guide in Health and Disease* (Des Moines, 1886), 293.

74. The idea of love as the gateway to lust is drawn from the enthusiast Orson Squire Fowler who, unlike restrictionists, welcomed the thought that "the gateway of lust itself lies through the pathway of pure, holy love." Fowler approved what restrictionists feared—love develops passion—in *Offspring, and Their Hereditary Endowment* (Boston, 1869), 156.

75. Anonymous, *Satan in Society*, 213.

76. Ibid., 143, 141, 145–46.

77. James Foster Scott, *The Sexual Instinct: Its Use and Dangers as Affecting Heredity and Morals* (New York, 1899), 20, 28.

78. John Todd, *The Young Man: Hints Addressed to the Young Men of the United States* (Northampton, 1846), 316.

79. Ibid., 315–16.

80. One secret sin on which medical advisers generally lavished their most intense images of organic as well as divine hell-fire and damnation (particularly after 1860) was masturbation. This is a commonplace in the story of Victorian repression. See, for example, John S. Haller, Jr., and Robin M. Haller, *The Physician and Sexuality in Victorian America* (Urbana, 1974), 105–6, 195–211; Ronald G. Walters, ed., *Primers for Prudery: Sexual Advice to Victorian America* (Englewood Cliffs, N.J., 1974), 35–42.

The anti-masturbation mania of 19th-century culture should, however, be treated with the same sense of historical relativity and respect for a different belief system that historians now routinely apply to the 17th-century belief in witches. Victorian scare tactics are more understandable if masturbation is interpreted as a dominant symbol of the secrecy and opportunity for non-conformity which the ethic of privacy allowed.

Historians have analyzed the vital connection between anti-masturbation rhetoric and social control, but insufficient attention has been paid to social control as an expression of anxiety over privacy itself. For examples of a sophisticated treatment of masturbation literature, see Smith-Rosenberg, "Sex as Symbol in Victorian Purity," in *Turning Points*, S212–45; and Rosenberg, "Sexuality, Class and Role," 133–37, 145–47.

81. [Samuel Bayard Woodward], *Hints for the Young: On a Subject Relating to the Health of Body and Mind* (Boston, 1838), 6. See also Storer, *Is It I?*, 63; Augustus K. Gardner, M.D., *Conjugal Sins Against the Laws of Life and Health* (New York, 1870; rpt., 1974), 74–75.

82. While medical and moral advice books were fighting to control private space, another kind of 19th-century guidebook, as Chapter 1 documents, was relinquishing control of the backroom and bedroom. For the etiquette adviser, the unviewed private family space was a haven to relax in and be less guarded, consequently more one's "true" self. Samuel Robert Wells, in *How to Behave*, admitted "Stiff formality and cold ceremoniousness are repulsive anywhere, and are particularly so in the family circle. . . ." Following the social geography of expression, he urged, "make no public exhibition of your endearments." *How to Behave; or A Pocket Manual of Republican Etiquette* (New York, 1858), 60, 113; also see 51–52 for a discussion of the sacredness of privacy. In *Good Behavior for Young Gentlemen: Founded on Principles of Common Sense, and the Usages of Good American Society* (Rochester, N.Y., 1848), 45, an American Gentleman urged, "Conjugal intimacy, it is true, dispenses with the etiquette established by politeness, but it does not dispense with amiability or attentions." Ironically, medical and moral advisers were attempting to extend social control to exactly those areas of the home for which etiquette books were sanctioning escape.

83. William Andrus Alcott, *The Young Wife* (Boston, 1837), 201, 203.

84. Gardner, *Conjugal Sins*, 78.

85. Anonymous, *Satan in Society*, 37.

86. Scott, *The Sexual Instinct*, 31.

87. Ibid., 35–36.

88. R[ussell] T[hacher] Trall, *Sexual Physiology: A Scientific and Popular Exposition of the Fundamental Problems in Sociology* (New York, 1870), 202.

89. Edmund S. Morgan, *The Puritan Family: Religion and Domestic Relations in Seventeenth-Century New England* (Boston, 1944; rev. ed., 1966), 61–63, 163–64, 166; Edmund S. Morgan, "The Puritans and Sex," *New England Quarterly* 15 (Dec. 1942), 591–607.

90. The process which established this public-private boundary awaits further research and explication.

91. Storer, *Is It I?*, 10.

92. Anonymous, *Satan in Society,* 41, 44.

93. Ibid.

94. Napheys, *Physical Life,* "Preface to the First Edition," published in 1869.

95. Foote, *Plain Home Talk,* 623.

96. Henry N. Guernsey, M.D., *Plain Talks on Avoided Subjects* (Philadelphia, 1882; rpt., 1907).

97. Cowan, *New Life,* 116–17.

98. Ibid., 171–72 and 136–87.

99. Ibid., 97.

100. Ibid., 99.

101. Ibid., 117. While it is important to note that his stance was unrepresentative, it is also significant that even this most extreme restrictionist adviser helped to bring the discussion of sex (405 pages) into the public sphere.

102. Anonymous, *Satan in Society,* 19, 23.

103. Ibid., 81.

104. Kellogg, *Plain Facts,* 178, 184–85.

105. Ibid., 185–203, 204.

106. I am indebted to Perry Miller's insights into the Puritan jeremiad in *The New England Mind: From Colony to Province* (Cambridge, Mass., 1953; rpt., 1961), 19–51; see also Sacvan Bercovitch, *The American Jeremiad* (Madison, 1978), 3–30. Though Bercovitch insists that his interpretation of the jeremiad diverges from Miller's, a significant portion of Bercovitch's understanding is congruent with Miller's mature analysis of the jeremiad in *Colony.* Bercovitch compares his interpretation with Miller's formulation in "Errand into the Wilderness," a popular essay but not Miller's fullest or most complex treatment of the jeremiad form. See Perry Miller, *Errand into the Wilderness* (Cambridge, Mass., 1956), 1–15.

107. Miller, *From Colony to Province,* 34–38.

108. Kellogg, *Plain Facts,* 116.

109. Gardner, *Conjugal Sins,* 181.

110. William Alcott, *The Young Husband* (Boston, 1839), 359.

111. Ibid., 264.

112. See Miller, *From Colony to Province,* 31–33.

113. Anonymous, *Satan in Society,* 37.

114. Gardner, *Conjugal Sins,* 195.

115. Kellogg, *Plain Facts,* 216.

116. J[ames] R[ush] Black, M.D., *The Ten Laws of Health; or, How Diseases Are Produced and Prevented* (Philadelphia, 1885), 238.

117. Ibid., 253; Miller, *From Colony to Province,* 33.

118. Kellogg, *Ladies' Guide,* 332.

119. Elizabeth Blackwell, M.D., *The Laws of Life with Special Reference to the Physical Education of Girls* (New York, 1852), 25.

120. Ibid., 28–32.

121. Gardner, *Conjugal Sins,* 17–23.

122. Miller, *From Colony to Province,* 52. He observes: "The sins paraded in

the sermons were not so much those of the notoriously scandalous but such as were bound to increase among good men. They thus had to be all the more vigorously condemned because they were incurable . . . ."

123. Cowan, *New Life,* 124–25.

124. Guernsey, *Avoided Subjects,* 89.

125. Trall, *Sexual Physiology,* 232, 248, 245.

126. Napheys, *The Physical Life,* 56, 98, 99–103.

127. See, for example, Mrs. Alice B. Stockham, *Tokology, A Book for Every Woman* (Chicago, 1883), 142–44; Trall, *Sexual Physiology,* 205; Duffey, *What Women Should Know,* 99.

128. Ibid.

129. Ibid., 91, 96.

130. Ibid., 113. Duffey was a popular feminist author, according to William Leach; she was no sexual enthusiast. See *True Love and Perfect Union,* 88, 94, 115.

131. Scott, *The Sexual Instinct,* 302; also see 104, 107.

132. Stockham, *Tokology,* 141–44.

133. Marion Harland (pseud.), *Eve's Daughters; or, Common Sense for Maid, Wife, and Mother* (New York, 1882), 403–4.

134. Bayley, *Marriage,* 112–13.

135. The Rev. Geo [rge] W. Hudson, *The Marriage Guide for Young Men: A Manual of Courtship and Marriage* (Ellsworth, Me., 1883), 125.

136. Ibid., 126.

137. Ibid.

138. Ibid., 126–27.

139. Ibid., 128.

140. Sometimes moderates might echo sentiments that were being espoused by the sex radicals or enthusiasts of their day. For example, free-lover Moses Hull insisted: "Thus marriage, as we have before intimated, is a union of spirits. Where the spirits are truly united, there is marriage; nowhere else." Moses Hull, *That Terrible Question* (Boston, 1874), 17, as quoted by Michael Gordon, "The Ideal Husband as Depicted in the Nineteenth-Century Marriage Manual," in *The American Man,* Elizabeth H. Pleck and Joseph H. Pleck, eds. (Englewood Cliffs, N.J., 1980), 145–157, esp. 153. The difference between moderates and 19th-century free lovers or sex radicals was psychologically wide and deep but substantively much narrower than one might expect, given the hostility the sexual radicals evoked from the Victorian mainstream. Often applying the logic of romantic love with a consistency that frightened the moderates, sex radicals were willing to abandon outward forms of propriety and the legal supports of marriage for the sake of emotional purity. See Hal D. Sears, *The Sex Radicals: Free Love in High Victorian America* (Lawrence, Kan., 1977), 3–27; Jesse F. Battan, "The Politics of Eros: Sexual Radicalism in Nineteenth-Century America" (Ph.D. diss., University of California, Los Angeles, 1988); John Spurlock, "The Free Love Network in America, 1850 to 1860," *Journal of Social History* 21 (Summer 1988), 765–79.

141. Degler, *At Odds,* 262–65.

142. Kathryn Ruth Hamilton, "Villains and Cultural Change: Aaron Burr and Victorian America" (M.A. thesis, California State University, Fullerton, 1985), 118–54.

143. Anonymous, *Satan in Society,* 146; Napheys, *The Physical Life,* 19; O[rson] S[quire] Fowler, *Sexual Science; or Manhood and Womanhood, including Perfect Husbands, Wives, Mothers, and Infants* (Boston, 1870), 63.

144. Foote, *Plain Talk,* 798.

145. Storer, *Is It I?,* 111.

146. Anonymous, *Satan in Society,* 147.

147. Duffey, *What Women Should Know,* 63.

148. Fowler, *Offspring,* 26–29.

149. Michael McGiffert, "American Puritan Studies in the 1960's," *William and Mary Quarterly* 3rd ser. 27 (Jan. 1970), 42–45; Robert G. Pope, *The Half-Way Covenant: Church Membership in Puritan New England* (Princeton, 1969); Gerald F. Moran, "Religious Renewal, Puritan Tribalism, and the Family in Seventeenth-Century Milford, Connecticut," *William and Mary Quarterly* 3rd ser. 36 (April 1979), 236–54.

150. Miller, *From Colony to Province,* 34.

151. R[ay] V[aughn] Pierce, *The People's Common Sense Medical Adviser in Plain English* (Buffalo, 1882), 196.

152. Foucault, *The History of Sexuality,* 21.

153. On the suppression of sexuality for capitalistic expansion, see, for example, Howard Gadlin, "Private Lives and Public Order: A Critical View of the History of Intimate Relations in the U.S.," *Massachusetts Review* 17 (Summer 1976), 304, 316–17; Regina Markell Morantz, "The Scientist As Sex Crusader: Alfred C. Kinsey and American Culture," *American Quarterly* 29 (Winter 1977), 586–87; Peter Cominos, "Late-Victorian Sexual Respectability and the Social System," *International Review of Social History* 8, part 2 (1963), 216. Charles Rosenberg questions the linkage of capitalism and sexual frugality on other grounds in a brief discussion in "Sexuality, Class and Role," 151–52.

154. Haller and Haller, *The Physician and Sexuality,* 184–87.

Chapter 5. Blurring Separate Spheres

1. William Perry Nebeker to Mrs. Mary McKean, July 24, 1880, Mormon File, Henry E. Huntington Library.

2. Ibid.

3. Nancy F. Cott, *The Bonds of Womanhood: "Woman's Sphere" in New England, 1780–1835* (New Haven, 1977); Kathryn Kish Sklar, *Catharine Beecher: A Study in American Domesticity* (New Haven, 1973); Ann Douglas, *The Feminization of American Culture* (New York, 1977); Carl N. Degler, *At Odds: Women and the Family in America From the Revolution to the Present* (New York, 1980); Linda Gordon, *Woman's Body, Woman's Right: A Social History of Birth Control in America* (New York, 1976); Mary P. Ryan, *Cradle of the Middle Class: The Family in*

Oneida County, New York 1790–1865 (New York, 1981); Barbara Welter, "The Cult of True Womanhood, 1820–1860," *American Quarterly* 18 (Summer 1966), 151–74; Barbara Leslie Epstein, *The Politics of Domesticity: Women, Evangelism and Temperance in Nineteenth-Century America* (Middletown, Conn., 1981); Sheila Rothman, *Woman's Proper Place: A History of Changing Ideals and Practices, 1870 to the Present* (New York, 1978); Rosalind Rosenberg, *Beyond Separate Spheres: Intellectual Roots of Modern Feminism* (New Haven, 1982). For an insightful review of the historiography of "separate spheres," see Linda K. Kerber, "Separate Spheres, Female Worlds, Woman's Place: The Rhetoric of Women's History," *Journal of American History* 75 (June 1988), 9–39.

4. As the previous two chapters document, the concept of purity does not necessarily imply an absence of sexuality in 19th-century usage. "Pure" Victorians acted within the limits of moral sexual expression. The "true woman" must be pure, but this did not necessarily mean that she must be a prude.

5. Charles Rosenberg, "Sexuality, Class and Role in 19th-Century America," *American Quarterly* 25 (May 1973), 131–53; Peter N. Stearns, *Be a Man! Males in Modern Society* (New York, 1979); Joseph H. Pleck and Jack Sawyer, eds., *Men and Masculinity* (Englewood Cliffs, N.J., 1974); G. J. Barker-Benfield, *The Horrors of the Half-Known Life: Male Attitudes Toward Women and Sexuality in Nineteenth-Century America* (New York, 1976); Peter Filene, *Him/Her/Self: Sex Roles in Modern America* (New York, 1975); Elizabeth H. Pleck and Joseph H. Pleck, eds., *The American Man* (Englewood Cliffs, N.J., 1980); Anthony E. Rotundo, "Body and Soul: Changing Ideals of American Middle-Class Manhood, 1770–1920," *Journal of Social History* 16 (Summer 1983), 23–38; Joseph F. Kett, *Rites of Passage: Adolescence in America, 1790 to the Present* (New York, 1977); Robert L. Griswold, *Family and Divorce in California, 1850–1890: Victorian Illusions and Everyday Realities* (Albany, 1982); Rupert Wilkinson, *American Tough: The Tough-Guy Tradition and American Character* (Westport, Conn., 1984); John Mack Faragher, *Women and Men on the Overland Trail* (New Haven, 1979).

6. Rosenberg, "Sexuality, Class and Role," 139–40.

7. John G. Cawelti, *Apostles of the Self-Made Man: Changing Concepts of Success in America* (Chicago, 1965); Irvin G. Wyllie, *The Self-Made Man in America: The Myth of Rags to Riches* (New Brunswick, N.J., 1954); Daniel T. Rodgers, *The Work Ethic in Industrial America, 1850–1920* (Chicago, 1978); John William Ward, *Andrew Jackson: Symbol for an Age* (New York, 1953).

8. Richard Slotkin, *Regeneration Through Violence: Mythology of the American Frontier, 1600–1860* (Middletown, Conn., 1973), 268–312.

9. Robert Griswold outlines the newer image of masculinity in *Family and Divorce,* 92–139.

10. Cott, *Bonds,* 172.

11. Ibid., 163–70.

12. Ibid., 161–62.

13. See Carroll Smith-Rosenberg, "The Female World of Love and Ritual: Relations Between Women in Nineteenth-Century America," *Signs: Journal of Women in Culture and Society* 1 (Autumn 1975), 1–29; while I respect the ground-

breaking insights of Carroll Smith-Rosenberg on 19th-century *female-female* relationships, her evidence of women's diaries and female-to-female correspondence did not provide the most solid foundation from which to generalize to *male-female* relationships. Thus when she says that rigid gender-role differentiations led to "the emotional segregation of women and men" (9), I must demur. The value and significance in 19th-century America of a specifically female world of intimacy and emotional richness are undeniable, but I would argue that women's emotional relationships with other women did not necessarily preclude their intensely emotional relationships with men. Male-female relationships were not characterized by "severe social restrictions on intimacy" (9), nor was the intimate contact between men and women frequently formal and stiff. The evidence of this study suggests that "closeness, freedom of emotional expression, and uninhibited physical contact" (27) were not exclusively the province of women's relationships with each other, but also might characterize women's romantic relationships with men. In her study of northeastern courtship, Rothman also found evidence and drew conclusions that contradicted Smith-Rosenberg. See Rothman, *Hands and Hearts,* 319, fn. 14.

14. Cott, *Bonds,* 168.

15. Marion Harland (pseud.), *Eve's Daughters; or, Common Sense for Maid, Wife, and Mother* (New York, 1882), 406.

16. Charles Godwin to Harriet Russell, Feb. 2[1], 1862, Strong Collection. The [1] has been supplied because I believe the letter to be misdated and the logic of development in the correspondence makes the 21st a more likely date.

17. E. Simkins to E. Trescot, Nov. 2, 1863, Simkins Collection.

18. S. F. Smith to M. Smith, March 20, 1834, Smith Collection.

19. E. Simkins to E. Trescot, Sept. 20, 1862, Simkins Collection.

20. Ibid.

21. A. Janin to V. Blair, Nov. 6, 1871, Janin Collection.

22. J. Marquis to N. Haile, Sept. 17, 1863, Marquis Collection.

23. Linda Kerber has demonstrated that in the late 18th century the Enlightenment, the Revolution, and Republican rhetoric furthered the acceptance of the educated woman. See *Women of the Republic: Intellect and Ideology in Revolutionary America* (Chapel Hill, 1980), 189–231.

24. S. F. Smith to M. Smith, March 6, 1834, Smith Collection.

25. N. Wheeler to C. Bradley, Jan. 10, 1875, Burdette Collection. Although the last digit in the date is difficult to read, I am certain from the context that the year is 1875.

26. Ibid., June 18, 1876.

27. Griswold, *Family and Divorce,* 1–17, 170–79; Degler, *At Odds,* 315. Preparation for motherhood was also a prominent justification for women's education in other sources, but it was not a predominant theme in the correspondence between men and women. See Cott, *Bonds,* 104–8; and Sklar, *Catharine Beecher,* 151–67.

28. The reader will find a more detailed discussion of William Harbert's attitudes toward women and education later in the chapter.

29. J. Bell to A. Hallock, Oct. 31, 1860, Bell Collection.

30. A. Janin to V. Blair, March 10, 1874, Janin Collection.

31. E. Simkins to E. Trescot, Nov. 2, 1863, Simkins Collection.

32. Degler, *At Odds,* 307–15; Cott, *Bonds,* 101–25.

33. Erving Goffman, *The Presentation of Self in Everyday Life* (Garden City, N.Y., 1959); Erving Goffman, *Encounters: Two Studies in the Sociology of Interaction* (Indianapolis, 1961); Ralf Dahrendorf, *"Homo Sociologicus:* On the History, Significance, and Limits of the Category of Social Role," in *Essays in the Theory of Society* (Stanford, 1968), 19–87.

34. Stearns, *Be a Man,* 47.

35. C. Strong to H. Russell, Dec. 20, 1862, Strong Collection.

36. J. Burnett to W. Burnett, June 3, 1857, Burnett Collection.

37. D. Lummis to C. Lummis, April 17, 1884, Moore Collection.

38. L. Clark to J. Clark, Dec. 24, 1848, Clark Collection.

39. J. Blair to M. Blair, March 19, 1848, Janin Collection.

40. Ibid., Sept. 30, 1850.

41. Ibid.

42. Ibid.

43. *The National Cyclopaedia of American Biography,* vol. 23 (New York, 1933), 164.

44. By 19th-century standards, Hague's courtship was a late one. He was therefore more established in his career than younger courting men. This may help to account for the fact that for this relationship, work was an especially divisive issue.

45. J. Hague to M. W. Foote, July 9, 1871, Hague Collection.

46. Ibid., July 20, 1871.

47. Ibid.

48. Mary's words as reported in ibid., April 3, 1872.

49. Ibid.

50. J. W. Foster to M. Appleton, April 23, 1824, Appleton-Foster Collection, Henry E. Huntington Library.

51. See Chapter 6 on courtship testing for a fuller treatment of women's greater insecurity about romantic love.

52. N. Hawthorne to S. Peabody, March 15, 1840, Hawthorne Collection.

53. Ibid., Aug. 21, 1839.

54. J. Bell to A. Hallock, March 29, 1863, Bell Collection.

55. Ibid., May 30, 1862; see also May 17, 1862, and March 6, 1863.

56. Ibid., May 2, 1858.

57. E. Simkins to E. Trescot, Aug. 8, 1864, Simkins Collection.

58. Ibid., [undated, c. Sept. 10–12, 1864].

59. L. Hodge to M. Granger, Jan. 31, 1869, Hodge Collection. Lyman had been waxing lyrical on romantic subjects up to this point; this letter marks a significant shift in tone and subject on the eve of his wedding.

60. J. Marquis to N. Haile, Jan. 20, 1866, Marquis Collection.

61. Ibid., Sept. 17, 1865.

62. C. Watts to E. Watts, Feb. 5, 1899, Charles William Watts Collection, Henry E. Huntington Library, hereafter cited as Watts Collection.

63. Priscilla Knuth, Research Associate for the Oregon Historical Society, to Mr. Lee Rohrbough, June 7, 1963; Lee Rohrbough to the editor of the Dawson paper and/or the Right Honorable Mayor of Dawson, Alaska, *Klondike Korner* (a mimeographed newsletter), June 27, 1963, Watts Collection.

64. C. Watts to E. Watts, Jan. 30, 1898.

65. There is a massive literature on the Protestant ethic. The classic statement is Max Weber's *The Protestant Ethic and the Spirit of Capitalism,* trans. Talcott Parsons (New York, 1956). Another classic use of the concept is Perry Miller's *The New England Mind: From Colony to Province* (Cambridge, Mass., 1953; rpt., 1961), 40–52; Daniel Rodgers provides an excellent summary of the changing cultural formulations of the work ethic in *Work Ethic,* 1–29.

66. Ibid., 9–11.

67. In fact, it may be that powerful romantic love actually weakened male work commitment or at least efficiency, just as powerful religious commitment may have also absorbed people's time and energy away from their normal work routines. The analogy from the Protestant ethic to romantic love is most appropriate at the level of ideology and belief.

68. S. F. Smith to M. Smith, March 6, 1834, Smith Collection. He also included intellectual interests.

69. Faragher, *Women and Men on the Overland Trail,* 59–63, 109–19, 144, 162–66, 174.

70. Suzanne Lebsock, *Free Women of Petersburg: Status and Culture in a Southern Town, 1784–1860* (New York, 1984), 32–34.

71. Degler, *At Odds,* 3–110, 279–327; Griswold, *Family and Divorce,* 10–17, 77–91; Cott, *Bonds,* 101–59, 197–206; Sklar, *Catharine Beecher,* 132–37, 151–67.

72. J. Burnett to W. Burnett, Sept. 8, 1866, Burnett Collection.

73. A. Hallock to J. Bell, April 11, 1863, Bell Collection. In a Jan. 13, 1863, letter to Gusta, James was enraged by southern women who did not perform their prescribed jobs of cooking, cleaning, etc. He too was tied to the traditional gender-specific symbolism of work.

74. J. Marquis to N. Haile, Jan. 9, 1865, Marquis Collection. This letter is misdated; it is actually 1866. Often in early January, letters will be dated incorrectly as correspondents inadvertently use the previous year.

75. A. Janin to V. Janin, May 24, June 19, 1875, Janin Collection.

76. A. Baldwin to F. Baldwin, Oct. 21, 1869, Baldwin Collection.

77. Ibid., Jan. 1, 1871.

78. Ibid., Jan. 19, 1871.

79. Ibid.

80. Ibid., Dec. 26, 1874.

81. This is an example of testing behavior which will be illustrated at length in courtship interactions in the next chapter. Testing diminished in marital interactions, except in the troubled relationship. Alice is testing her husband through

the standard medium of self-criticism: inviting him to reassure her that she is beautiful and has a nice figure.

82. A. Baldwin to F. Baldwin, Jan. 16, 1875, Baldwin Collection.

83. D. Lummis to C. Lummis [March 31, 1884], Moore Collection.

84. A. Baldwin to F. Baldwin, Oct. 8, 1876, Baldwin Collection.

85. Ibid., Jan. 3, 1877.

86. D. Lummis to C. Lummis [Jan. 21–23, 1884], Moore Collection.

87. E. Trescot to E. Simkins, Jan. 6, [1864], Simkins Collection.

88. Ibid.

89. Ibid., Sept. 14 [1864].

90. E. Lovell to M. Lovell, Nov. 23, 1862, Lovell Collection.

91. This courtship tension was manifested by almost all women in the study for whom adequate premarital evidence exists and revolved around the testing and re-testing of a man's emotional commitment by the woman who was considering marriage. See Chapter 6.

92. Reframing is a term I first heard used by Lucy Steiner, M.F.C.C., to refer to one therapeutic strategy of reorienting the mind-set of a patient in psychotherapy. See Paul Watzlawick, John H. Weakland, and Richard Fisch, *Change: Principles of Problem Formation and Problem Resolution* (New York, 1974); William Hudson O'Hanlon, *Taproots: Underlying Principles of Milton Erickson's Therapy and Hypnosis* (New York, 1987). I have borrowed the word but changed its context and meaning for my own analytical purposes.

93. N. Wheeler to C. Bradley, April 4, 1878, Burdette Collection.

94. Ibid.

95. Theoretically, the opposite could also be true. Women might reframe when they renounced power which they were supposed to assume in their sphere, and men might reframe when they assumed power in the feminine sphere which they were supposed to renounce. I found almost no examples of either kind of reframing but it is certainly possible.

96. N. Wheeler to C. Bradley, Jan. 17, 1875, Burdette Collection.

97. E. Harbert to W. Harbert, March 18, 1886, Harbert Collection.

98. E. Lovell to M. Lovell, Oct. 29, 1862, Lovell Collection.

99. L. Clark to his parents, Sept. 22, 1846, typescript in "American Scene— Clark Family Narrative," pp. 138–42, Clark Collection. I could not find this letter in manuscript form; it only appeared as a typescript copy in L. Clark's unpublished narrative.

100. L. Clark to J. Clark, April 4, 1847.

101. Ibid. All the quotations in the preceding two paragraphs are from the April 4, 1847, letter.

102. J. Clark to L. Clark, April 20, 1847.

103. L. Clark to J. Clark, May 7, 1847.

104. Ibid., May 21, 1847. Julia herself may have taken the lead in reframing this experience sometime after her April 20 letter. From the extant evidence, it is impossible to draw a clear inference. It appears that she at least gently nudged Lincoln in the direction of claiming back the decision.

105. J. Clark to L. Clark, June 25, 1864.

106. See Mary Beth Norton, *Liberty's Daughters: The Revolutionary Experience of American Women, 1750–1800* (Boston, 1980); Anne Firor Scott, *The Southern Lady: From Pedestal to Politics, 1830–1930* (Chicago, 1970), 81–133; Glenda Riley, *Frontierswomen: The Iowa Experience* (Ames, Iowa, 1981), 110–35; Michael Fellman, *Inside War* (New York, 1989).

107. Degler notes that "the woman's movement throughout the 19th century left untouched the great mass of women, married and unmarried." See Degler, *At Odds,* 306; and Blanche Glassman Hersh, *The Slavery of Sex: Feminist-Abolitionists in America* (Urbana, 1978), 256. Ellen Carol DuBois argues for the wider cultural significance of suffrage on feminist grounds in *Feminism and Suffrage: The Emergence of an Independent Women's Movement in America, 1848–1869* (Ithaca, 1978), 16–18, 201–2.

108. D. Lummis to C. Lummis [Feb. 28–March 2, 1884], Moore Collection.

109. A. Hallock to J. Bell, Oct. 11, 1861; J. Bell to A. Hallock, Oct. 3, 1861, Bell Collection.

110. Perhaps there are critiques of the male role in female-female correspondence. This awaits further research.

111. N. Wheeler to C. Wheeler, July 21, 1880, Burdette Collection. This was Clara Burdette's first husband. See also Clara Burdette, *The Answer to "Clara, what are you going to do with your life?"* (Pasadena, 1951), 67–71.

112. E. Trescot to E. Simkins, Dec. 13, 1863, Simkins Collection.

113. E. Simkins to E. Trescot, Dec. 17, 1863.

114. J. Marquis to N. Haile, April 5, 1863, Marquis Collection.

115. Ibid.

116. Ibid.

117. J. Hague to M. W. Foote, July 1, 1871, Hague Collection. All the Anthony-Stanton conversation and commentary are from this letter.

118. Ibid. See Sklar, *Catharine Beecher,* 258–73. Hague is most likely referring to Beecher's *Woman's Profession As Mother and Educator with Views in Opposition to Woman Suffrage* (1872).

119. Cott, *Bonds,* 201.

120. J. Hague to M. W. Foote, Aug. 20, 1871, Hague Collection. This was written from the mines.

121. Ibid., Aug. 11, 1871.

122. Ibid., Sept. 11, 1871.

123. W. Harbert to E. Boynton, Oct. 15, 1869, Harbert Collection.

124. Ibid., Feb. 17, 1869.

125. Cott, *Bonds,* 202–6.

126. W. Harbert to E. Boynton, Feb. 17, 1869, Harbert Collection.

127. Ibid. Elizabeth Boynton [Harbert] was prominent in the woman's suffrage movement and continued to participate after her marriage. See Steven M. Buechler, "Elizabeth Boynton Harbert and the Woman Suffrage Movement, 1870–1896," *Signs: Journal of Women in Culture and Society* 13 (Autumn 1987), 78–97.

Chapter 6. Testing Romantic Love

1. Theocritus, Jr. (pseud.), *Dictionary of Love* (New York, 1858), 73.

2. Examples of courtship testing appear throughout the book. I usually do not identify testing behaviors outside Chapter 6, but see, for example, Chapter 7, 195–96; Chapter 8, 243; Chapter 2, 39–40, 47–48. Testing was a pervasive pattern in middle-class premarital courtship letters but dropped off dramatically after marriage, though the behavior might recur sporadically or even proceed unabated in the "troubled" marital relationship. See Chapter 7, 216; Chapter 5, 140–41.

Though there need be no consistency within culture or between culture and social structure, Victorian courtship testing was, in fact, congruent with a system of mate selection that had just replaced external parental restraints with the lovers' own internal appraisal of romantic love. The development of some alternative method of emotional "control" in partner selection was culturally symmetrical to this change.

The length of a middle-class courtship could obviously vary considerably, depending on individual circumstances. Longer courtships allowed for a lengthier period of testing. But whether lengthier testing was always more effective in creating a stronger romantic bond for the marriage is unclear. Intensity of testing may operate somewhat independently of the duration of a courtship.

3. There is a paucity of secondary materials on pre-20th-century American courtship. The best study and only book-length treatment of the subject is Ellen K. Rothman's *Hands and Hearts: A History of Courtship in America* (New York, 1984); working with different letter collections and thus offering some opportunity for cross-checking the validity of generalization, Rothman's conclusions for the 1830–80 period often support my own. We diverge in our interpretations of the last twenty years of the 19th century as well as in our basic chronological scope, topical focus, and analytic approach. On the subjects of the wedding ceremony and "honeymoon" ritual, readers should consult Rothman. I found too little data in these areas to make meaningful generalizations. Her discussion of both topics is sensitive to change over time, informative, and enlightening. While Rothman's evidence does not appear to contradict my own interpretation of middle-class Victorian courtship, she makes little mention of the testing rituals which I believe form the central structure and defining principle of middle-class courtship from 1830 to 1900.

Though there is a vast and growing literature on sexuality and sex roles and scattered works on romantic love (see references in Chapters 2, 3, and 5), scholarship on 19th-century courtship outside these three areas is not numerous. However, works of varied relevance to the topic include: Steven M. Stowe, " 'The Thing, Not Its Vision': A Woman's Courtship and Her Sphere in the Southern Planter Class," *Feminist Studies* 9 (Spring 1983), 113–30; Daniel Scott Smith, "Parental Power and Marriage Patterns: An Analysis of Historical Trends in Hingham, Massachusetts," *Journal of Marriage and the Family* 35 (Aug. 1973), 419–28; Michael Grossberg, *Governing the Hearth: Law and the Family in Nineteenth-Century America* (Chapel Hill, 1985); Sondra Herman, "Loving Courtship or the Marriage Market? The Ideal and Its Critics, 1871–1911," *American Quarterly* 25 (May 1973), 235–52; Karen Halttunen, *Confidence Men and Painted Women: A Study of Middle-Class Culture in America, 1830–1900* (New Haven, 1982).

4. M. Smith to S. F. Smith, April 7, 1834, Smith Collection.

5. Mr. and Mrs. Jonathan K. Smith (Mary's parents) to S. F. Smith, April 9, 1834, Smith Collection.

6. Sarah Smith (unrelated to Mary) to L. Clark, Aug. 25, 1836, Clark Collection.

7. Ibid.

8. J. Bell to A. Hallock, Aug. 27, 1863, Bell Collection.

9. C. Strong to H. Russell, Jan. 1, 1863, Strong Collection.

10. Ibid., Jan. 11, 1863.

11. Ibid.

12. Rothman, *Hands and Hearts,* 160, 214–15.

13. E. Simkins to E. Trescot, Dec. 4, 1863, Simkins Collection.

14. E. Trescot to E. Simkins, Jan. 14, 1864.

15. Ibid.

16. E. Simkins to E. Trescot, Feb. 1, 1864.

17. N. Wheeler to C. Bradley, Jan. 11, 1877, Burdette Collection. I could document only one instance in which a man participated in a ritual of consent with his own father. See. J. Blair to M. Jesup, Sept. 22, 1845, Janin Collection. Also refer to Rothman, *Hands and Hearts,* 115–17.

18. N. Hawthorne to S. Peabody, May 27, 1842, Hawthorne Collection. Hawthorne's inability to tell his family of his engagement was unusual, but cannot be attributed to any belief that his mother had a right to interfere with his choice of a mate. Hawthorne blamed his reticence on emotional constrictions he learned very early in family life: "We are conscious of one another's feelings, always; but there seems to be a tacit law, that our deepest heart-concernments are not to be spoken of. I cannot gush out in their presence—I cannot take my heart in my hand, and show it to them." N. Hawthorne to S. Peabody, Feb. 27, 1842. See also Arlin Turner, *Nathaniel Hawthorne: A Biography* (New York, 1980), 140–42.

19. E. Shepard to N. Shepard, Aug. 23, 1869, Raphael Pumpelly Collection, Henry E. Huntington Library.

20. Ibid.

21. Ibid.

22. Ibid.

23. E. Shepard to K. Shepard, Sept. 1, 1869.

24. See Rothman, *Hands and Hearts,* 119–20 for corroboration of this conclusion.

25. J. Marquis to N. Haile, Jan. 1, 1866, Marquis Collection.

26. Ibid.

27. Ibid.

28. In 19th-century American culture, the disruptive and socially alienating

potential of romantic love was offset by the rough social structural separation of class, race, and ethnic groups. Of course, this separation of associational networks might break down in transitional geographies and social situations.

29. L. Hodge to M. Granger, Jan. 9, 1866, Hodge Collection.

30. D. Lummis to C. Lummis [Feb. 28–March 2, 1884], Moore Collection.

31. M. Smith (unmarried) to S. F. Smith, March 17, 1834. She repeats these sentiments Aug. 4, 1834, Smith Collection.

32. M. Smith (married) to S. F. Smith, April 5, 1835.

33. C. Watts to E. Watts, [no date], envelope postmarked Feb. 5, 1900, Watts Collection.

34. M. Hague to J. Hague, June 1, 1892, Hague Collection.

35. Ibid. Mary Hague assumed that there were no alternatives available to the unhappy parents, once the inappropriate choice was made by their daughter.

36. Outside a certain range, differences of class, race, ethnicity, and education may be undesirable to the participants themselves.

37. Rothman, *Hands and Hearts,* 25–28, 120–22.

38. Ibid., 22–23, 287–88. Thomas P. Monahan, *The Pattern of Age at Marriage in the United States,* vol. 2 (Philadelphia, 1957), 276–77, 280–84.

39. M. Granger to L. Hodge, Jan. 28, 1866, Hodge Collection.

40. C. Strong to H. Russell, Dec. 20, 1862, Strong Collection.

41. N. Wheeler to C. Bradley, Nov. 1, 1874, Burdette Collection.

42. Ibid., Jan. 7, 1878.

43. J. Hague to M. W. Foote, June 25, 1871, Hague Collection.

44. E. Simkins to E. Trescot, Aug. 14, 1864, Simkins Collection.

45. Ibid. The antebellum southern courtship correspondence which Steven M. Stowe examines in " 'The *Thing,* Not Its Vision' " appears similar to southern correspondence in this study. Though I question Stowe's overall interpretation, it appears from his description that the southern couple under investigation were involved in a standard courtship-testing ritual, with a dramatic female crisis of doubt precipitated after the engagement. See Stowe, 122–28, and 130, fn. 28: "As their marriage drew near, Bessie's and Thomas's doubts involved their family and friends in some rather partisan arguments about whether the wedding should take place." Stowe calls this experience a "courtship crisis." Daniel Blake Smith, *Inside the Great House: Planter Family Life in Eighteenth-Century Chesapeake Society* (Ithaca, 1980), 130–50, 154–59 argues that for the middle to upper-middle strata of southern society, romantic love and companionate marriage grew increasingly significant after 1750. Jan Lewis's study, *The Pursuit of Happiness: Family and Values in Jefferson's Virginia* (New York, 1983), covers the period from the early 18th century to 1830 and essentially supports this point of view; see 188–208. While Catherine Clinton in *The Plantation Mistress: Woman's World in the Old South* (New York, 1982), 59–68, does not rule out romantic motives, she seems to give primacy to wealth and property in southern mate selection; Clinton's study applies to the period between 1780 and 1830. Less tentative generalizations on pre- and post-Civil War southern courtship practices await more intensive research on mid-19th century southern materials.

46. A. Hallock to J. Bell, Aug. 6, 1855, Bell Collection. While group socializing may have been significant in the early courtship initiation process, it was unclear what role group activities played in the later stages of romantic involvement. See Rothman, *Hands and Hearts,* 23–25, 162–63.

47. A. Hallock to J. Bell, Feb. 1, 1857, Bell Collection. From the context I have concluded Gusta was not referring to 3 o'clock in the afternoon, though no specific meridian time was given.

48. Ibid.

49. A major dimension of 19th-century courtship was the effort to resolve doubt before marriage. Ironically, success meant that significant dramatic aspects of a relationship were eclipsed. See Chapter 7 for a discussion of the calm, quiet domestic conceptions of marriage expressed by courting couples, particularly men.

50. As I have discussed in Chapter 5, men were also anxious, but over their ability to be depended upon, particularly to be a woman's sole economic support.

51. J. Marquis to N. Haile, Aug. 1, 1862, Marquis Collection.

52. Ibid.

53. Ibid., April 9, 1865.

54. Ibid., Sept. 17, 1865.

55. Ibid.

56. Ibid. All quotations in this paragraph are taken from this letter.

57. Ibid.

58. Ibid., and Oct. 10, 1865.

59. Ibid., Sept. 17, 1865.

60. Ibid., Oct. 10, 1865.

61. Ibid.

62. A. Janin to V. Blair, Nov. 23, 1871, Janin Collection.

63. Ibid., Jan. 15, 1872. See Chapter 2 for a more detailed description of the incident.

64. Ibid., March 2, 1872.

65. V. Blair to A. Janin, Dec. 11, 1871.

66. A. Janin to V. Blair, Jan. 1, 1872.

67. Ibid., Nov. 13, 1872.

68. Ibid.

69. Ibid., Nov. 13, 1872. A pendant is a companion piece or a counterpart.

70. V. Blair to A. Janin, April 2, 1874; March 1, March 12, Nov. 20, 1873.

71. Ibid., April 22, 1874.

72. Ibid.

73. A. Janin to V. Blair, April 28, 1874.

74. Ibid., July 3, 1874.

75. Ibid., Aug. 30, 1874.

76. N. Hawthorne to S. Peabody, March 6, 1839, Hawthorne Collection.

77. Ibid.

78. Ibid., July 3, 1839. He describes her query in his letter.

79. Ibid.

80. Ibid., July 24, 1839.

81. Ibid., July 30, 1839. Her reaction is described in this letter.

82. Ibid.

83. Ibid., Oct. 3, and Dec. 1, 1839.

84. Ibid., March 15, 1840.

85. J. Hague to M. W. Foote, July 20, 1871, Hague Collection.

86. N. Hawthorne to S. Peabody, Aug. 16, 1841, Hawthorne Collection.

87. J. Hague to M. W. Foote, July 9, 1871, Hague Collection.

88. Ibid., July 17, 1871.

89. Mary Ward Foote's side of the transaction has been read from James's letters, as only her later correspondence to her husband was found.

90. J. Hague to M. W. Foote, Aug. 6, 1871, Hague Collection.

91. Ibid., Aug. 25, 1871.

92. Ibid., Aug. 31, 1871.

93. Ibid., Sept. 5, 1871.

94. Ibid., Sept. 11 and 14, 1871.

95. Ibid., Sept. 25, 1871. After recovering from a bout of ill-health, Mary expressed herself "ready" to marry James. Seeming to expect the immediate fulfillment of his promise to return East to marry her when her health permitted, the severity of Mary's courtship test may have been a reaction to James's delay.

96. Ibid.

97. Ibid.

98. Ibid., Sept. 27, 1871. The letter James wrote to her on Sept. 25 probably did not arrive until approximately Oct. 5, 1871.

99. Ibid., Oct. 14 and 17, 1871.

100. Reported in J. Hague's letter to M. W. Foote, Nov. 27, 1871.

101. R. Burdette to C. Baker, Feb. 24, 1898, Burdette Collection.

102. Ibid.

103. Ibid., Feb. 28, 1898.

104. Ibid.

105. Ibid., March 4, 1898.

106. Ibid., March 14, 1898.

107. Ibid., March 15, 1898.

108. Ibid., Chicago Wednesday night [March 16, 1898].

109. Ibid., April 3, 1898.

110. Ibid., April 25, 1898.

111. Ibid., Aug. 3, 1898.

112. Ibid.

113. Ibid., Aug. 25, 1898. These are quotations which Robert cites from a letter Clara wrote to him. I could not find her letter in the collection. As Clara was a more rigorous censor of her own letters than of his, it may have been destroyed. Robert placed all this material in quotations, occasionally interspersing his own remarks. I have excluded all his interjections.

114. Ibid.

115. Ibid., Aug. 27, 1898.

116. C. Baker to R. Burdette, Aug. 21, 1898.

117. R. Burdette to C. Baker, Sept. 1, 1898.

118. Telegram message from C. Baker to R. Burdette, Sept. 1, 1898 as quoted in a letter of R. Burdette to C. Baker, Sept. 2, 1898.

119. R. Burdette to C. Baker, Sept. 2, 1898.

120. Ibid., Tuesday night, Sept. 6, 1898.

121. C. Baker to R. Burdette, Oct. 17, 1897 [actually 1898].

122. R. Burdette to C. Baker, Nov. 22, 1898.

123. Ibid., Nov. 27, 1898.

124. N. Wheeler to C. Bradley, Feb. 3, 1878, Burdette Collection.

125. A. Hallock to J. Bell, May 28, 1862, Bell Collection.

126. Ibid., April 30, 1863.

127. J. Bell to A. Hallock, Sept. 19, 1863.

128. Theocritus, Jr. (pseud.), *Dictionary of Love,* 94–96.

129. E. Simkins to E. Trescot, July 6, 1863, Simkins Collection.

130. E. Trescot to E. Simkins, July 11, 1861.

131. Ibid.

132. E. Simkins to E. Trescot, July 16 [19], 1863.

133. Ibid., July 19, 1863.

134. Ibid.

135. Ibid., Sept. 4, 1863.

136. Ibid., Sept. 11, 1863.

137. Ibid. "Moments of indifference" is Eldred's term.

138. Ibid.

139. E. Trescot to E. Simkins, Oct. 26, 1863.

140. E. Simkins to E. Trescot, Oct. 17, 1863.

141. Ibid., July 16 [19], 1863.

142. Ibid.

143. This is his report of her letter. Ibid., Sept. 4, 1863.

144. E. Trescot to E. Simkins, Oct. 26, 1863.

145. Ibid., Jan. 14, 1864.

146. Ibid.

147. See Rothman, *Hands and Hearts,* 161–76 on 1830–1880 wedding ritual.

148. Occasionally men delayed a marriage, but this was unusual. See Rothman, *Hands and Hearts,* 157–58.

149. Don Graham to Margaret Collier, Oct. 8, 1873, Margaret Collier Graham Collection, Henry E. Huntington Library, hereafter cited as Graham Collection.

150. J. W. North to A. Loomis, July 8, 1848, North Collection.

151. E. Simkins to E. Trescot, Dec. 4, 1863, Simkins Collection.

152. Ibid., Aug. 14, 1864.

153. C. Godwin to H. Russell, undated [perhaps written around Nov. 10, 1861], Strong Collection. Harriet's fiancé was also a Charles (Strong), not to be confused with the spurned suitor, Charles Godwin.

154. Ibid.

155. Ibid., Jan. 5, 1861. (The letter is misdated, which is not unusual for correspondence around the beginning of a new year. It was written in 1862.)

156. Ibid., Jan. 12, 1862.

157. Ibid., Jan. 18, 1862.

158. Ibid.

159. Ibid., Feb. 5, 1862.

160. Ibid.

161. Ibid.

162. Ibid., March 13, 1862.

163. Ibid., April 16, 1862.

164. Ibid., April 26, 1862.

165. Ibid., April 30, 1862.

166. Ibid.

167. The problem for the historian is that couples often saw no reason to dissect the motivating forces in their choice of a lifetime partner. They believed that love was a mysterious force. See Chapter 7 for further discussion of the consequences of this idea.

168. The balance of personal criteria which tipped the scales in someone's favor was among the most elusive aspects of the entire 19th-century courtship process for both participants and historian. Individual proclivities were no doubt shaped by cultural forces such as novels, schooling, parenting, and so on. Nonetheless, they reflected an interest in someone else for individual attributes. More or less non-romantic motives—wealth, family lineage, political expediency, and social status—primarily involved qualities of experience outside the self of the potential partner. Though non-romantic motives did not necessarily exclude the development of romantic love, they also did not necessitate personal attraction or even interest.

169. N. Wheeler to C. Bradley, Nov. 1, 1874, Burdette Collection.

170. A. Hallock to J. Bell, Jan. 14–15, 1861, Bell Collection.

171. E. Trescot to E. Simkins, undated [Aug. 5, typescript/Sept. 5, 1860, manuscript], Simkins Collection.

172. A. Hallock to J. Bell, Jan. 10, 1858, Bell Collection.

173. J. H. Meteer to E. Boynton, Oct. 21, 1867, Harbert Collection.

174. Ibid., Dec. 20, 1867.

175. Male cousin to J. Cleveland [Burnett], June 19, 1855, Burnett Collection.

176. E. Simkins to E. Trescot, Jan. 1, 1864, Simkins Collection.

177. E. Trescot to E. Simkins, Jan. 6, 1864.

178. Maude Hulbert to Jack Horn, undated [Dec. 27 postmark on envelope, 1897], private collection. See Nora Beale Jacob, "Biography of Maude Hulbert Horn, Editor of *The Georgetown* (California) *Gazette,* 1891–1924" (M.A. thesis, California State University, Fullerton, 1980.)

179. Ibid.

180. Ibid.

181. S. Child to P. Newcomb, Dec. 22, [c. 1817–1825], Newcomb-Johnson Collection, Henry E. Huntington Library.

182. N. Wheeler to C. Bradley, Nov. 1, 1874, Burdette Collection.

183. J. Bell to A. Hallock, Oct. 2, 1857, Bell Collection.

184. Ibid.

185. J. Hague to M. W. Foote, Sept. 10, 1871, Hague Collection.

186. Barbara Myerhoff, *Number Our Days* (New York, 1978), 32. "Enacted beliefs have a capacity for arousing belief that mere statements do not. 'Doing is believing,' hence ritual and ceremony generate conviction when reason and thought may fail." I am also indebted to Myerhoff for the phrase "definitional ceremonies."

187. J. Hague to M. W. Foote, Nov. 8, 1871, Hague Collection.

Chapter 7. Husbands and Wives

1. The secondary literature on 19th-century American marriage is not vast, but there is more work of direct relevance on marriage than courtship. Though not a study of American couples, Phyllis Rose's *Parallel Lives: Five Victorian Marriages* (New York, 1984) is exceptionally cogent; Blanche Glassman Hersh in *The Slavery of Sex: Feminist-Abolitionists in America* (Urbana, 1978) has a neglected chapter on feminist marriages (218–51); Carl N. Degler studies the workings of the companionate marriage in *At Odds: Women and the Family in America from the Revolution to the Present* (New York, 1980), 26–65; Robert L. Griswold effectively utilizes divorce data to study normal cultural expectations of spousal roles in *Family and Divorce in California, 1850–1890: Victorian Illusions and Everyday Realities* (Albany, 1982); Suzanne Lebsock questions women's power in the companionate marriage in *The Free Women of Petersberg: Status and Culture in a Southern Town, 1784–1860* (New York, 1984). See also Steven Mintz, *A Prison of Expectations: The Family in Victorian Culture* (New York, 1983), 103–46; William L. O'Neill, *Divorce in the Progressive Era* (New Haven, 1967); Elaine Tyler May, *Great Expectations: Marriage and Divorce in Post-Victorian America* (Chicago, 1980); John Mack Faragher, *Women and Men on the Overland Trail* (New Haven, 1979); Julie Roy Jeffrey, *Frontier Women: The Trans-Mississippi West, 1840–1880* (New York, 1979); Sidney Herbert Ditzion, *Marriage, Morals and Sex in America: A History of Ideas* (New York, 1953; rev. ed., 1969); George Elliott Howard, *A History of Matrimonial Institutions*, 3 vols. (Chicago, 1904; rpt., 1964); Mary P. Ryan, *Cradle of the Middle Class: The Family in Oneida County, New York, 1790–1865* (New York, 1981); Edward Shorter, *The Making of the Modern Family* (New York, 1975); Michael Grossberg, *Governing the Hearth: Law and the Family in Nineteenth-Century America* (Chapel Hill, 1985).

2. An American Gentleman, *Good Behavior for Young Gentlemen: Founded on Principles of Common Sense, and the Usages of Good American Society* (Rochester, N.Y., 1848), 43.

3. T[imothy] S[hay] Arthur, *Advice to Young Ladies on Their Duties and Conduct in Life* (Philadelphia, [1860]), 214.

4. S[amuel] R. Wells, *Wedlock; or, the Right Relations of the Sexes: Disclosing the Laws of Conjugal Selection, and Showing Who May, and Who May Not Marry* (New York, 1879), 52.

5. Ibid., 51.

6. I use the term "troubled," not as an objective description of a Victorian marriage, but as a reflection of the dissatisfaction of at least one spouse. The categorization of 19th-century marriages in this chapter is strictly limited to the subjective perceptions of the husband or wife.

7. William L. O'Neill, "Divorce in the Progressive Era," *American Quarterly* 17 (Summer 1965), 203–17; O'Neill, *Divorce in the Progressive Era,* 33.

8. N. Hawthorne to S. Peabody, Dec. 24, 1839, Hawthorne Collection. "And nights" is an inked-over passage in the manuscript that has been restored by the Centenary edition of Hawthorne's letters. See Chapter 4, fn. 2.

9. A. Janin to V. Blair, March 26, 1874, Janin Collection. These sentiments invoke the imagery of home and family as chronicled by Christopher Lasch in *Haven in a Heartless World: The Family Besieged* (New York, 1977).

10. N. Hawthorne to S. Peabody, March 30, 1840, Hawthorne Collection.

11. J. Hague to M. W. Foote, Nov. 15, 1871, Hague Collection.

12. E. Simkins to E. Trescot, Nov. 2, 1863, Simkins Collection.

13. A. Hallock to J. Bell, July 8, 1858, Bell Collection.

14. Ibid., Dec. 23, 1858.

15. E. Trescot to E. Simkins, Aug. 31, 1864, Simkins Collection.

16. E. Elliot to A. Hallock, Dec. 30, 1860, Bell Collection.

17. D. Lummis to C. Lummis [Dec. 13–16, 1883], Moore Collection.

18. M. Smith to S. F. Smith, Aug. 25, 1834, Smith Collection.

19. Ibid.

20. J. Hague to M. W. Foote, July 17, 1871, Hague Collection.

21. Nephew to A. Lake, Nov. 6, 1852, Ann (Getz) Lake Collection, Henry E. Huntington Library.

22. V. Blair to A. Janin, Nov. 6, 1872, Janin Collection.

23. J. Hague to M. W. Foote, Oct. 27, 1871, Hague Collection.

24. J. Bell to A. Hallock, Jan. 8, 1859, Bell Collection.

25. Ibid., Oct. 31, 1860.

26. A. Hallock to J. Bell, Nov. 18, 1860.

27. Ibid.

28. A girlfriend to Anna E. Dane, April 6, 1879, in the "unsigned and uni-dentified letter file," Dane Family Collection, Henry E. Huntington Library, hereafter cited as Dane Collection. The unidentified letter writer was apparently attracted to a young man who did not reciprocate her interest.

29. M. Smith to S. F. Smith, Sept. 1, 1834, Smith Collection.

30. Ibid., April 14, 1834.

31. J. Marquis to N. Marquis, March 13, 1870, Marquis Collection.

32. A. Janin to V. Janin, July 3, 1874, Janin Collection.

33. Ibid., June 10, 1875.

34. V. Janin to A. Janin, August 13, 1874.

35. M. Smith to S. F. Smith, April 5, 1835, Smith Collection.

36. Ibid., April 8, 1835.

37. Ibid., April 12, 1835.

38. N. Hawthorne to S. Hawthorne, June 10, 1844, Hawthorne Collection.

39. Ibid., April 26, 1850.

40. Ibid., March 18, 1856.

41. These avowals of love should not be dismissed as merely the application of ritualized formula. First, they were varied by both style, content, and the writer's particular fluency. Second, they contained an intensity of tone that communicated emotion far beyond the necessities of polite exchange. Third, they were private communications, written with no other audience in mind than husband or wife. Fourth, they were voluntarily offered, and not compelled by any public ritual or ceremony. Nonetheless, courtship letters were more frequent, continuous, and usually more revealing than marriage letters. Marital correspondence usually offers the historian more sporadic contact with the wedded couple.

42. E. Harbert to W. Harbert, May 11, 1879, Harbert Collection.

43. E. Lovell to M. Lovell, Nov. 15, 1862, Lovell Collection.

44. J. North to A. North, July 6, 1861, North Collection.

45. M. Graham to D. Graham, Aug. 25, 1881. Graham Collection.

46. D. Graham to M. Graham, September 28, 1889.

47. L. Dane to her mother, Dec. 7, 1864, Dane Collection.

48. L. Dane to E. Dane, May 28, 1876.

49. E. Dane to L. Dane, June 4, 1876. Ezra routinely capitalized all his Ys, which I have altered.

50. J. Burnett to W. Burnett, Oct. 1, 1865, Burnett Collection. The house she refers to is his parents'.

51. R. Burdette to C. Burdette, June 20, 1914, Burdette Collection. Though their marriage exceeded the time limits of the study, the Burdettes courted within the 19th-century framework and were thoroughly imbued with Victorian mores and attitudes. He was born July 20, 1844, and her birthday was July 22, 1855.

52. C. Burdette to R. Burdette, June 21, 1914.

53. L. Clark to J. Clark, Oct. 23, 1841, typescript in "American Scene—Clark Family Narrative," p. 91, Clark Collection.

54. Ibid., Jan. 22, 1843, typescript, pp. 98–99.

55. Ibid., Oct. 31, 1846. Unless indicated, the Clark letters are in original manuscript form.

56. Ibid., Dec. 20, 1846.

57. Ibid., Aug. 4, 1848.

58. Ibid., Dec. 27, 1848.

59. Ibid., May 15, 1849.

60. Ibid., Feb. 5, 1852.

61. Ibid., April 10, 1852.

62. Ibid., Jan. 29, 1858.

63. N. Hawthorne to S. Peabody, Jan. 3, 1840, Hawthorne Collection.

64. Griswold, *Family and Divorce,* esp. 137–40, but also 46–48, 66–70, 72–74, 96–105, 110–13, and 131–34. Admittedly the introduction of mental cruelty as a criterion of marital dissolution was a step in the direction of greater parallelism between reasons to get married and reasons to remain married.

65. The testing rituals of 19th-century courtship were one response to the tensions created by these contradictions within the concept of marriage. Another response was the effort to stay as close as possible to a kinship network that could give economic assistance in time of marital difficulty and might also diffuse emotional problems with a spouse by offering other emotional ties. See Marilyn Ferris Motz, *True Sisterhood: Michigan Women and Their Kin, 1820–1929* (Albany, 1983). A third response to the tensions within the 19th-century view of marriage was to liberalize divorce laws and to begin the process of legitimizing divorce as an acceptable alternative to suffering in a bad marriage. See May, *Great Expectations,* 75–163; O'Neill, *Divorce in the Progressive Era.*

66. The relatively "satisfied" marriages in this study provided less evidence of the marital contradictions outlined above. More research is needed in the area of normative marital tensions, where these contradictions may be played out in more subtle forms.

67. D. Lummis to C. Lummis, Oct. 9 [actually Oct. 8, 1883], Moore Collection.

68. Ibid., [Oct. 18–21, 1883].

69. Ibid., Oct. 9 [actually Oct. 8, 1883].

70. Ibid., [Oct. 9, 1883].

71. Ibid., [Oct. 26–28, 1883].

72. Ibid., [Oct. 29, 1883].

73. Griswold, *Family and Divorce,* 63–119; also fn. 64 above; May, *Great Expectations,* 23–48.

74. It is only in the 20th century that wifely and husbandly role performance itself becomes a function of love on the level of cultural ideals, and ideal marital relations are seen in wholly voluntaristic terms.

75. D. Lummis to C. Lummis [Nov. 5–7, 1883], Moore Collection.

76. Ibid., [Nov. 12–13, 1883].

77. Ibid., [Nov. 26–28, 1883].

78. Ibid., [Feb. 21–24, 1884].

79. Ibid., [March 27–30, 1884].

80. Ibid.

81. Grossberg, *Governing the Hearth,* 9, 18–23, 98, 104, 119–20; O'Neill, *Divorce in the Progressive Era,* 12, 33, 72–74.

82. Charles Watts was a journalist and his letters are full of vivid and incisive reporting on his experiences in the Yukon. C. Watts to Emma Watts, June 19, 1898, Watts Collection.

83. C. Watts to Emma and Erma Watts, July 10?, 1898. Emma is his wife and Erma his daughter. The number of the date has been inked over as a 10 but the letter was written in pencil.

84. Ibid., Nov. 15, 1898.

85. C. Watts to Erma Watts, Jan. 14, 1900.

86. C. Watts to Emma Watts, Feb. 5, 1899.

87. C. Watts to Erma Watts, May 28, 1899.

88. Ibid.

89. Ibid., Sept. 6, 1899.

90. Ibid., Nov. 29, 1899. He began the New Year by writing to both mother and daughter, but generally continued to address his letters to his daughter.

91. Ibid., Jan. 21, 1900.

92. A note typed on the bottom of a letter from Mr. Lee Rohrbough to Mrs. Jane Warner Zumwalt, May 30, 1963. Mrs. Zumwalt was Charles Watts's granddaughter.

93. See Mary Richardson's 1833–1837 Diary (read in typescript), esp. 72–75, and her May 9, 1837–June 9, 1838, Diary in manuscript, the last portion of which was written after she married Elkanah Walker on March 5, 1838, Walker Collection.

94. M. Walker, Diary, April 24, 1838.

95. Ibid., June 18, 1838.

96. Ibid., June 27, 1839. Though this entry has been almost completely crossed out, it can be read.

97. Ibid.

98. Ibid.

99. Ibid., Thursday evening (June 27, 1839). This is a separate entry.

100. Ibid., "To Rev. E. Walker," entry that opens the July 24, 1839, to Jan. 14, 1840, Diary. Each of Mary's diaries is written in a small journal-book and contains varying amounts of material.

101. Ibid., Sept. 14, 1839.

102. Ibid., Dec. 12 and 13, 1839.

103. Ibid., Dec. 29, 1839. Mary's diaries never again address her husband or appear in any way to deliberately and directly communicate with him.

104. E. Walker, Diary, Feb. 5, 1841.

105. M. Walker, Diary, Feb. 13, 1841.

106. See Chapter 8, where both supportive evidence is presented and my reasoning is explored at length.

107. J. Hague to M. Hague, May 12, 1875, Hague Collection.

108. Ibid., July 22, 1875.

109. M. Hague to J. Hague, June 18, 1892.

110. Ibid.

111. Ibid., [June or early July 1892].

112. Ibid.

113. Ibid., July 11, 1892. This is Mary's summary of her husband's "ideas."

114. Ibid.

115. Ibid.

116. Ibid., Aug. 6, 1892.

117. Ibid.

118. Ibid., Aug. 3, 1892.

119. Ibid., Nov. 30, 1894.

120. Ibid., Feb. 8, 1898.

121. Ibid; also Jan. 31, 1898.

122. Ibid., Feb. 8, 1898.

123. Ibid., Feb. 10, 1898.

124. Ibid. Her husband responded to this jibe in a letter which she never opened.

125. J. Hague to M. Hague, Feb. 11, 1898. Mary died of complications from her surgery, and this letter remained sealed until Mrs. Virginia Rust opened it some 70 years later in the vaults of the Huntington Library.

126. Ibid.

127. See Degler, *At Odds,* 375–85, for a cogent analysis of 19th-century married women's economic position. On the issue of 19th-century female political autonomy see Ellen Carol DuBois, *Feminism and Suffrage: The Emergence of An Independent Women's Movement in America, 1848–1869* (Ithaca, 1978), 42–47.

128. H. Strong to C. Strong, May 22, 1865, Strong Collection.

129. C. Strong to H. Strong, June 12, 1865.

130. H. Strong to C. Strong, Feb. 26, 1864.

131. C. Strong to H. Strong, Nov. 16 and Dec. 29, 1874.

132. H. Strong to C. Strong, Sept. 4, 1876. The stamp was not cancelled and the envelope included with this letter showed no signs of a postmark.

133. Ibid.

134. H. Russell to C. Strong, Jan. 19, 1863.

135. H. Strong to C. Strong, Sept. 4, 1876.

136. Ibid.

137. Ibid.

138. Edmund S. Morgan, *The Puritan Family: Religion and Domestic Relations in Seventeenth-Century New England* (Boston, 1944; rev. ed., 1966), 47–54, 59.

139. Sondra Herman, "Loving Courtship or the Marriage Market? The Ideal and Its Critics, 1871–1911," *American Quarterly* 25 (May 1973), 237.

140. Ibid., 240.

141. C. Strong to H. Strong, Feb. 2, 1877, Strong Collection.

142. See, for example, ibid., and May 31, 1877.

143. Ibid., Jan. 15, 1877.

144. Ibid., Aug. 28, 1877.

145. Rose, *Parallel Lives,* 7.

146. C. Strong to H. Strong, Feb. 26, 1873, Strong Collection.

147. Ibid., June 14 and July 10, 1877.

148. Ibid., July 19 and July 20, 1877.

149. Ibid., Feb. 2, 1877.

150. Ibid., March 20, 1877. Charles had a history of "nervous disorders" including severe headaches and may have suffered, in modern parlance, from depression. Apparently Harriet worried for years about her husband's mental and emotional stability. See Mamie Strong to H. Strong, Feb. 18, 1883.

151. Ibid., March 27, and April 10, 1877.

152. Ibid., Nov. 20, 1877.

153. Even Harriet's ill-health, which might be considered an external impediment to their reunion, was probably more an effect of their marital strain than the cause of their separation. See H. Strong to C. Strong, Sept. 4, 1876. In this

letter Harriet bluntly attributed her physical problems to the emotional disaffection and tension of their marriage. Both Harriet and Charles appeared to suffer symptoms of neurasthenia. Professor Susan Albertine is currently engaged in a study of Harriet Strong that will intensively examine issues of health and finance in her life. After Charles's death, Harriet was transformed from sickly housewife to highly successful businesswoman and inventor—in the forefront of water control in Southern California and a leading Los Angeles clubwoman.

154. Notes of Harriet Strong, written on the back of a letter from Charles to her dated January 14, 1877 (actually 1878), Strong Collection.

155. Ibid., dated Jan. 12, 1877 (actually 1878). Though this segment was written on an earlier letter, it finishes the thoughts she expressed on the back of the Jan. 14 epistle from Charles.

156. Ibid.

157. Ibid.

158. Ibid., Aug. 18, 23, 24, 1878.

159. Ibid., Oct. 26, 1878.

160. See a discussion of this issue in Chapter 5.

161. C. Strong to H. Strong, Jan. 31, 1883.

Chapter 8. Not for God Only

1. Robert L. Griswold, *Family and Divorce in California, 1850–1890: Victorian Illusions and Everyday Realities* (Albany, 1982), 5.

2. Daniel Blake Smith, "The Study of the Family in Early America: Trends, Problems, and Prospects," *William and Mary Quarterly* 3rd ser. 39 (Jan. 1982), 3–28, esp. 18. This entire issue of the *WMQ* was devoted to "The Family in Early America."

3. Carl N. Degler, *At Odds: Women and the Family in America from the Revolution to the Present* (New York, 1980), 8–25.

4. Mary Beth Norton, "The Myth of the Golden Age," in *Women of America: A History,* Carol Ruth Berkin and Mary Beth Norton, eds. (Boston, 1979), 37–47, 45. See also Linda Kerber, *Women of the Republic: Intellect and Ideology in Revolutionary America* (Chapel Hill, 1980), esp. 158–231, 269–93; Daniel Scott Smith, "Parental Power and Marriage Patterns: An Analysis of Historical Trends in Hingham, Massachusetts," *Journal of Marriage and the Family* 35 (Aug. 1973), 419–428; Ann D. Gordon, "The Young Ladies Academy of Philadelphia," in *Women of America,* 69–87.

5. Nancy F. Cott, "Divorce and the Changing Status of Women in Eighteenth-Century Massachusetts," *William and Mary Quarterly* 3rd ser. 33 (Oct. 1976), 613, and *The Bonds of Womanhood: "Woman's Sphere" in New England, 1780–1835* (New Haven, 1977).

6. Jan Lewis, *The Pursuit of Happiness: Family and Values in Jefferson's Virginia* (New York, 1983), 216, 209–30.

7. See Gerda Lerner, *The Creation of Patriarchy* (New York, 1986), 238–39,

for an argument against a restricted definition. For another perspective, see Randolph Trumbach, *The Rise of the Egalitarian Family: Aristocratic Kinship and Domestic Relations in Eighteenth-Century England* (New York, 1978), 3–14. Trumbach differentiated patriarchy in patterns of household organization from patterns of kinship. He found that while household organizations in Western culture had been patriarchal for thousands of years, kinship ties in England, except for the very wealthy, operated according to principles which he labeled cognatic. By this he meant that kinship was traced from the mother, father, and often from relationships by marriage; there was some attempt to distribute (*not* necessarily evenly divide) the family "wealth" among both sons and daughters, and marriage was forbidden to any close relation.

Though closely intertwined, kinship patterns may be distinguished from household organization in America as well as England. Excluding a few very wealthy families in early America who followed a more patriarchal kinship pattern in which descent was traced from the father and only the first son inherited the bulk of the family estate, American kinship was also organized along cognatic lines.

American household organization, however, had patriarchal roots. Trumbach describes the patriarchal household's original presumption as follows: "at the head of each household stood a man who in his roles as master, father, and husband owned his wife, his children, his slaves, his animals and his land. . . . Many men were the property of other men; and all women and children, of some man." This pattern of subjection was first questioned in Western culture, Trumbach believes, in the 17th century.

See also Mary Beth Norton, "The Evaluation of White Women's Experience in Early America," *American Historical Review* 89 (June 1984), 603. Norton, by contrast, emphasizes the patriarchal tendencies of kinship ties in early America, particularly embodied in inheritance laws and practices. Also see Keith Thomas, "The Double Standard," *Journal of the History of Ideas* 20 (April 1959), 195–216; Friedrich Engels, *The Origin of the Family, Private Property, and the State,* trans. Ernest Untermann (Chicago, 1905); Annette Kuhn, "Structures of Patriarchy and Capital in the Family," in *Feminism and Materialism: Women and Modes of Production,* Annette Kuhn and AnnMarie Wolpe, eds. (London, 1978), 42–67.

The concept of female chastity as the property of men is, I believe, the foundation of patriarchal household organization. By the early 19th-century, the concept of romantic love was shifting the ownership of female chastity from father and husband to daughter and wife. This was a basic challenge to patriarchal power. See Chapter 3.

8. Smith, "The Study of the Family in Early America," 18.

9. Trumbach, *Egalitarian Family,* 3.

10. Mary Beth Norton, *Liberty's Daughters: The Revolutionary Experience of American Women, 1750–1800* (Boston, 1980); Linda Kerber, *Women of the Republic,* 189–231, 269–93; Lonna M. Malmsheimer, "Daughters of Zion: New England Roots of American Feminism," *New England Quarterly* 50 (Sept. 1977), 484–504; Jay Fliegelman, *Prodigals and Pilgrims: The American Revolution Against*

Patriarchal Authority, 1750–1800 (New York, 1982); Carl N. Degler, *At Odds,* 298–327; Cott, *Bonds;* Ann Douglas, *The Feminization of American Culture* (New York, 1977); Kathryn Kish Sklar, *Catharine Beecher: A Study in American Domesticity* (New Haven, 1973).

11. Simone de Beauvoir's great chapter, "The Woman in Love," in *The Second Sex,* trans. H. M. Parshley (New York, 1953; rpt., 1974), 712–43, argues that the idea of romantic love has been a dream of self-abnegation for women and resulted in their failure to assume full responsibility for their lives. In de Beauvoir's view, romantic love has been the existential siren song of women in European and American culture, leading them to abandon their selfhood to men in a doomed attempt to become sovereign subjects. While de Beauvoir's insights on the woman in love are profound and may explain the situation of some American women after World War II, her argument reflects neither the direction of change from 17th- to 19th-century male-female relationships nor the character of middle-class 19th-century American romantic relationships. Again, one must guard against an essentialist or ahistorical treatment of the historical reality of romantic love.

12. Rhys Isaac, *The Transformation of Virginia, 1740–1790* (Chapel Hill, 1982), 5–138; Lewis, *The Pursuit of Happiness,* 1–39. On the northern side, the Puritan attention to feeling between spouses and their conception of spiritual equality was already compromising the experience of female submission, but reformed Protestants nonetheless maintained a firm commitment to civil inequality in the family. See Laurel Thatcher Ulrich, *Good Wives: Image and Reality in the Lives of Women in Northern New England, 1650–1750* (New York, 1982), 106–25.

13. Dichotomized gender roles had a firm grip on the 19th-century family. Nonetheless, as Chapter 5 illustrates, Victorian men and women were beginning to see each other as more alike in terms of both head and heart. This is a relative judgment, however, based upon comparison with earlier cultural conceptions of gender roles.

14. Obviously, the rate of change in this cultural pattern for any particular class and ethnic group in America varied widely, with different estimations of the relative decline of patriarchy even in the mainstream family.

15. J. Marquis to N. Haile, Nov. 11, 1865, Marquis Collection.

16. Ibid., Feb. 11, 1870.

17. Another aspect of this gift of self—attributed to wives—were children who were also coming to be seen as precious selves to be nurtured and developed. When children were seen in part as "selves," they also became more equal in the household political economy. The encouragement of self-expression and self-formation in 19th-century childrearing practices is discussed in William E. Bridges, "Family Patterns and Social Values in America 1825–1875," *American Quarterly* 17 (Spring 1965), 3–11; and Richard L. Rapson, "The American Child as Seen by British Travelers, 1845–1935," *American Quarterly* 17 (Fall 1965), 520–34. See also Philip J. Greven, Jr., *The Protestant Temperament: Patterns of Child-rearing, Religious Experience, and the Self in Early America* (New York, 1977); Bernard Wishy, *The Child and the Republic: The Dawn of Modern American Child Nurture* (Philadelphia, 1968); Joseph M. Hawes and N. Ray Hiner, *American Childhood: A*

Research Guide and Historical Handbook (Westport, Conn., 1985); Joseph F. Kett, *Rites of Passage: Adolescence in America, 1790 to the Present* (New York, 1977). On the invention of the self, see John O. Lyons, *The Invention of the Self: The Hinge of Consciousness in the Eighteenth Century* (Carbondale, Ill., 1978).

18. J. Marquis to N. Haile, March 10, 1865, Marquis Collection.

19. V. Blair to A. Janin, April 9, 1874, Janin Collection.

20. A. Janin to V. Blair, Dec. 21, 1871.

21. J. Marquis to N. Marquis, March 6, 1870, Marquis Collection. Of course, sometimes the lines between men's solicitude for women's happiness and men's concern over failure to fulfill the masculine role demands of provider may blur. There is no necessary reason that the goal of good provider must be linked to masculine responsibility for female happiness and emotional fulfillment. But it so happens that the work and love ethic were joined in American Victorian culture. See Chapter 5.

22. Susan Strasser, *Never Done: A History of American Housework* (New York, 1982), 263–81; Ruth Schwartz Cowan, *More Work for Mother: The Ironies of Household Technology from the Open Hearth to the Microwave* (New York, 1983), 100–101. Cowan explains: "Modern labor-saving devices eliminated drudgery, not labor." Maxine L. Margolis, *Mothers and Such: Views of American Women and Why They Changed* (Berkeley, 1984), 108–47; Susan M. Strasser, "An Enlarged Human Existence? Technology and Household Work in Nineteenth-Century America," in *Women and Household Labor,* Sarah Fenstermaker Berk, ed. (Beverly Hills, 1980), 29–51. Also middle-class women's access to servants was limited by both finances and the labor supply. See Strasser, *Never Done,* 162–64; and Cowan, *More Work for Mother,* 121–26. Alice Kessler-Harris seems to suggest that servants were somewhat more common in "comfortable, but by no means affluent" homes before the Civil War in *Out to Work: A History of Wage-Earning Women in the United States* (New York, 1982), 54–55.

23. The data of this study revealed attitudes, intentions, and emotions and included as well a number of behavioral reports on men's efforts to please the women they loved. Letters were themselves a concrete behavioral manifestation of the effort to please.

24. L. Clark to J. Clark, March 3, 1838, Clark Collection.

25. Ibid., Feb. 2, 1847.

26. J. Bell to A. Hallock, April 21, 1861, Bell Collection.

27. W. Burnett to J. Burnett, May 2, 1857, Burnett Collection.

28. J. Blair to M. Blair, March 21, 1847, Janin Collection.

29. Ibid., April 20, 1849.

30. Ibid., April 20 and Sept. 26, 1849. James sent his wife a large sum of money five months later to spend as she wished. He suggested she might employ a portion of the funds to build a cottage.

31. Ibid., Feb. 25, 1847.

32. C. Strong to H. Russell, Dec. 17, 1862, Strong Collection.

33. W. Burnett to J. Burnett, April 29, 1857, Burnett Collection.

34. Ibid., June 7, 1864.

35. E. Simkins to E. Trescot, Nov. 2, 1863, Simkins Collection.

36. J. Blair to M. Blair, Sept. 30, 1850, emphasis added, Janin Collection.

37. A. Janin to V. Blair, Feb. 15, 1872.

38. V. Blair to A. Janin, April 9, 1874.

39. A. Janin to V. Blair, Nov. 3, 1873.

40. See Griswold, *Family and Divorce,* 120–40.

41. For example, see A. Janin to V. Blair, Nov. 7 and 17, Dec. 24, 1871; Jan. 16, 1872, Janin Collection.

42. Ibid., Nov. 23, 1871. (The courtship testing by women is discussed at length in Chapter 6.)

43. Ibid., Sept. 4, 1874.

44. V. Janin to A. Janin, Nov. 23, 1874.

45. Ibid., June 27, 1874.

46. For example, V. Blair to A. Janin, Nov. 16, 1872.

47. Ibid., March 9, 1874.

48. Ibid., March 12, 1873.

49. The situation was more complex after marriage. Masculine sex roles clearly gave men broader rights of self-assertion than women, though power relationships in the 19th-century household were somewhat ameliorated as long as the married couple were in love.

50. N. Hawthorne to S. Peabody, May 26, 1839, Hawthorne Collection.

51. Ibid., March 15, 1840.

52. L. Clark to J. Clark, April 4, 1847, Clark Collection.

53. N. Wheeler to C. Wheeler, July 26, 1880, Burdette Collection.

54. The incident surrounding Clara Wheeler's resignation is found in Chapter 5.

55. N. Wheeler to C. Wheeler, July 31, 1881, Burdette Collection.

56. E. Simkins to E. Trescot, Dec. 4, 1863, Simkins Collection.

57. Ibid., Dec. 12, 1863.

58. Ibid., Aug. 8, 1864.

59. G. Bomford to C. Bomford, June 14, 1846, Baldwin Family Collection, Henry E. Huntington Library (not to be confused with the Frank Dwight Baldwin Collection).

60. Ibid.

61. Marital relationships in which love faltered and/or died sometimes presented evidence of husbandly power abuse, yet as Robert Griswold demonstrates in his sample of 19th-century divorce documents, "it was clear that a so-called Victorian patriarch was not an ideal husband." See his *Family and Divorce,* 11, also 92–140. Of course, both sexes expected minimum sex-role competence from the other. For women, this included performance of household tasks which were a service to men and children. But men who worked in jobs they disliked or barely tolerated in order to support their family might also be said to be performing a service for women and children.

62. W. Burnett to J. Burnett, April 30, 1857, Burnett Collection.

63. J. Bell to A. Hallock, Dec. 26–28, 1862, Bell Collection.

64. J. Blair to M. Blair, March 21, 1847, Janin Collection.

65. Edward Taylor as quoted and summarized in Lyle Koehler, *A Search for Power: The "Weaker Sex" in Seventeenth-Century New England* (Urbana, 1980), 17.

66. Thomas H. Johnson, ed., *The Poetical Words of Edward Taylor* (New York, 1939; rpt., 1966), "Meditation One," 123.

67. Ulrich, *Good Wives*, 108.

68. Edmund S. Morgan, *The Puritan Family: Religion and Domestic Relations in Seventeenth-Century New England* (Boston, 1944; rev. ed., 1966), 61.

69. Cotton Mather, *The Mystical Marriage* (Boston, 1728), 6. See also Ulrich, *Good Wives*, 108.

70. John Davenport to Lady Mary Vere, July 21, 1635, in *Letters of John Davenport: Puritan Divine*, Isabel Macbeath Calder, ed. (New Haven, 1937), 58.

71. M. Winthrop to J. Winthrop, Feb. 2, 1630, in *The Puritans: A Sourcebook of Their Writings*, Perry Miller and Thomas H. Johnson, eds., 2 vols. (New York, 1938; rev. ed., 1963), vol. 2, 468–69.

72. Peter Bulkeley, *The Gospel-Covenant; or, the Covenant of Grace Opened* (London, 1651), 103. See also Morgan, *Puritan Family*, 162.

73. John Cotton, *Christ the Fountaine of Life* (London, 1651), 36–37. See also Morgan, *Puritan Family*, 163–64.

74. Cott, *Bonds*, 170. For a general overview, see Sydney E. Ahlstrom, *A Religious History of the American People* (New Haven, 1972), 295–313.

75. Jonathan Edwards, *The Works of Jonathan Edwards*, Perry Miller, gen. ed.; vol. 2, *Religious Affections*, John E. Smith, ed. (New Haven, 1959), 106–8.

76. Ibid.

77. Ibid., 111–12.

78. Ibid., 113.

79. Ahlstrom, *A Religious History*, 287. Nancy Cott suggests that Jonathan Edwards's conception of the religious affections "fostered female religious friendships by illuminating the 'heart,' at the same time that it induced men to approve and share similiar religious affections." Cott, *Bonds*, 170. On the differing views of Edwards's impact on American non-elite culture, see William J. Scheick, *Critical Essays on Jonathan Edwards* (Boston, 1980), xviii–xxi.

80. Steven Mintz discusses the application of evangelical language to the vocabulary of love in the personal papers of Harriet Beecher Stowe, Robert Louis Stevenson, and George Eliot, in *A Prison of Expectations: The Family in Victorian Culture* (New York, 1983), 118–28.

81. Lewis O. Saum, *The Popular Mood of Pre-Civil War America* (Westport, Conn., 1980), esp. 3–54, 107–42; Saum's work suggests this secular drift in world view did not affect the lower-middle classes until at least the Civil War.

82. John Cotton, *A Practical Commentary, or An Exposition with Observations, Reasons and Uses Upon the First Epistle Generall of John* (London, 1656), 126. See also Morgan, *Puritan Family*, 48.

83. Ibid. See also Ulrich, *Good Wives*, 109. William Haller and Malleville Haller, "The Puritan Art of Love," *Huntington Library Quarterly* 5 (Jan. 1942), 257.

84. Morgan, *Puritan Family,* 50–51.

85. J. Bolton to A. Harrison, July 16, 1838, Harrison Collection.

86. Ibid.

87. Saum, *The Popular Mood,* 17.

88. As quoted in ibid., 117–18.

89. J. Marquis to N. Marquis, Jan. 30, 1870, Marquis Collection.

90. Richard D. Mosier, *The American Temper: Patterns of Our Intellectual Heritage* (Berkeley, 1952), 201, 181–203; also quoted in Saum's *The Popular Mood,* 108.

91. M. Smith to S. F. Smith, Jan. 10, 1834, Smith Collection.

92. S. F. Smith to M. Smith, Jan. 22, 1834.

93. M. Smith to S. F. Smith, Jan. 27, 1834.

94. S. F. Smith to M. Smith, Jan. 30, 1834.

95. M. Smith to S. F. Smith, Feb. 3, 1834.

96. Ibid., Feb. 10, 1834.

97. S. F. Smith to M. Smith, Feb. 20, 1834. In her Feb. 10 letter, Mary asked if she could call Samuel by his middle name, a symbolic request for more intimacy which he grumpily rejected in a letter of Feb. 13. Mary calmly retaliated on Feb. 17 by referring to him as "friend." This moved Samuel, in response, to openly avow his affection. He entreated her on Feb. 20 to address him more intimately in her next letter.

98. Ibid., Feb. 27, 1834.

99. M. Smith to S. F. Smith, March 10, 1834; S. F. Smith to M. Smith, March 13, 1834.

100. S. F. Smith to M. Smith, March 14, 1834.

101. M. Smith to S. F. Smith, March 17, 1834.

102. Ibid.

103. S. F. Smith to M. Smith, March 20, 1834.

104. Ibid., March 27, 1834.

105. Ibid., April 3, 1834.

106. M. Smith to S. F. Smith, April 7, 1834.

107. S. F. Smith to M. Smith, April 10, 1834.

108. M. Smith to S. F. Smith, May 6, 1834.

109. Ibid., May 28, 1834.

110. Ibid., June 9, 1834.

111. S. F. Smith to M. Smith, June 12, 1834.

112. Ibid., June 20, 1834.

113. Ibid.

114. Ibid., July 3, 1834.

115. Ibid., July 18 and July 25, 1834.

116. Ibid., Aug. 1, 1834.

117. Ibid., Aug. 26, 1834.

118. N. Hawthorne to S. Peabody, March 6, 1839, Hawthorne Collection.

119. Ibid., May 26, 1839.

120. Ibid., July 24, 1839.

121. Ibid., Oct. 4, 1840.

122. W. Burnett to J. Burnett, Jan. 16, 1857, Burnett Collection.

123. Ibid., April 30, 1857.

124. N. Wheeler to C. Bradley, Sept. 13, 1874, Burdette Collection.

125. Ibid., Feb. 3, 1878.

126. D. Lummis to C. Lummis [April 16, 1884], Moore Collection.

127. Ibid., [Nov. 12–13 and Nov. 26–28, 1883].

128. R. Burdette to C. Baker, April 3, 1898, Burdette Collection.

129. Ibid., "On the Burlington Eastbound, Tuesday evening." This letter was located in the 1914 letter file but was certainly written before Burdette was married to Clara.

130. E. Simkins to E. Trescot, Aug. 8, 1864, Simkins Collection.

131. E. Trescot to E. Simkins, Aug. 4, 1864.

132. E. Simkins to E. Trescot, Aug. 8, 1864.

133. Ibid., Sept. 2, 1864.

134. Annie Fields, "The Canticles of Married Love," number four, Sunday, Nov. 15, 1863. Fields Addenda, Annie Adams Fields Collection, Henry E. Huntington Library.

135. C. Baker to R. Burdette, Oct. 17, 1897, Burdette Collection. I believe the letter was actually written in 1898.

136. Ibid., Aug. 14, 1898.

137. Ibid., Sunday evening, Jan. 22, 1899.

138. N. Hawthorne to S. Peabody, April 2, 1839, Hawthorne Collection; L. Clark to J. Clark, Dec. 4, 1836, Clark Collection; L. Hodge to M. Granger, Feb. 10, 1867, Hodge Collection.

139. R. Burdette to C. Baker, Nov. 18, 1898, Burdette Collection.

140. The secular "religious" elevation of sexuality in the context of romantic love is perhaps the most significant non-mythic legacy of Victorian attitudes toward sex.

141. C. Strong to H. Russell, Dec. 14, 1862, Strong Collection.

142. Ibid.; and Jan. 8, 1863.

143. The First Commandment: "Thou shalt have no other gods before me." Exodus 20:3.

144. C. Strong to H. Russell, Dec. 14, 1862, Strong Collection.

145. In this sample, the sense of conflict between divine and secular love was concentrated in the period before 1870.

146. L. Clark to J. Clark, Oct. 15, 1836, Clark Collection. Lincoln is evidently referring to Christ.

147. Ibid., June 6–7, 1841, in "American Scene—Clark Family Narrative," p. 84. This letter is in typescript. I could not find the original manuscript in the collection.

148. Ibid., June 1, 1843.

149. Ibid., Oct. 25, 1846.

150. Ibid., April 25, 1847.

151. M. Garland to Mr. Garland, Aug. 10, 1823, Garland Family Papers, Robert Alonzo Brock Collection, Henry E. Huntington Library.

152. M. Walker, Diary, Sunday, Oct. 17, 1847 (copied by J. E. Walker in Nov. 1896), Walker Collection.

153. E. Lovell to M. Lovell, Nov. 15, 1862, Lovell Collection.

154. J. Bell to A. Hallock, May 27, 1858, Bell Collection.

155. Ibid., Nov. 20, 1858.

156. A. Hallock to J. Bell, Dec. 23, 1858.

157. J. Bell to A. Hallock, Jan. 8, 1859.

158. A. Hallock to J. Bell, Feb. 14, 1859.

159. Ibid.

160. J. Bell to A. Hallock, Feb. 23, 1859.

161. Ibid., April 3, 1859.

162. A. Hallock to J. Bell, March 27 and April 3, 1859.

163. J. Bell to A. Hallock, Jan. 25, 1861.

164. Ibid., Jan. 5, 1862.

165. Ibid., Dec. 26, 1862.

166. Ibid., Jan. 27, 1863.

167. A. Hallock to J. Bell, April 11–12, 1863.

168. Ibid., April 30, 1863.

169. J. Bell to A. Hallock, May 13, 1863.

170. A. Hallock to J. Bell, May 19, 1863.

171. Ibid., Aug. 8, 1863.

172. James Bell's dying words presumably recorded by Augusta Hallock, Stanton Hospital, Washington, D. C., Monday Evening Oct. 5 and Tues. morn [Oct. 6], 1863.

173. A. Hallock to I. Cooper, Aug. 29, 1865 (rough draft).

174. Secularization is a treacherous term, for what is defined as secular depends to a large extent on how one conceptualizes religion. For a non-sectarian definition with a wide range of applicability, see Clifford Geertz, "Religion as a Cultural System," in *The Interpretation of Cultures* (New York, 1973), 87–125. Steven Mintz suggests that secularization is best understood as "a transfer, or displacement, of religious needs and aspirations onto other secular objects" rather than as a "lapse of religious belief." See *Prison of Expectations,* 145.

175. Ralph H. Gabriel, "Evangelical Religion and Popular Romanticism in Early Nineteenth-Century America," *Church History* 19 (March 1950), 34–47.

176. David Leverenz, *The Language of Puritan Feeling: An Exploration in Literature, Psychology, and Social History* (New Brunswick, N.J., 1980), 79–87. On the pervasiveness of Milton's influence in early American culture, see George F. Sensabaugh, *Milton in Early America* (Princeton, N.J., 1964).

MANUSCRIPT COLLECTIONS

Henry E. Huntington Library, San Marino, California

Collections that proved most helpful are marked with an asterisk.

Adams, Phinehas
Anderson, Richard Clough
*Appleton-Foster
*Baldwin Family
*Baldwin, Frank Dwight
*Bell, James Alvin
Bissell, Mary Eleanor
Bouvier, John
*Brock, Robert Alonzo
   Subcollections consulted:
   Cabell Family
   Chevallie Family
   Daniel Family
   Garland Family
   Harrison Family
   Minor Family
   Pleasants Family
   Tinsley Family
   Tucker Family
*Burdette, Clara Bradley
*Burnett, Wellington Cleveland
Carr, Jeanne Caroline Smith
*Clark, Lincoln
*Dane Family
Dodge, Mary
Eldridge, James William
Field, Eugene
Fields, Annie Adams

Galloway, Joseph
Gardner, William Bunker
Graham, John Lorimer
Graham, Margaret Collier
*Hague, James Duncan
*Harbert, Elizabeth Morrisson Boynton
Harper, Ida Husted
*Hawthorne, Nathaniel
*Hodge, Benjamin
*Janin Family
Lake, Ann Getz
*Lovell, Mansfield
*Marquis, Neeta
Megquier, Mary Jane Cole
Mellish, George H.
*Moore, Dorothea Rhodes Lummis
*Mormon File
Muir, John
Newcomb-Johnson
Newman, Emma E.
*North, John Wesley
Park, Alice Locke
Peck, Orrin M.
*Pumpelly, Raphael
Rhees, William Jones
Rust, Horatio Nelson
*Simkins, Eldred
*Smith, Samuel Francis
*Spence-Lowell
*Strong, Harriet Williams Russell
*Walker, Elkanah Bartlett
*Watts, Charles William
Weinland, William Henry
Wood, Charles Erskine Scott

VICTORIAN ADVICE-BOOK BIBLIOGRAPHY

A Physician. [Graham, Thomas John.] *Sure Methods of Improving Health, and Prolonging Life: or, A Treatise on the Art of Living Long and Comfortably, by Regulating the Diet and Regimen.* . . . Philadelphia: Carey, Lea and Carey, 1828 (first American ed., with additions).

Acton, William. *The Functions and Disorders of the Reproductive Organs in Childhood, Youth, Adult Age, and Advanced Life Considered in Their Physiological, Social and Moral Relations.* Philadelphia: Lindsay and Blakiston, 1867 (second American from the fourth London edition).

Alcott, William [Andrus]. *The Young Husband; or, Duties of Man in the Marriage Relation.* Boston: George W. Light, 1839.

Alcott, William [Andrus]. *The Young Man's Guide.* Boston: Lilly, Wait, Colman, and Holden, 1833.

Alcott, William [Andrus]. *The Young Wife; or, Duties of Woman in the Marriage Relation.* Boston: George W. Light, 1837.

An American Gentleman. *Good Behavior for Young Gentlemen: Founded on Principles of Common Sense, and the Usages of Good American Society: Containing . . . also Instructions in Letter Writing.* Rochester, N.Y.: D. M. Dewey, 1848.

Anonymous. *A New Letter-Writer, For the Use of Gentlemen; Embodying Letters on the Simplest Matters of Life, and on Various Subjects.* . . . Philadelphia: Porter and Coates, 186 .

Anonymous. *A New Letter-Writer, For the Use of Ladies; Embodying Letters on the Simplest Matters of Life.* . . . Philadelphia: Porter and Coates, 186-.

Anonymous [Cooke, Nicholas Francis]. *Satan in Society.* By a Physician. Cincinnati: Edward F. Hovey, 1882.

Anonymous. *The Art of Good Behavior; and Letter Writer on Love, Courtship, and Marriage: A Complete Guide for Ladies and Gentlemen, Particularly Those Who Have Not Enjoyed the Advantages of Fashionable Life.* . . . New York: C. P. Huestis, 1846.

Anonymous. *The Ladies' and Gentleman's Model Letter-Writer: A Complete Guide to Correspondence on All Subjects with Household and Commercial Forms.* New York: Frederick Warne, 1871.

Arthur, T[imothy] S[hay]. *Advice to Young Ladies on Their Duties and Conduct in Life*. Philadelphia: John E. Potter, [1860].

Arthur, T[imothy] S[hay]. *Advice to Young Men on Their Duties and Conduct in Life*. Philadelphia: J[ohn] E. Potter [1847?].

Bayley, Rev. John. *Marriage As It Is and As It Should Be*. New York: M. W. Dodd, 1857.

Bean, Rev. James. *The Christian Minister's Affectionate Advice to a Married Couple*. New York: American Tract Society, 18–.

Black, M.D., J[ames] R[ush]. *The Ten Laws of Health; or, How Diseases Are Produced and Prevented: And Family Guide to Protection against Epidemic Diseases and Other Dangerous Infections*. Philadelphia: J. B. Lippincott, 1885.

Blackwell, M.D., Elizabeth. *The Laws of Life with Special Reference to the Physical Education of Girls*. New York: George P. Putnam, 1852.

Brigham, M.D., Amariah. *Observations on the Influence of Religion Upon the Health and Physical Welfare of Mankind*. Boston: Marsh, Capen, and Lyon, 1835.

Brigham, M.D., Amariah. *Remarks on the Influence of Mental Cultivation and Mental Excitement Upon Health*. Philadelphia: Lea and Blanchard, 1845.

Brown, Eli F. *Sex and Life: The Physiology and Hygiene of the Sexual Organization*. Chicago: F. J. Schulte, 1891.

Capp, M.D., William M. *The Daughter: Her Health, Education and Wedlock: Homely Suggestions for Mothers and Daughters*. Philadelphia: F. A. Davis, 1891.

Chavasse, M.D., P[ye] Henry. *Physical Life of Man and Woman; or Advice to Both Sexes . . . Advice to Wife and Mother . . . Advice to a Maiden, Husband, and Son, from the Most Recent French and German Works. . . .* Cincinnati: National Publishing Co., 1871.

Cowan, M.D., John. *The Science of a New Life*. New York: Cowan and Co., 1873.

Culverwell, Robert James. *PorneioPathology: A Popular Treatise On Venereal and Other Diseases of the Male and Female Genital System; With Remarks on Impotence, Onanism, Sterility, Piles, and Gravel, and Prescriptions for Their Treatment*. New York: J. S. Redfield, 1844.

Duffey, Mrs. E[liza] B[isbee]. *What Women Should Know. A Woman's Book about Women. Containing Practical Information for Wives and Mothers*. Philadelphia: J. M. Stoddart, 1873.

[Farrar, Mrs. Eliza]. By a Lady. *The Young Lady's Friend*. Boston: American Stationers' Co. 1837.

Finck, Henry T[heophilus]. *Romantic Love and Personal Beauty; Their Development, Causal Relations, Historic and National Peculiarities*. New York: Macmillan and Co., 1887.

Foote, M.D., Edward B. *Medical Common Sense; Applied to the Causes, Prevention and Cure of Chronic Diseases and Unhappiness in Marriage*. New York: Edward B. Foote, 1867.

Foote, M.D., Edward B. *Plain Home Talk About the Human System—the Habits of Men and Women . . . Embracing Medical Common Sense Applied to Causes, Prevention, and Cure of Chronic Diseases. . . .* New York: Wells and Coffin, 1870; rev. ed., Chicago: Thompson and Thomas, 1896.

Fowler, O[rson] S[quire]. *The Family; or, Man's Social Faculties and Domestic Relations, including Man, Woman, Love, Selection, Courtship, Marriage, Paternity, Maternity, Infancy, Childhood, Home, Country, etc., as Taught by Phrenology and Kindred Sciences.* Boston: H. O. Houghton and Co., 1869.

Fowler, O[rson] S[quire]. *Offspring, and Their Hereditary Endowment: Or, Paternity, Maternity, and Infancy; Including Sexuality.* . . . Boston: H. O. Houghton and Co., 1869.

Fowler, O[rson] S[quire]. *Sexual Science; or Manhood and Womanhhood, including Perfect Husbands, Wives, Mothers, and Infants. Together with Maternity . . . as Taught by Phrenology and Physiology* Boston: H. O. Houghton and Co., 1870.

Gardner, M.D., Augustus K[insley]. *Conjugal Sins Against the Laws of Life and Health [and Their Effects upon the Father, Mother and Child].* New York: J. S. Redfield, 1870; reprint ed. New York: Arno Press, 1974.

Gray, George Zabriskie. *Husband and Wife or The Theory of Marriage and Its Consequences.* Boston: Houghton, Mifflin, 1885.

Guernsey, M.D., Henry N[ewell]. *Plain Talks on Avoided Subjects.* Philadelphia: F. A. Davis, 1882; reprint ed., 1907.

Hanchett, Henry G[ranger]. *Sexual Health . . . A Plain and Practical Guide for the People in All Matters Concerning the Organs of Reproduction in Both Sexes and All Ages.* New York: Charles T. Hurlburt, 1887.

Harland, Marion [pseud.] [Terhune, Mrs. Mary Virginia (Hawes)]. *Eve's Daughters; or, Common Sense for Maid, Wife, and Mother.* New York: John R. Anderson and Henry S. Allen, 1882.

Hersey, Thomas, Physician. *The Midwife's Practical Directory; or, Woman's Confidential Friend; Comprising Extensive Remarks on the Various Casualities and Forms of Disease.* . . . Columbus, Ohio: Clapp, Gillett, 1834.

Hudson, Rev. Geo[rge] W. *The Marriage Guide for Young Men. A Manual of Courtship and Marriage.* Ellsworth, Me.: Geo. W. Hudson, 1883.

Kellogg, M.D., J[ohn] H[arvey]. *Ladies' Guide in Health and Disease. Girlhood, Maidenhood, Wifehood, Motherhood.* Des Moines: W. D. Condit and Nelson, 1886.

Kellogg, M.D., J[ohn] H[arvey]. *Plain Facts for Old and Young.* Burlington, Iowa: I. F. Segner, 1882.

Lewis, M.D., Dio. *Chastity; or, Our Secret Sins.* Philadelphia: George Maclean, 1875.

Meigs, M.D., Charles D[elucena]. *Woman: Her Diseases and Remedies: A Series of Letters to His Class.* Philadelphia: Blanchard and Lea, 1854.

Napheys, George H[enry]. *The Physical Life of Woman: Advice to the Maiden, Wife and Mother.* Philadelphia: H. C. Watts and Co., 1884.

Peyre-Ferry, François. *The Art of Epistolary Composition; or Models of Letters, Billets, Bills of Exchange . . . to Which are Added, a Collection of Fables.* Middletown, Conn.: E. and H. Clark, 1826.

Pierce, R[ay] V[aughn]. *The People's Common Sense Medical Adviser in Plain En-*

glish; or, Medicine Simplified. Buffalo, N. Y.: World's Dispensary Printing Office and Bindery, 1882.

Ray, M.D., I[saac]. *Mental Hygiene.* Boston: Ticknor and Fields, 1863.

Scott, M.D., James Foster. *The Sexual Instinct; Its Use and Dangers as Affecting Heredity and Morals. Essentials to the Welfare of the Individual and the Future of the Race.* New York: E. B. Treat, 1899.

Stall, Sylvanus. *What a Man of Forty-five Ought to Know.* Philadelphia: Vir Publishing Co., 1901.

Stockham, M.D., Mrs. Alice B[unker]. *Tokology, A Book for Every Woman.* Chicago: [no pub. given], 1883.

Storer, M.D., Horatio R[obinson], and Franklin Fiske Heard. *Criminal Abortion: Its Nature, Its Evidence, and Its Law.* Boston: Little Brown, 1868.

Storer, Dr. H[oratio] R[obinson]. *Is It I? A Book for Every Man.* Boston: Lee and Shepard, 1868.

Theocritus, Junior [pseud.]. *Dictionary of Love: Containing a Definition of All the Terms Used in the History of the Tender Passion, with Rare Quotations from the Ancient and Modern Poets of All Nations; Together with Specimens of Curious Model Love Letters. . . .* New York: Dick and Fitzgerald, 1858.

Thornwell, Emily. *The Lady's Guide to Perfect Gentility, in Manners, Dress, and Conversation . . . also a Useful Instructor in Letter Writing. . . .* New York: Derby and Jackson, 1856.

Titcomb, Timothy, Esquire. [pseud.] [Holland, Josiah Gilbert]. *Titcomb's Letters to Young People, Single and Married.* New York: Charles Scribner, 1858.

Todd, Rev. John. *Woman's Rights.* Boston: Lee and Shepard, 1867.

Todd, John. *The Young Man. Hints Addressed to the Young Men of the United States.* Northampton: J. H. Butler, 1846.

Trall, M.D., R[ussell] T[hacher]. *Sexual Physiology: A Scientific and Popular Exposition of the Fundamental Problems in Sociology.* New York: Wood and Holbrook, 1870.

Weaver, Rev. G[eorge] S[umner]. *The Ways of Life, Showing the Right Way and the Wrong Way; Contrasting the High Way and the Low Way; the True Way and the False Way; the Upward Way and the Downward Way; the Way of Honor and the Way of Dishonor.* New York: Fowler and Wells, 1855.

[Wells, Samuel Roberts]. *How to Behave; or A Pocket Manual of Republican Etiquette, and Guide to Correct Personal Habits.* New York: Fowler and Wells, 1858.

Wells, S[amuel] R[oberts]. *Wedlock; or, the Right Relations of the Sexes: Disclosing the Laws of Conjugal Selection, and Showing Who May, and Who May Not Marry.* New York: S. R. Wells, 1879.

Wood-Allen, M.D., Mrs. Mary. *What A Young Girl Ought to Know.* rev. ed. Philadelphia: Vir Publishing Co., 1905.

[Woodward, Samuel Bayard]. *Hints for the Young: On a Subject Relating to the Health of Body and Mind.* Boston: Weeks, Jordan, 1838.

Index